Max-Plus Linear Stochastic Systems and Perturbation Analysis

T0205646

THE INTERNATIONAL SERIES
ON DISCRETE EVENT DYNAMIC SYSTEMS

Series Editor

Yu-Chi Ho
Harvard University

MAX-PLUS LINEAR STOCHASTIC SYSTEMS AND PERTURBATION ANALYSIS

by

Bernd Heidergott
Vrije Universiteit, Amsterdam, The Netherlands

 Springer

Bernd Heidergott
Vrije Universiteit
Faculty of Economics and Business Administration
Department of Econometrics and Operations Research
De Boelelaan 1105
1081 HV Amsterdam
Email: bheidergott@feweb.vu.nl

Max-Plus Linear Stochastic Systems and Perturbation Analysis
by Bernd Heidergott

e-ISBN-13: 978-0-387-38995-0
ISBN-13: 978-1-4419-4198-5   e-ISBN-10: 0-387-38995-4

Printed on acid-free paper.

9 8 7 6 5 4 3 2 1

springer.com

Preface

This monograph presents perturbation analysis for max-plus linear stochastic systems. Max-Plus algebra has been successfully applied to many areas of stochastic networks. For example, applying Kingman's subadditive ergodic theorem to max-plus linear queuing networks, one can establish ergodicity of the inverse throughput. More generally, applying backward coupling arguments, stability results for max-plus linear queuing systems follow. In addition to that, stability results for waiting times in open queuing networks can be obtained.

Part I of this book is a self-contained introduction to stochastic max-plus linear systems. Chapter 1 provides an introduction to the max-plus algebra. More specifically, we introduce the basic algebraic concepts and properties of max-plus algebra. The emphasis of the chapter is on modeling issues, that is, we will discuss what kind of discrete event systems, such as queueing networks, can be modeled by max-plus algebra. Chapter 2 deals with the ergodic theory for stochastic max-plus linear systems. The common approaches are discussed and the chapter may serve as a reference to max-plus ergodic theory.

Max-Plus algebra is an area of intensive research and a complete treatment of the theory of max-plus linear stochastic systems is beyond the scope of this book. An area of applications of max-plus linearity to queuing systems not covered in this monograph is the generalization of Lindley–type results for the GI/G/1 queue to max-plus linear queuing networks. For example, in [1, 2] Altman, Gaujal and Hordijk extend a result of Hajek [59] on admission control to a GI/G/1 queue to max-plus linear queuing networks. Furthermore, the focus of this monograph is on stochastic systems and we only briefly present the main results of the theory of deterministic max-plus systems. Readers particularly interested in deterministic theory are referred to [10] and the more recent book [65]. For this reason, network calculus, a min-plus based mathematical theory for analyzing the flow in deterministic queueing networks, is not covered either and readers interested in this approach are referred to [79]. Various approaches that are extensions of, or, closely related to max-plus algebra are not addressed in this monograph. Readers interested in min-max-plus systems are referred to [37, 72, 87, 98]. References on the theory of non–expansive maps are [43, 49, 58], and for **MM** functions we refer to [38, 39]. For applications of max-plus methods to control theory, we refer to [85].

Part II studies perturbation analysis of max-plus linear systems. Our approach to perturbation analysis of max-plus linear systems mirrors the hierarchical structure inherited by the structure of the problem. More precisely, the individual chapters will have the following internal structure:

Random variable level: We set off with carefully developing a concept of differentiation for random variables and distributions, respectively.

Matrix level: For the kind of applications we have in mind, the dynamic of a system is modeled by random matrices, the elements of which are (sums of) simple random variables. Our theory will provide sufficient conditions such that (higher–order) differentiability or analyticity of the elements of a matrix in the max-plus algebra implies (higher–order) differentiability or analyticity of the matrix itself.

System level: For (higher–order) differentiability or analyticity we then provide product rules, that is, we will establish conditions under which the (random) product (or sum) of differentiable (respectively, analytic) matrices is again differentiable (respectively, analytic). In other words, we establish sufficient conditions for (higher–order) differentiability or analyticity of the state–vector of max-plus linear systems.

Performance level: The concept of differentiability is such that it allows statements about (higher–order) derivatives or Taylor series expansions for a predefined class of performance functions applied to max-plus linear systems. We will work with a particular class of performance functions that covers many functions that are of interest in applications and that is most suitable to work with in a max-plus environment.

The reason for choosing this hierarchical approach to perturbation analysis is that we want to provide conditions for differentiability that are easy to check. One of the highlights of this approach is that we will show that if a particular service time in a max-plus linear queuing network is differentiable [random variable level], then the matrix modeling the network dynamic is differentiable [matrix level] and by virtue of our product rule of differentiation the state–vector of the system is differentiable [system level]. This fact can then be translated into expressions for the derivative of the expected value of the performance of the system measured by performance functions out of a predefined class [performance level]. We conclude our analysis with a study of Taylor series expansions of stationary characteristics of max-plus linear systems.

Part II is organized as follows. Chapter 3 introduces our concept of weak differentiation of measures, called $\mathcal{D}$–differentiation of measures. Using the algebraic properties of max-plus, we extend this concept to max-plus matrices and vectors and thereby establish a calculus of unbiased gradient estimators. In Chapter 4, we extend the $\mathcal{D}$–differentiation approach of Chapter 3 to higher–order derivatives. In Chapter 5 we turn our attention to Taylor series expansions of max-plus systems. This area of application of max-plus linearity has been initiated by Baccelli and Schmidt who showed in their pioneering paper [17] that waiting times in max-plus linear queuing networks with Poisson–λ–arrival stream can be obtained via Taylor expansions w.r.t. λ, see [15]. For certain classes of open queuing networks this yields a feasible way of calculating the waiting time distribution, see [71]. Concerning analyticity of closed networks, there are promising first results, see [7], but a general theory has still to be developed. We provide a unified approach to the aforementioned results on Taylor series expansions and new results will be established as well.

A reader interested in an introduction to stochastic max-plus linear systems will benefit from Part I of this book, whereas the reader interested in perturbation analysis, will benefit from Chapter 3 and Chapter 4, where the theory of $\mathcal{D}$-differentiation is developed. The full power of this method can be appreciated when studying Taylor series expansions, and we consider Chapter 5 the highlight of the book.

Notation and Conventions

This monograph covers two areas in applied probability that have been disjoint until now. Both areas (that of max-plus linear stochastic systems and that of perturbation analysis) have developed their own terminology independently. This has led to notational conventions that are sometimes not compatible. Throughout this monograph we stick to the established notation as much as possible. In two prominent cases, we even choose for ambiguity of notation in order to honor notational conventions. The first instance of ambiguity will be the symbol θ. More specifically, in ergodic theory of max-plus linear stochastic systems (in the first part of this monograph) the shift operator on the sample space Ω, traditionally denoted by θ, is the standard means for analysis. On the other hand, the parameter of interest in perturbation analysis is typically denoted by θ too, and we will follow this standard notation in the second part of the monograph. Fortunately, the shift operator is only used in the first part of the monograph and from the context it will always be clear which interpretation of θ is meant. The second instance of ambiguity will be the symbol λ. More specifically, for ergodic theory of max-plus linear stochastic systems we will denote by λ the Lyapunov exponent of the system and λ will also be used to denote the intensity of a given Poisson process. Both notations are classical and it will always be clear from the context which interpretation of λ is meant.

Throughout this monograph, we assume that an underlying probability space $(\Omega, \mathcal{A}, P)$ is given and that any random variable introduced is defined on $(\Omega, \mathcal{A}, P)$. Furthermore, we will use the standard abbreviation 'i.i.d.' for 'independent and identically distributed,' and 'a.s.' for 'almost surely.' To avoid an inflation of subscripts, we will suppress in Part II the subscript θ when this causes no confusion. In addition to that, we will write $\mathbb{E}_\theta$ in order to denote the expected value of a random variable evaluated at θ. Furthermore, let $a < b$, for $a, b \in \mathbb{R}$, and let $f : (a, b) \to \mathbb{R}$ be n times differentiable with respect to θ on (a, b), then we write $\frac{d^n}{d\theta^n}\big|_{\theta=\theta_0} f(\theta)$ for the nth derivative of f evaluated at θ_0. We will frequently work with the set $\mathbb{R} \cup \{-\infty\}$ and we introduce the following convention: for any $x \in \mathbb{R}$ we set $x + (-\infty) = -\infty + x = -\infty = -\infty - x$, $x - (-\infty) = \infty$, and $-\infty + (-\infty) = -\infty$ and $-\infty - (-\infty) = 0$.

Acknowledgements

This work was supported in parts by EC–TMR project ALAPEDES under Grant ERBFMRXCT960074 and in parts by Deutsche Forschungsgemeinschaft Grant He 3139/1-1.

The author wants to thank François Baccelli and Dohy Hong for many fruitful discussions. Furthermore, we are grateful to Bruno Gaujal for bringing the problem addressed in Remark 1.5.2 and Remark 1.5.3 in Section 1.5 to our attention. The presentation of max-plus ergodic theory benefitted from our discussions with Stephane Gaubert and Jean Mairesse. We are also grateful to Arie Hordijk and Haralambie Leahu for many insightful discussions on the topic of measure-valued differentiation. Finally, the author wants to thank TU Delft, EURANDOM, TU Eindhoven, and Vrije Universiteit Amsterdam for giving him the opportunity to work and continue working on this project during the various stages of his professional career.

The author   Amsterdam, The Netherlands, May 2006

Contents

II   Perturbation Analysis   117

3   A Max-Plus Differential Calculus   119

4   Higher-Order $\mathcal{D}$-Derivatives   151

5   Taylor Series Expansions   179

Part I

Max-Plus Algebra

Chapter 1

Max-Plus Linear Stochastic Systems

In this chapter we introduce max-plus algebra. The basic properties of max-plus algebra are discussed in Section 1.1. A first example of a max-plus linear system is presented in Section 1.2. A variant of the basic max-plus model best suited for studying the asymptotic behavior of max-plus linear systems is presented in Section 1.3. In general, the type of system that can be analyzed through max-plus techniques is best described in terms of Petri nets which are introduced in Section 1.4. To make the modeling aspects involved more transparent, we present in Section 1.5 a characterization of max-plus linearity in terms of queueing networks. This section also contains many examples of max-plus linear systems. Finally, we discuss in Section 1.6 properties of max-plus algebra that are of importance when max-plus linear recurrences are studied. In particular, we present approaches with which the growth rate of a max-plus linear system can be measured and we discuss various ways of making max-plus algebra a metric space.

1.1 The Max-Plus Algebra

In this section we introduce max-plus algebra. For an extensive discussion of the max-plus algebra and similar structures we refer to [10, 65]. An early reference is [37]. A historical overview on the beginnings of the max-plus theory can be found in [47].

Max-Plus algebra is usually introduced as follows. Let $\varepsilon = -\infty$, $e = 0$ and denote by $\mathbb{R}_{\max}$ the set $\mathbb{R} \cup \{\varepsilon\}$. For elements $a, b \in \mathbb{R}_{\max}$ we define the operations $\oplus$ and $\otimes$ by

$$a \oplus b = \max(a, b) \quad \text{and} \quad a \otimes b = a + b, \tag{1.1}$$

where we adopt the convention that for all $a \in \mathbb{R}$: $\max(a, -\infty) = \max(-\infty, a) = a$ and $a + (-\infty) = -\infty + a = -\infty$. The set $\mathbb{R}_{\max}$ together with the operations $\oplus$

and $\otimes$ is called max-plus algebra and is denoted by $\mathcal{R}_{\max} = (\mathbb{R}_{\max}, \oplus, \otimes, \varepsilon, e)$. In particular, ε is the neutral element for the operation $\oplus$ and *absorbing* for $\otimes$, that is, for all $a \in \mathbb{R}_{\max}$: $a \otimes \varepsilon = \varepsilon \otimes a = \varepsilon$. The neutral element for $\otimes$ is $e = 0$.

The max-plus algebra is an example of an algebraic structure, called *semiring*, that is introduced below.

Definition 1.1.1 *A semiring is a nonempty set R endowed with two binary relations, $\oplus_R$ and $\otimes_R$, so that $\oplus_R$ is associative and commutative with identity element ε_R; $\otimes_R$ distributes over $\oplus_R$, is associative, has identity element e_R and ε_R is absorbing for $\otimes_R$. Such a semiring is denoted by $\mathcal{R} = (R, \oplus_R, \otimes_R, \varepsilon_R, e_R)$. We call $\mathcal{R}$ commutative if $\otimes_R$ is commutative and we call it idempotent if $a \oplus_R a = a$ for all $a \in R$. To simplify the notation, the relation $\otimes_R$ precedes $\oplus_R$.*

The following example provides interpretations of $\mathcal{R}$ which are of interest in applications.

Example 1.1.1

- *If we identify $\oplus_R$ with conventional addition and $\otimes_R$ with conventional multiplication, then the neutral elements are $\varepsilon_R = 0$ and $e_R = 1$. We call $\mathcal{R}_{st} = (\mathbb{R}, \cdot, +, 0, 1)$ the standard model of $\mathcal{R}$. Since conventional multiplication is commutative, the standard model of $\mathcal{R}$ is a commutative semiring. Note that the standard model is not idempotent.*

- *The structure $\mathcal{R}_{\max}$ is an idempotent semiring and we call $\mathcal{R}_{\max}$ the max-plus model of $\mathcal{R}$. Note that $\otimes$ is commutative. Hence, the max-plus model of $\mathcal{R}$ is an idempotent, commutative semiring.*

- *In the same way as for the max-plus model of $\mathcal{R}$ we find the min-plus model $\mathcal{R}_{\min} = (\mathbb{R}_{\min} = \mathbb{R} \cup \{\infty\}, \min, +, \infty, 0)$ of $\mathcal{R}$. Note that $\mathcal{R}_{\min}$ is an idempotent, commutative semiring.*

- *Let S be a non-empty set. Denote the power set of S by R, then $(R, \cup, \cap, \emptyset, S)$ is a commutative, idempotent semiring.*

For more examples of semirings we refer to the excellent overview in [47]. To keep the notation simple, we will in the following suppress the subscript R when referring to a semi-ring.

An element $A \in R^{I \times J}$ is called *matrix* and its elements are denoted by A_{ij} for $1 \leq i \leq I$, $1 \leq j \leq J$. A matrix $A \in R^{I \times J}$ is called *regular* if A contains at least one element different from ε in each row. The *transpose* of a matrix A, denoted by $A^{\top}$, is defined in the usual way: $(A^{\top})_{ij} = A_{ji}$ for all i, j. For matrices $A \in R^{I \times J}$ and $B \in R^{J \times K}$ the matrix product $A \otimes B$ is defined in the usual way as follows:

$$(A \otimes B)_{ik} = \bigoplus_{j=1}^{J} A_{ij} \otimes B_{jk} \overset{\text{def}}{=} A_{i1} \otimes B_{1k} \oplus \cdots \oplus A_{iJ} \otimes B_{Jk}, \qquad (1.2)$$

for $1 \leq i \leq I$, $1 \leq k \leq K$, and for $A(k) \in R^{J \times J}$, $1 \leq k \leq m$,

$$\bigotimes_{k=1}^{m} A(k) \stackrel{\text{def}}{=} A(m) \otimes A(m-1) \otimes \cdots \otimes A(1) .$$

Specifically, for $A \in R^{I \times J}$ and $\alpha \in R$, scalar multiplication is defined by

$$(\alpha \otimes A)_{ij} = \alpha \otimes A_{ij} , \tag{1.3}$$

for $1 \leq i \leq I$, $1 \leq j \leq J$. Addition of matrices $A \in R^{I \times J}$ and $B \in R^{I \times J}$, denoted by $A \oplus B$, is defined through

$$(A \oplus B)_{ij} = A_{ij} \oplus B_{ij} , \tag{1.4}$$

for $1 \leq i \leq I$, $1 \leq j \leq J$, and for $A(k) \in R^{I \times J}$, $1 \leq k \leq m$,

$$\bigoplus_{k=1}^{m} A(k) \stackrel{\text{def}}{=} A(m) \oplus A(m-1) \oplus \cdots \oplus A(1) .$$

Let $\mathcal{E}(I, J)$ denote the $I \times J$ matrix with all elements equal to ε and $E(I, J)$ the matrix with e on the diagonal and ε elsewhere. For $R^{I \times J}$, the $\oplus$-sum, as defined in (1.4), is associative, commutative and has zero element $\mathcal{E}(I, J)$, and for $R^{J \times J}$ the $\otimes$-product is associative, distributive with respect to $\oplus$, has identity element $E(J, J)$ and $\mathcal{E}(J, J)$ is absorbing for $\otimes$. Idempotent semirings are called *dioids* in [10]. Note that if $\oplus$ is idempotent, then the addition of matrices in (1.4) is idempotent. Thus, if $\oplus$ is idempotent, then $\mathcal{R}^{J \times J} = (R^{J \times J}, \oplus, \otimes, \mathcal{E}(J, J), E(J, J))$ is a dioid. For example, both the max-plus model and the min-plus model of $\mathcal{R}$ are dioids, cf. Example 1.1.1. Observe that generally $\mathcal{R}^{J \times J}$ fails to be commutative even if $\mathcal{R}$ is a commutative semiring, which is due to the definition of the matrix product in (1.2).

Remark 1.1.1 *The elements of $R^J \stackrel{\text{def}}{=} R^{J \times 1}$ are called vectors. In the following we will carefully distinguish R^J (the set of J-dimensional vectors in R), $R^{I \times J}$ (the set of $I \times J$-dimensional matrices in R), and $R^{J \times J}$ (the set of square matrices in R). Note that for $A \in R^{I \times J}$ and $x \in R^J$ the product $A \otimes x$ is defined in (1.2), whereas $A \otimes A$ is only defined for $A \in R^{J \times J}$.*

For the kind of applications we will study in this monograph, we focus on the max-plus semiring. Roughly speaking $\mathcal{R}_{\max}$ is used to model departure times, called *daters*, in a class of discrete event dynamic systems which will be introduced in Section 1.4 and Section 1.5. In what follows formulas have thus to be interpreted in max-plus algebra.

Let A be a $J \times J$ dimensional matrix. We denote the *communication graph* of A by $\mathcal{G}(A) = (\mathcal{N}(A), \mathcal{D}(A))$, where $\mathcal{N}(A) = \{1, \ldots, J\}$ denotes the set of nodes and $\mathcal{D}(A) \subset \{1, \ldots, J\} \times \{1, \ldots, J\}$ the set of arcs where $(i, j) \in \mathcal{D}(A)$ if and only if $A_{ji} \neq \varepsilon$. For any two nodes i, j, a sequence of arcs $\rho = ((i_n, j_n) : 1 \leq n \leq m)$, so that $i = i_1$, $j_n = i_{n+1}$ for $1 \leq n < m$ and $j_m = j$, is called a *path* from

i to j. In case $i = j$, ρ is also called a *circuit*. For any arc (i, j) in $\mathcal{G}(A)$, we call A_{ji} the *weight* of arc (i, j). The *weight of a path* in $\mathcal{G}(A)$ is defined by the sum of the weights of all arcs constituting the path; more formally, let $\rho = ((i_n, j_n) : 1 \leq n \leq m)$ be a path from i to j of length m, then the weight of ρ, denoted by $|\rho|_w$, is given by

$$|\rho|_w = \sum_{n=1}^{m} A_{j_n i_n} = \bigotimes_{n=1}^{m} A_{j_n i_n}.$$

Let A^n denote the nth power of A, or, more formally, set $A(k) = A$, for $1 \leq k \leq n$, and

$$A^n \overset{\text{def}}{=} A^{\otimes n} = \bigotimes_{k=1}^{n} A(k), \tag{1.5}$$

where $A^0 = E$. With these definitions it can be shown that A_{ji}^n is equal to the maximal weight of paths of length n (that is, consisting of n arcs) from node i to node j, and $A_{ji}^n = \varepsilon$ refers to the fact that there is no path of length n from i to j, see [10] or [37].

Some remarks on the particularities of max-plus algebra seem to be in order here. Idempotency of $\oplus$ implies that $\oplus$ has no inverse. Indeed, if $a \neq \varepsilon$ had an inverse element, say b, w.r.t. $\oplus$, then $a \oplus b = \varepsilon$ would imply $a \oplus a \oplus b = a \oplus \varepsilon$. By idempotency, the left-hand side equals $a \oplus b$, whereas the right-hand side is equal to a. Hence, we have $a \oplus b = a$, which contradicts $a \oplus b = \varepsilon$. For more details on idempotency, see [43]. For this reason, $\mathcal{R}_{\max}$ is by no means an algebra in the classical sense. The name 'max-plus algebra' is only historically justified and the correct name for $\mathcal{R}_{\max}$ would be 'idempotent semiring' or 'dioid' (which may explain why the name 'max-plus algebra' is still predominant in the literature). The structure $\mathcal{R}_{\max}$ is richer than that of a dioid since $\otimes$ is commutative and has an inverse. However, in what follows we will work with matrices in $\mathbb{R}_{\max}$ and thereby lose, like in conventional algebra, commutativity and general invertibility of the product.

In the following we will study matrix-vector recurrence relations defined over a semiring. With respect to applications this means that we study systems whose dynamic can be described in such a way that the state-vector of the system, denoted by $x(k)$, follows the linear recurrence relation

$$x(k + 1) = A(k) \otimes x(k) \oplus B(k), \quad k \geq 0,$$

with $x(0) = x_0$, where $\{A(k)\}$ is a sequence of matrices and $\{B(k)\}$ a sequence of vectors of appropriate size. The above recurrence relation is said to be *inhomogeneous*. As we will see below, many systems can be described by *homogeneous* recurrence relation of type

$$x(k + 1) = A(k) \otimes x(k), \quad k \geq 0,$$

with $x(0) = x_0$, where $\{A(k)\}$ is a sequence of square matrices. See Section 1.4.3 for more details. As explained above, examples of this kind of systems are conventional linear systems, that is, $\otimes$ represents conventional matrix-vector multiplication and $\oplus$ conventional addition of vectors, max-plus linear and min-plus linear systems. Since max-plus linear systems are of most interest in applications, we will work with max-plus algebra in the remainder of this book. Hence, the basic operations ⊕ and ⊗ are defined as in (1.1) and extended to matrix operations as explained in (1.4) and (1.2).

1.2 Heap of Pieces

In this section, we present a first example of a max-plus linear system. The type of system studied in this section is called *heap models*. In a heap model, solid blocks are piled up according to a 'Tetris game' mechanism. More specifically, consider the blocks, labeled 'α', 'β' and 'γ', in Figure 1.1 to Figure 1.3.

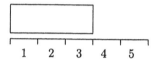

Figure 1.1: Block α

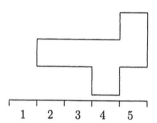

Figure 1.2: Block β

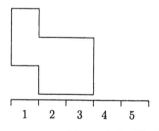

Figure 1.3: Block γ

The blocks occupy columns out of a finite set of columns $\mathcal{R}$, in our example given by the set $\{1, 2, \ldots, 5\}$. When we pile these blocks up according to a fixed sequence, like, for example, 'α β α γ β', this results in the heap shown in Figure 1.4. Situations like the one pictured in Figure 1.4 typically arise in scheduling problems. Here, blocks represent tasks that compete for a limited number of resources, represented by the columns. The extent of an individual block over a particular column can be interpreted as the time required by the task of this resource. See [28, 51, 50] for more on applications of heap models in scheduling.

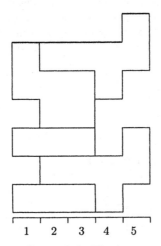

Figure 1.4: The heap $w = \alpha\,\beta\,\alpha\,\gamma\,\beta$.

Before we can continue, we have to introduce some notation. Let $\mathcal{A}$ denote the finite set of blocks, in our example $\mathcal{A} = \{\alpha, \beta, \gamma\}$. We call a sequence of blocks out of $\mathcal{A}$ a *heap*. For example, $w = \alpha\,\beta\,\alpha\,\gamma\,\beta$ is a heap. We denote the *upper contour* of a heap w by a vector $x_{\mathcal{H}}(w) \in \mathbb{R}^{\mathcal{R}}$, where $(x_{\mathcal{H}}(w))_r$ is the height of the heap on column r, for example, $x_{\mathcal{H}}(\alpha\,\beta\,\alpha\,\gamma\,\beta) = (3, 4, 4, 3, 3)$, where we started from ground level. The upper contour of the heap $\alpha\,\beta\,\alpha\,\gamma\,\beta$ is indicated by the boldfaced line in Figure 1.4.

A piece $a \in \mathcal{A}$ is characterized by its lower contour, denoted by $l(a)$, and its upper contour, denoted by $u(a)$. Denote by $\mathcal{R}(a)$ the set of resources required by a. The upper and lower contour of a piece a enjoy the following properties: $l(a), u(a) \in \mathbb{R}_{\max}^{\mathcal{R}}$, $l_r(a) \leq u_r(a)$ for $r \in \mathcal{R}(a)$, and $l_r(a) = u_r(a) = -\varepsilon$ for $r \notin \mathcal{R}(a)$. We associate a matrix $M(a)$ with piece a through

$$(M(a))_{rs} = \begin{cases} 0 & \text{for } s = r,\, r \notin \mathcal{R}(a)\,, \\ u_r(a) - l_s(a) & \text{for } r, s \in \mathcal{R}(a)\,, \\ -\infty & \text{otherwise.} \end{cases}$$

The matrices corresponding to the blocks α, β and γ are as follows:

$$M(\alpha) = \begin{pmatrix} 1 & 1 & 1 & -\infty & -\infty \\ 1 & 1 & 1 & -\infty & -\infty \\ -\infty & -\infty & 0 & -\infty & -\infty \\ -\infty & -\infty & -\infty & 0 & -\infty \\ -\infty & -\infty & -\infty & -\infty & 0 \end{pmatrix},$$

$$M(\beta) = \begin{pmatrix} 0 & -\infty & -\infty & -\infty & -\infty \\ -\infty & 1 & 1 & 2 & 1 \\ -\infty & 1 & 1 & 2 & 1 \\ -\infty & 1 & 1 & 2 & 1 \\ -\infty & 2 & 2 & 3 & 2 \end{pmatrix}$$

and

$$M(\gamma) = \begin{pmatrix} 2 & 3 & 3 & -\infty & -\infty \\ 1 & 2 & 2 & -\infty & -\infty \\ 1 & 2 & 2 & -\infty & -\infty \\ -\infty & -\infty & -\infty & 0 & -\infty \\ -\infty & -\infty & -\infty & -\infty & 0 \end{pmatrix}.$$

For a heap w and a block $\eta \in \mathcal{A}$, we write $w\,\eta$ for the heap constituted out of piling block η on heap w. It is easily checked that the upper contour follows the recurrence relation:

$$(x_{\mathcal{H}}(w\,\eta))_r = \max\{(M(\eta))_{rs} + (x_{\mathcal{H}}(w))_s : s \in \mathcal{R}\},$$

with initial contour $x_{\mathcal{H}}(\emptyset) = (0, \ldots, 0)$. Elaborating on the notational power of the max-plus semiring, the above recurrence relation reads:

$$(x_{\mathcal{H}}(w\,\eta))_r = \bigoplus_{s \in \mathcal{R}} (M(\eta))_{rs} \otimes (x_{\mathcal{H}}(w))_s, \quad r \in \mathcal{R},$$

or, in a more concise way,

$$x_{\mathcal{H}}(w\,\eta) = M(\eta) \otimes x_{\mathcal{H}}(w).$$

For a given sequence η_k, $k \in \mathbb{N}$, of pieces, set $x_{\mathcal{H}}(k) = x_{\mathcal{H}}(\eta_1\,\eta_2 \cdots \eta_k)$ and $M(k) = M(\eta_k)$. The upper contour follows the recursive relation:

$$x_{\mathcal{H}}(k+1) = M(k) \otimes x_{\mathcal{H}}(k), \quad k \geq 0,$$

where $x_{\mathcal{H}}(0) = (0, \ldots, 0)$. For given schedule η_k, $k \in \mathbb{N}$, the *asymptotic growth rate* of the heap model, given by

$$\lim_{k \to \infty} \frac{1}{k} x_{\mathcal{H}}(k),$$

provided that the limit exists, describes the speed or efficiency of the schedule. Limits of the above type are studied in Chapter 2.

1.3   The Projective Space

On $\mathbb{R}_{max}^J$ we introduce an equivalence relation, denoted by $\cong$, as follows: for $Y, Z \in \mathbb{R}_{max}^J$, $Y \cong Z$ if and only if there is an $\alpha \in \mathbb{R}$ so that $Y = \alpha \otimes Z$, that is, $Y_i = \alpha + Z_i$, $1 \leq i \leq J$, or, in a more concise way,

$$Y \cong Z \quad \Longleftrightarrow \quad \exists \alpha \in \mathbb{R} : \quad Y = \alpha \otimes Z.$$

If $Y \cong Z$, we say that Y and Z are *linear dependent*, and if $Y \not\cong Z$, we say that Y and Z are *linear independent*. For example, $(1,0)^\top \cong (0,-1)^\top$ and the vectors $(1,0)^\top$ and $(0,-1)^\top$ are linear dependent; and $(1,0)^\top \not\cong (0,0)^\top$ which implies that the vectors $(1,0)^\top$ and $(0,0)^\top$ are linear independent.

For $Z \in \mathbb{R}_{max}^J$, we write $\overline{Z}$ for the set $\{Y \in \mathbb{R}_{max}^J : Y \cong Z\}$. Let $\mathbb{PR}_{max}^J$ denote the quotient space of $\mathbb{R}_{max}^J$ by equivalence relation $\cong$, or, more formally,

$$\mathbb{PR}_{max}^J = \{\overline{Z} : Z \in \mathbb{R}_{max}^J\}.$$

$\mathbb{PR}_{max}^J$ is called the *projective space* of $\mathbb{R}_{max}^J$ with respect to $\cong$. The bar-operator is the canonical projection of $\mathbb{R}_{max}^J$ onto $\mathbb{PR}_{max}^J$. In the same vein, we denote by $\mathbb{PR}^J$ the quotient space of $\mathbb{R}^J$ by the above equivalence relation.

For $x \in \mathbb{R}^J$, set $z(x) \stackrel{\text{def}}{=} (0, x_2 - x_1, x_3 - x_1, \ldots, x_J - x_1)^\top$. For example, $z((2,3,1)^\top) = (0,1,-1)^\top$. Consider $\overline{x} \in \mathbb{PR}^J$ with $x \in \mathbb{R}^J$. Then, $z(x)$ lies in $\overline{x}$, which stems from the fact that $x_1 \otimes z(x) = x$. Moreover, for any vectors $u, v \in \overline{x}$ it holds that $z(u) = z(v)$, which can be expressed by saying that z maps $\overline{x}$ onto a single element of $\mathbb{R}^J$ the first component of which is equal to zero. We may thus disregard the first element and set $\overline{z}(x) = (x_2 - x_1, x_3 - x_1, \ldots, x_J - x_1)^\top$. For example, $\overline{z}((2,3,1)^\top) = (1,-1)^\top$. Hence, $\overline{z}(\cdot)$ identifies any element in $\mathbb{PR}^J$ with an element in $\mathbb{R}^{J-1}$.

1.4   Petri Nets

In this section, we study discrete event systems whose sample path dynamic can be modeled by max-plus algebra. Section 1.4.1 introduces the modeling tool of Petri nets. In Section 1.4.2, we discuss max-plus linear recurrence relations for so called autonomous and non-autonomous Petri nets. In Section 1.4.3, we explain the relation between autonomous and non-autonomous representations of a discrete event system, such as, for example, a queueing network, and an algebraic property, called irreducibility, of the max-plus model. Eventually, Section 1.4.4 discusses some particular issues that arise when dealing with waiting times in non-autonomous systems.

1.4.1   Basic Definitions

Max-Plus algebra allows one to describe the dynamics of a class of networks, called *stochastic event graphs*, via vectorial equations. Before we are able to give a precise definition of an event graph, we have to provide a brief introduction to Petri nets. A Petri net is denoted by $\mathcal{G} = (\mathcal{P}, \mathcal{Q}, \mathcal{F}, \mathcal{M}_0)$, where

$\mathcal{P} = \{p_1, \dots, p_{|\mathcal{P}|}\}$ is the set of places, $\mathcal{Q} = \{q_1, \dots, q_{|\mathcal{Q}|}\}$ is the set of transitions (also called *nodes* for event graphs), $\mathcal{F} \subset \mathcal{Q} \times \mathcal{P} \cup \mathcal{P} \times \mathcal{Q}$ is the set of arcs and $\mathcal{M}_0 : \mathcal{P} \to \{0, 1, \dots, M\}^{|\mathcal{P}|}$ is the initial number of tokens in each place, called *initial marking*; M is called the *maximal marking*. For $(p_i, q_j) \in \mathcal{F}$ we say that p_i is an upstream place for q_j, and for $(q_j, p_i) \in \mathcal{F}$ we say that p_i is a downstream place for q_j and call q_j an upstream transition of p_i. We denote the set of all upstream places of transition j by $\pi^q(j)$, i.e., $i \in \pi^q(j)$ if and only if $(p_i, q_j) \in \mathcal{F}$, and the set of all upstream transitions of place i by $\pi^p(i)$, i.e., $j \in \pi^p(i)$ if and only if $(q_j, p_i) \in \mathcal{F}$. We denote by $\pi^{j\,l}$ the set of places having downstream transition q_j and upstream transition q_l.

Roughly speaking, places represent conditions and transitions represent events. A certain transition (that is, event) has a certain number of input and output places representing the pre-conditions and post-conditions of the event. The presence of a token in a place is interpreted as the condition associated with the place being fulfilled. In another interpretation, m_i tokens are put into a place p_i to indicate that m_i data items or resources are available. If a token represents data, then a typical example of transitions is a computation step for which these data are needed as an input. The marking of a Petri net is identified with the state. Changes occur according to the following rules: (1) a transition is said to be *enabled* if each upstream place contains at least one token, (2) a *firing* of an enabled transition removes one token from each of its upstream places and adds one token to each of its downstream places. A transition without predecessor(s) is called *source transition* or simply *source*. Similarly, a transition which does not have successor(s) is called *sink transition* or simply *sink*. A source transition is an input of the network, a sink transition is an output of the network. If there are no sources in the network, then we talk about an *autonomous* network and we call it *nonautonomous* otherwise. It is assumed that only transitions can be sources or sinks (which is no loss of generality, since one can always add a transition upstream or downstream to a place if necessary).

A Petri net is called an *event graph* if each place has exactly one upstream and one downstream transition, that is, for all $i \in \mathcal{P}$ it holds $|\pi^p(i)| = 1$ and $|\{j \in \mathcal{Q} : i \in \pi^q(j)\}| = 1$. Event graphs are sometimes also referred to as *marked graphs* or *decision free Petri nets*. Typical examples are the G/G/1-queue, networks of (finite) queues in tandem, Kanban systems, flexible manufacturing systems, fork/join queues or any parallel and/or series composition made by these elements.

The original theory of Petri nets deals with the ordering of events, and questions pertaining to when events take place are not addressed. However, for questions related to performance evaluation it is necessary to introduce time. This can be done in two basic ways by associating durations with either transition firings or with the sojourn times of tokens in places.

The *firing time* of a transition is the time that elapses between the starting and the completion of the firing of the transition. We adopt the convention that the tokens that are to be consumed by the transition remain in the preceding places during the firing time. Such tokens are called *reserved* tokens. Firing times can be used to represent production times in a manufacturing environment, where transitions represent machines, the length of a code in a computer science setting etc.

The *holding time* of a place is the time a token must spend in the place before contributing to the enabling of the downstream transitions. Firing times represent the actual time it takes to fire a transition, whereas holding times can be viewed as minimal time tokens have to spend in places. In practical situations, both types of durations may be present. However, it can be shown that for event graphs one can disregard durations associated with transitions without loss of generality (or vice versa). In what follows we associate durations with places and assume that the firing of transitions consumes no time.

A Petri net is said to be *timed* if such durations are given as data associated with the network. If these times are random variables defined on a common probability space, then we call the Petri net a *stochastic Petri net*.

A place p_i is said to be *first in first out* (FIFO) if the kth token to enter this place is also the kth token which becomes available in this place. In the same way, we call a transition q_j FIFO if the kth firing of q_j to start is also the kth firing to complete. If all places and transitions are FIFO, then the Petri net is said to be FIFO.

1.4.2 The Max-Plus Recursion for Firing Times

In what follows we study (stochastic) FIFO event graphs. We discuss the autonomous case in Section 1.4.2.1 and the non-autonomous case in Section 1.4.2.2.

1.4.2.1 The Autonomous Case

Let $\sigma_i(k)$ denote the kth holding time incurred by place p_i and let $X_j(k)$ denote the time when transition j fires for the kth time. We take the vector $X(k) = (X_1(k), \ldots, X_{|\mathcal{Q}|}(k))$ as state of the system.

To any stochastic event graph, we can associate matrices $A(0, k), \ldots, A(M, k)$, all of size $|\mathcal{Q}| \times |\mathcal{Q}|$, given by

$$(A(m,k))_{jl} = \bigoplus_{\{i \in \pi^{j\,l} \mid \mathcal{M}_0(i) = m\}} \sigma_i(k), \tag{1.6}$$

for $j, l \in \mathcal{Q}$, and in case the set on the right-hand side is empty, we set $(A(m,k))_{jl} = \varepsilon$. In other words, to obtain $(A(m,k))_{jl}$ we consider all places with downstream transition q_j and upstream transition q_l with initially m tokens, and we take as $(A(m,k))_{jl}$ the maximum of the kth holding time of these places.

If we consider the state variables $X_i(k)$, which denote the kth time transition i initiates firing, then the vector $X(k) = (X_1(k), \ldots, X_{|\mathcal{Q}|}(k))$ satisfies the following (linear) equation:

$$X(k) = A(0,k) \otimes X(k) \oplus A(1,k) \otimes X(k-1) \oplus \tag{1.7}$$
$$\cdots \oplus A(M,k) \otimes X(k-M),$$

see Corollary 2.62 in [10].

A Petri net is said to be *live* (for the initial marking $\mathcal{M}_0$) if for each marking $\mathcal{M}$ reachable from $\mathcal{M}_0$ and for each transition q, there exists a marking $\mathcal{N}$ which is reachable from $\mathcal{M}$ such that q is enabled in $\mathcal{N}$. For a live Petri net, any arbitrary transition can be fired an infinite number of times. A Petri net that is not live is called *deadlocked*. An event graph is live if and only if there exists a permutation P of the coordinates so that the matrix $P^T \otimes A(0, k) \otimes P$ is strictly lower triangular for all k.

We define the formal power series of $A(0, k)$ by

$$A^*(0, k) \stackrel{\text{def}}{=} \bigoplus_{i=0}^{\infty} A^i(0, k) .$$

If the event graph is live, then $A(0, k)$ is (up to a permutation) a lower triangular matrix, and a finite number p exists, such that

$$A^*(0, k) = \bigoplus_{i=0}^{p} A^i(0, k) . \tag{1.8}$$

Set

$$b(k) = \bigoplus_{i=1}^{M} A(i, k) \otimes X(k - i) ,$$

then (1.7) reduces to

$$X(k) = A(0, k) \otimes X(k) \oplus b(k) . \tag{1.9}$$

For fixed k, the above equation is of type $x = A \otimes x \oplus b$. It is well-known that $A^* \otimes b$ solves this equation, see Theorem 3.17 in [10] or Theorem 2.10 in [65]. Therefore, $X(k)$ can be written

$$X(k) = A^*(0, k) \otimes b(k) ,$$

or, more explicitly,

$$X(k) = A^*(0, k) \otimes A(1, k) \otimes X(k-1) \oplus \cdots \oplus A^*(0, k) \otimes A(M, k) \otimes X(k-M). \tag{1.10}$$

The difference between (1.7) and (1.10) is that the latter contains no 0th order recurrence relation, that is, $X(k)$ occurs only on the left-hand side of the equation.

As a next step we transform (1.10) into a first-order recurrence relation. In order to do so, we take as new state vector the $(|\mathcal{Q}| \times M)$-dimensional vector

$$x(k) = (X(k), X(k - 1), \ldots, X(k - M + 1))^T \tag{1.11}$$

and $(|\mathcal{Q}| \times M) \times (|\mathcal{Q}| \times M)$-dimensional matrices

$$
A(k{-}1) = \begin{pmatrix} A^*(0,k) \otimes A(1,k) & A^*(0,k) \otimes A(2,k) & \cdots \cdots & A^*(0,k) \otimes A(M,k) \\ E & \mathcal{E} & \cdots \cdots & \mathcal{E} \\ \mathcal{E} & E & \ddots & \mathcal{E} \\ \vdots & & \ddots & \vdots \\ \mathcal{E} & \mathcal{E} & \cdots \ E & \mathcal{E} \end{pmatrix}.
$$

$$(1.12)$$

Then (1.10) can be written as

$$
x(k) = A(k-1) \otimes x(k-1), \quad k \geq 1,
$$

or, equivalently,

$$
x(k+1) = A(k) \otimes x(k), \quad k \geq 0.
$$

We call the above equation the *standard autonomous equation*. Any live FIFO autonomous event graph can be modeled by a standard autonomous equation.

1.4.2.2   The Non-Autonomous Case

Let $\mathcal{I} \subset \mathcal{Q}$ denote the set of input transitions, set $\mathcal{Q}' = \mathcal{Q} \setminus \mathcal{I}$, and denote the maximal initial marking of the input places by M'. We let $\alpha_i(k)$ denote the kth firing time of input transition q_i. We now define $|\mathcal{Q}'| \times |\mathcal{I}|$ dimensional matrices $B(0,k), \ldots, B(M',k)$, so that

$$
(B(m,k))_{jl} = \bigoplus_{\{i \in \pi^j{}^l \mid \mathcal{M}_0(i)=m\}} \sigma_i(k),
$$

for $j \in \mathcal{Q}'$ and $l \in \mathcal{I}$, and in case the set on the right-hand side is empty, we set $(B(m,k))_{jl} = \varepsilon$. In words, to obtain $(B(m,k))_{jl}$ we consider all places with downstream transition q_j (being not an input transition) and upstream transition q_l (being an input transition) with initially m tokens. We take as $(B(m,k))_{jl}$ the maximum of the kth holding time of these places. Furthermore, we let $U(k)$ be a $|\mathcal{I}|$-dimensional vector, where $U_i(k)$ denotes the time of the kth firing of input transition i.

The vector of the kth firing times satisfies the following (linear) equation:

$$
\begin{aligned}
X(k) = {}& A(0,k) \otimes X(k) \oplus A(1,k) \otimes X(k-1) \oplus \cdots \oplus A(M,k) \otimes X(k-M) \\
& \oplus B(0,k) \otimes U(k) \oplus B(1,k) \otimes U(k-1) \oplus \cdots \\
& \cdots \oplus B(M',k) \otimes U(k-M'),
\end{aligned}
$$

$$(1.13)$$

where $X_j(k)$ and $U_j(k)$ are ε if $k \leq 0$, see Theorem 2.80 in [10]. Note that $X(k)$ is the vector of kth firing times of transitions q_i with $i \in \mathcal{Q}'$. Put differently, $X(k)$ models the firing times of all transitions which are not input transitions.

In what follows, we say that the non-autonomous event graph is live if the associated autonomous event graph is live (that is, if the event graph obtained from the non-autonomous one through deleting all input transitions is live). From now on we restrict ourselves to non-autonomous event graphs that are live.

Equation (1.13) is equivalent to

$$
\begin{aligned}
X(k) = A^*(0,k) \otimes A(1,k) \otimes X(k-1) \ \oplus \ \cdots \oplus A^*(0,k) \otimes A(M,k) \otimes X(k-M) \\
\oplus A^*(0,k) \otimes B(0,k) \otimes U(k) \ \oplus \ \cdots \\
\cdots \oplus A^*(0,k) \otimes B(M',k) \otimes U(k-M') \,,
\end{aligned}
\tag{1.14}
$$

compare recurrence relation (1.10) for the autonomous case; and we define $x(k)$ like in (1.11).

Define the $(|\mathcal{I}| \times (M'+1))$-dimensional vector

$$
u(k) = (U(k), U(k-1), \ldots, U(k-M'))^T
$$

and the $(|\mathcal{Q}'| \times M) \times (|\mathcal{I}| \times (M'+1))$ matrix

$$
B(k-1) = \begin{pmatrix}
A^*(0,k) \otimes B(0,k) & A^*(0,k) \otimes B(1,k) & \cdots & A^*(0,k) \otimes B(M',k) \\
\mathcal{E} & \mathcal{E} & \cdots & \mathcal{E} \\
\vdots & \vdots & \cdots & \vdots \\
\mathcal{E} & \mathcal{E} & \cdots & \mathcal{E}
\end{pmatrix} .
$$

Then (1.14) can be written as

$$
x(k) = A(k-1) \otimes x(k-1) \oplus B(k-1) \otimes u(k) \,, \quad k \geq 1 \,,
$$

with $A(k-1)$ as defined in (1.12) or, equivalently,

$$
x(k+1) = A(k) \otimes x(k) \oplus B(k) \otimes u(k+1) \,, \quad k \geq 0 \,.
\tag{1.15}
$$

We call the above equation the *standard non-autonomous equation*. Any live FIFO non-autonomous event graph can be modeled by a standard non-autonomous equation.

1.4.3 Autonomous Systems and Irreducible Matrices

So far we have distinguished two types of max-plus recurrence relations for firing times in event graphs: homogeneous recurrence relations of type

$$
x(k+1) = A(k) \otimes x(k)
\tag{1.16}
$$

that describe the firing times in an autonomous event graph and inhomogeneous recurrence relations of type

$$
x(k+1) = A(k) \otimes x(k) \oplus B(k) \otimes u(k+1)
\tag{1.17}
$$

that describe the firing times in a non-autonomous event graph.

In principle, the sample dynamic of a max-plus linear discrete event system can be modeled either by a homogeneous or an inhomogeneous recurrence relation. Indeed, recurrence relation (1.17) is easily transformed into a recurrence relation of type (1.16). To see this, assume, for the sake of simplicity, that there is only one input transition, i.e., $|I| = 1$, and that the initial marking of this input place is one, i.e., $M' = 1$. This implies, that $u(k)$ is a scalar, and denoting the kth firing time of the source transition by $\sigma_0(k)$ it holds that

$$u(k) = \sum_{i=1}^{k} \sigma_0(i).$$

In order to do transform (1.17) into an inhomogeneous equation, set

$$\tilde{x}(k) = \begin{pmatrix} u(k) \\ x(k) \end{pmatrix}$$

and

$$\tilde{A}(k) = \begin{pmatrix} \sigma_0(k+1) & \varepsilon \\ B(k) \otimes \sigma_0(k+1) & A(k) \end{pmatrix}.$$

Then, it is immediate that (1.17) can be rewritten as

$$\tilde{x}(k+1) = \tilde{A}(k) \otimes \tilde{x}(k). \tag{1.18}$$

This transformation is tantamount to viewing the input transition as a recycled transition where the holding times of the recycling place are given by the sequence $\sigma_0(k)$. In the following we study the difference between (1.17) and (1.18) more closely, which leads to the important notion of *irreducibility*.

We call a matrix $A \in \mathbb{R}_{max}^{J \times J}$ *irreducible* if its communication graph $\mathcal{G}(A)$ is strongly connected, that is, if for any two nodes i, j there is a sequence of arcs $((i_n, j_n) : 1 \le n \le m)$ so that $i = i_1$, $j_n = i_{n+1}$ for $1 \le n < m$ and $j_m = j$. This definition is equivalent to the definition of irreducibility that is predominant in algebra, namely, that a matrix is called irreducible if no permutation matrix P exists, such that $P^T \otimes A \otimes P$ has an upper triangular block structure, see [10] for more details. If a matrix is not irreducible, it is called *reducible*.

Remark 1.4.1 *If A is irreducible, then every row of A contains at least one finite element. In other words, an irreducible matrix is regular.*

The relation between the (algebraic) type of recurrence relation (1.18) and the type of system modeled can now be phrased as follows: If $\tilde{A}(k)$ is irreducible, then $\tilde{x}(k)$ models the sample path dynamic of an autonomous system, or, in terms of queueing, that of a closed network; see Section 1.5 for a description of closed queueing systems. If, on the other hand, $\tilde{A}(k)$ is of the above particular form (and thus not irreducible), then $\tilde{x}(k)$ models the sample path dynamic of an non-autonomous systems, or, in terms of queueing, that of an open queueing system; see Section 1.5 for a description of open queueing systems. Hence, a homogeneous equation can model either an autonomous or a non-autonomous system. However, given that $A(k)$ is irreducible, homogeneous equations are related to non-autonomous systems.

In order to define irreducibility for random matrices, we introduce the concept of fixed support of a matrix.

Definition 1.4.1 *We say that $A(k)$ has fixed support if the probability that $(A(k))_{ij}$ equals ε is either 0 or 1 and does not depend on k.*

With the definition of fixed support at hand, we say that a random matrix A is *irreducible* if (a) it has fixed support and (b) it is irreducible with probability one. For random matrices, irreducibility thus implies fixed support.

The following lemma establishes an important consequence of the irreducibility of a (random) matrix: there exists a power of the matrix such that all entries are different from ε.

Lemma 1.4.1 *Let $A(k) \in \mathbb{R}_{\max}^{J \times J}$, for $k \geq 0$, be irreducible such that (i) all finite elements are bounded from below by some finite constant δ and (ii) all diagonal elements are different from ε. Then,*

$$G(k) \stackrel{\text{def}}{=} \bigotimes_{j=k-J}^{k-1} A(j), \; \text{for } k \geq J,$$

satisfies $(G(k))_{ij} \geq J \cdot \delta$ for all $(i, j) \in J \times J$.

Proof: Without loss of generality assume that $\delta = 0$. Let $A_{ij} = 0$ if $A_{ij}(k) \neq \varepsilon$ with probability one and $A_{ij} = \varepsilon$ otherwise. For the proof of the lemma it suffices to show that $A_{ij}^J \neq \varepsilon$ for any i, j. Because $A(k)$ is irreducible, so is A. Hence, for any node i, j there exists a number m_{ij}, such that there is a path of length m_{ij} from i to j in the communication graph of A. Such a path contains each arc at most once and is hence of maximal length J. We have thus shown that for any i, j a $m_{ij} \leq J$ exists such that $(A^{m_{ij}})_{ij} \neq \varepsilon$. Since all diagonal elements of A are different from ε, this yields

$$\forall n \geq m_{ij} : \quad (A^n)_{ij} \neq \varepsilon,$$

for any i, j. Indeed, we can add arbitrarily many recycling loops (i, i) to the path. Using the fact that $\max(m_{ij} : i, j) \leq J$, completes the proof of the lemma. $\square$

1.4.4 The Max-Plus Recursion for Waiting Times in Non-Autonomous Event Graphs

We consider a non-autonomous event graph with one source transition denoted by q_0. We furthermore assume that the initial marking of this source is equal to one and that the maximal marking of the input place of the source transition equals one as well. For each transition q in $\mathcal{G}$ we consider the set $P(q)$ of all paths from q_0 to q. We denote by

$$M(\pi) = \sum_{p \in \pi} \mathcal{M}_0(p)$$

the total number of all initial tokens on path π, and set

$$L(q) = \min_{\pi \in P(q)} M(\pi) .$$

Lemma 1.4.2 *The $(k + L(q))^{th}$ firing of transition q of $\mathcal{G}$ consumes a token produced by the kth firing of transition q_0.*

Proof: Let s_q be the shortest path from q_0 to q with $L(q)$ tokens. The length of s_q is called the distance from q_0 to q. The proof holds by induction w.r.t. the length of s_q. If $s_q = 0$, then $q = q_0$ and the result is true. Suppose that the result is true for all transitions with distance $k - 1$ from q_0. Choose q at distance k, then the transition q' preceding q on path s_q is at distance $k - 1$ from q_0 and the induction applies to q'. Now the place p between q' and q contains m tokens. By definition of q', $L(q') = L(q) - m$, and, by induction, the $(k + L(q'))^{th}$ firing of transition q' uses token number k. Because the place between q' and q is FIFO, the $(k + L(q))^{th}$ firing of q will use that token. $\square$

For the sake of simplicity we assume that for every transition q in $\mathcal{G}$ there exists a path from q_0 to q that contains no tokens. For queueing networks this condition means that the network is initially empty. Note that the queueing network being empty does not mean that the initial marking in the Petri net model is zero in all places. This stems from the fact that tokens representing physical constrains, like limited buffer capacity, are still present in the Petri net model even though the queueing network is empty. We now set

$$W_q(k) = x_q(k) - u(k) , \quad 1 \le q \le \mathcal{Q}' , \tag{1.19}$$

for $k \ge 1$. In a queueing network interpretation, let the firing of transition q represent the beginning of service at a particular server j. Then Lemma 1.4.2 justifies the interpretation of $W_q(k)$ as the travel time of the kth customer between her/his entrance in the system and the beginning of her/his service at node j. Since we consider a non-autonomous system, the basic recurrence relation for the firing times is given as

$$x(k + 1) = A(k) \otimes x(k) \oplus B(k) \otimes u(k + 1) . \tag{1.20}$$

We have assumed that there is only one input transition, i.e., $|\mathcal{I}| = 1$, and that the initial marking of this input place is one, i.e., $M' = 1$. This implies, that $u(k + 1)$ is a scalar. If we consider only component $x_q(k)$, we can subtract $u(k + 1)$ on both sides of equation (1.20) and get

$$\begin{aligned} W_q(k + 1) &= x_q(k + 1) - u(k + 1) \\ &= (A(k) \otimes x(k))_q \otimes (-u(k + 1)) \oplus B_q(k) . \end{aligned}$$

Let $\sigma_0(k)$ denote the kth firing time of the source transition, that is,

$$u(k) = \sum_{i=1}^{k} \sigma_0(i) ,$$

then

$$(A(k) \otimes x(k))_q \otimes (-u(k+1)) = \bigoplus_{r \in Q'} A_{qr}(k) \otimes (-\sigma_0(k+1)) \otimes (x_r(k) - u(k))$$

$$= \bigoplus_{r \in Q'} A_{qr}(k) \otimes (-\sigma_0(k+1)) \otimes W_r(k).$$

Let $C(h)$ denote a diagonal matrix with $-h$ on the diagonal and ε else. Then, we can write the expression on the right-hand side of the above equation as follows

$$\bigoplus_{r \in Q'} A_{qr}(k) \otimes (-\sigma_0(k+1)) \otimes W_r(k) = \left(A \otimes C(\sigma_0(k+1)) \otimes W(k) \right)_q.$$

Combining the above formulas, we obtain the following vectorial form of the recurrence relation for $W(k+1)$:

$$W(k+1) = A(k) \otimes C(\sigma_0(k+1)) \otimes W(k) \oplus B(k). \qquad (1.21)$$

Lemma 1.4.3 Let $x_0 = (0, \ldots, 0)$ in (1.20). If $W(0) = x_0$ in (1.21), then $W(1) = B(0)$.

Proof: We have assumed that the Petri net is live. There exists thus a path from the source q_0 to any transition q. The element $A_{qr}(0)$ is the time it takes for the first firing of transition r to trigger the first firing of transition q, and $A_{qr}(0) = \varepsilon$ if transition r has no influence on transition q. Moreover, the time it takes for the first firing of the source transition to trigger a firing of transition q is $B_q(0)$. Because any transition r that can possibly trigger a firing of q lies on a path from q_0 to q, it holds that

$$A_{qr}(0) \leq B_q(0), \quad r \in Q',$$

or, equivalently,

$$\bigoplus_{r \in Q'} A_{qr}(0) \leq B(0),$$

which yields

$$A(0) \otimes x_0 \leq B(0).$$

The above inequality implies

$$A(0) \otimes C(\sigma_0(1)) \otimes x_0 \leq B(0).$$

By equation (1.21), it follows $W(1) = B(0)$, which concludes the proof of the lemma. $\square$

By using elementary matrix operations in the max-plus algebra, Equation (1.21) can be rewritten as

$$W(k+1) = \bigotimes_{i=0}^{k} A(i) \otimes C(\sigma_0(i+1)) \otimes W(0)$$

$$\oplus \bigoplus_{i=0}^{k} \bigotimes_{j=i+1}^{k} A(j) \otimes C(\sigma_0(j+1)) \otimes B(i), \qquad (1.22)$$

with $W(0) = x_0$.

When it comes to queueing networks, we obtain from (1.22) a closed-form expression for the vector of $(k+1)^{st}$ waiting/sojourn times in an open queueing network that is initially empty and whose sequence of interarrival times is given by $\{\sigma_0(k)\}$. More precisely, depending on whether we model beginning of service or departure times by $x(k)$, $W_j(k)$ models the time the kth arriving customer spends in the system until her/his service at server j starts, or until she/he departs from server j; see also Section 1.5.3.3. Equation (1.22) is called the *forward construction* of waiting times.

1.5   Queueing Systems and Timed Event Graphs

Petri net models of queueing networks heavily depend on the initial population. In particular, for a given timed event graph we cannot tell whether a token represents a physical restriction, like a finite buffer capacity, or a moving item, like a customer. This violates the queueing theorist's intuition that physical aspects of the system and items/customers moving through the network represent different levels of information. In other words, Petri-net theory is not (yet) a standard tool for queueing theorists, and the characterization of max-plus linearity via subclasses of Petri-nets does not contribute to understanding. We therefore provide a purely queueing theoretic characterization of max-plus linearity.

As we will explain in Chapter 2, in order to obtain stability results for queueing networks via Kingman's subadditive ergodic theorem, a max-plus linear model has to satisfy structural conditions. The most important of these conditions is that the matrices, which govern the transitions in a max-plus linear system, have fixed support. See Definition 1.4.1. In other words, besides identifying max-plus linear queueing networks, we have to find conditions that imply that the corresponding max-plus models have fixed support. In order to do so, we first obtain a recurrence relation for departure times in a general queueing network, called *general sample path formula* (GSPF). Using the GSPF we derive the standard max-plus linear model for the departure times. We then identify structural conditions guaranteeing that the max-plus linear model has fixed support. In a last step, we show that these structural limitations can be summarized in a simple condition. This condition is based on the flow of items/customers through the network. We introduce the notion of *distance* of customers, where the distance of two customers, say k and k', at a certain node, say j, measures the difference between the position of k and k' in the arrival stream at j: If k triggers the mth arrival at j whereas k' triggers the nth one, then the distance between k and k' is $|m - n|$. We then prove that a queueing network is max-plus linear with fixed support if and only if distances between customers are invariant, that is, (1) customers enter nodes in the same order as they leave them, (2) if k is n 'customers ahead' of k' at j, then k is always exactly n customers ahead, and (3) if two customers visit the same node, then they have the same route through the network.

Stability analysis of queueing networks via Kingman's subadditive ergodic theorem requires that the matrices, which govern the transitions in a max-plus linear system, have fixed support. From the above it is clear that this imposes a severe restriction on the class of queueing systems that can be treated. However, elaborating on backward coupling arguments, Mairesse developed a different approach to stability analysis of max-plus linear queueing systems, which does not require that the system dynamic has fixed support. Here, the key property is that the max-plus linear system has a *pattern*, see [84]. A more detailed description of this approach and a discussion of the modeling issues will be given in Section 2.5.

The present section is organized as follows. In Section 1.5.1 we describe the class of queueing networks whose max-plus linearity we will study. Examples of max-plus models of queueing networks are given in Section 1.5.2. Section 1.5.3 provides an analysis of the sample path behavior of the queueing networks under consideration. In Section 1.5.4 we derive a simple structural condition for a queueing network to be max-plus linear. Section 1.5.5 studies possible extensions of our results to other types of queueing networks. Finally, Section 1.5.6 discusses modeling issues when the fixed support assumption is dropped. The material put forward in this section is based on [61].

1.5.1   Queueing Networks

Roughly speaking, a queueing network is a system consisting of nodes, which are connected through routes. Items circulate through the system via the routes and are delayed on their way at the nodes. A node consists of two kinds of places: service and buffer places. On a service place an item is delayed for a predefined (stochastic) time, called 'service time.' When an item arrives at a node and receives no service place, it has to wait for service on a buffer place. In the following we give a precise description of the dynamics of a generic queueing network.

We consider a queueing network with J nodes. If items arrive at the network from the outside and leave the network, we call the network *open*, otherwise we call it *closed*. To facilitate considering both the open and the closed case, we assume that there is only one stream of arrivals. We include a fictitious node 0, which is never idle, from which all arrivals to the system originate and to which all departures from the system go. Typically, items are divided into several distinct classes. However, in what follows we assume that all items belong to one class. As will become clear later, this restriction is necessary to obtain a max-plus linear model, see Section 1.5.5.1.

The interactions between the nodes, carried out through the items, are governed by the following phenomena:

- **Fork:** A departure from a node may generate arrivals at more than one node, that is, items may split up into several (sub) items. More precisely, the kth departure from node j generates arrivals at the nodes out of the set $\mathcal{B}(j,k)$. For example, $\mathcal{B}(j,k) = \{j_1, j_2\}$ means that the item which triggers the kth departure from j splits up into two new items: one moving to node j_1, the other to node j_2. We call node j a *fork node* if $|\mathcal{B}(j,k)| > 1$ for some k.

- **Blocking:** Upon service completion, an item finds no place at the next node. Therefore, the item is forced to stay at the current node and can only move on if a place becomes available at the next node.

  Due to a fork mechanism an item may have to wait for buffer places at several nodes. In this case we assume that the fork operation takes place *after* the item has left the node and *before* it reaches the next (ones). In particular, the kth item departing from j can only be blocked by the nodes in $\mathcal{B}(j,k)$. If the network is open, we assume that the source cannot be blocked, that is, we assume $\mathcal{B}(0,k) = \emptyset$ for all k.

- **Join:** If a node can only commence service if one item from each of the upstream nodes has arrived, we call this node a *join node*. Service of an item at a join node consumes one item of each of the upstream nodes. The join operation is tantamount to synchronizing arrival streams. More precisely, let the kth item departing from j originate from an arrival from each of the nodes $i \in \mathcal{A}(j,k)$. For example, $\mathcal{A}(j,k) = \{j_1, j_2\}$ means that the kth item departing from j originates from joining two items: one arriving from j_1, the other from j_2. We say that j is a join node if $|\mathcal{A}(j,k)| > 1$ for some k. In particular, we assume that the join mechanism is applied only to the newly arriving items and that the items initially present at j have already been 'joined.' If the network is open, we set $\mathcal{A}(j,k) = \emptyset$, which expresses the fact that no arrivals occur at the source.

  Another frequently used join mechanism is called *batching*. Here, several items originating from *one* node are grouped together to form one new item. However, this join mechanism is ruled out by max-plus linearity, as will be demonstrated in Section 1.5.5.4.

- **Variable Origins:** We say that a node admits *variable origins* if an arrival to the node may originate from different nodes. If there are no variable origins at j, then $\mathcal{A}(j,k) = \mathcal{A}(j)$ for all k. In other words, a node j has no variable origins if j is either (1) a join node, so that for each item present, exactly one item must arrive from each node out of the set $\mathcal{A}(j)$, or, (2) there is exactly one node from which items can directly reach j.

- **Variable Destinations:** After completing service at a node, say j, an item may split up according to a fork mechanism. If the set of nodes receiving an (sub) item upon a departure from j varies over time, then this phenomenon is called *variable destinations*. On the other hand, if there are no variable destinations, then $\mathcal{B}(j, k)$ does not depend on time, i.e., $\mathcal{B}(j, k) = \mathcal{B}(j)$, for all k. In other words, a node j does not admit variable destinations if j is either (1) a fork node, so that each departure at j always leads to an arrival of an item at each node out of the set $\mathcal{B}(j)$, or, (2) there is exactly one node to which all items go directly from j.

  A particular node may admit variable origins but no variable destinations, or the node may admit variable destinations and no variable origins. In what follows we say that a *queueing network admits no routing* if all nodes admit neither variable origins nor variable destinations.

- **Internal Overtaking:** In general, the order in which items leave a node is different from the order in which they enter the node. Internal overtake freeness can be forced by a so-called resequencing mechanism. A resequencing queue is such that an item whose service is completed remains on its service place until the service of all items that entered the node before this particular item is finished. The resequencing mechanism is of importance in computer communication systems where the flow of packets or messages entering a communication system in chronological order from the same port or from different ports may be disordered, see [9] for more details. We call a node where items are reordered according to a resequencing mechanism a *resequencing node*.

The way in which the items are processed at the nodes is called the *queueing discipline*. The most prominent example is the first come, first served (FCFS) queueing discipline. If one node simultaneously blocks several other nodes, then the order in which this blocking is resolved is determined via a *blocking discipline*, like, for example, first blocked, first unblocked (FBFU). If an item is blocked, we assume that the item is blocked at the *end* of service and remains on its service place until a free place at the next node becomes available. This is referred to as blocking after service (of manufacturing type), see eg. [31]. To keep the presentation simple, we postpone the discussion of other possible blocking schemes to Section 1.5.5.3.

Remark 1.5.1 *Internal overtake-freeness at a single-server node implies FCFS. At multi-server nodes, one can have more sophisticated queueing disciplines, such as processor sharing or exchangeable items, see [40]. However, this implies that the service times are state dependent which rules out max-plus linearity, see Section 1.5.5.2.*

For ease of reference, we summarize our assumptions:

(A) The queueing network under consideration has only one class of items, no state-dependent service times, all queues are FCFS with blocking after service (of manufacturing type) and blocking is resolved according to FBFU.

Although it seems a bit awkward to reduce our analysis *a priori* to networks satisfying (**A**), it will turn out that (**A**) identifies the class of networks for which we can derive necessary and sufficient conditions for max-plus linearity. In other words, condition (**A**) actually imposes no restriction with respect to the generality of our results. This will be discussed in more detail in Section 1.5.5. The reason for postponing this discussion is that it requires more background on the modeling of queueing networks via max-plus recurrence relations (which will be provided in the next sections).

Remark 1.5.2 *In the above description we associated fork and join operations with nodes, that is, a join operation can only take place immediately before a node (i.e., this particular node is a join node) and a fork operation can only take place immediately after a node (i.e., this particular node is fork node). However, one may want to model an isolated fork/join operation that is not attached to a node. Figure 1.5 shows a sample network with four nodes and an isolated join operation.*
Items arrive from the outside at node 1. After finishing service at node 1, items

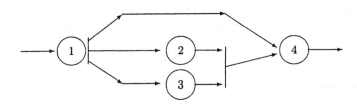

Figure 1.5: A network with an isolated join operation.

are split up into three (sub) items proceeding either to node 4 or to node 2 and 3, respectively. Items finishing their service at node 2 and 3, respectively, are joined to form a new (super) item and this new item proceeds to node 4. This join operation is not attached to a node and, therefore, this network does not fall into the class of queueing networks we introduced so far. However, we may include an fictitious node so that the join operation takes place immediately before this node, that is, the node is a join node. Letting this particular node be a single-server node with infinite buffer capacity and setting the service times equal to zero, we obtain a network that is equivalent to that in Figure 1.5 but that falls into the class of queueing networks introduced above. Figure 1.6 shows the modified network.

1.5.2 Examples of Max-Plus Linear Systems

This section provides a series of examples of max-plus linear queueing systems.

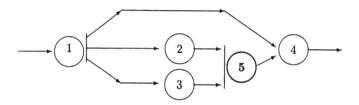

Figure 1.6: The network in Figure 1.5 with a fictitious node included.

Example 1.5.1 *Consider a closed system of J single-server queues in tandem, with infinite buffers. In the system, customers have to pass through the queues consecutively so as to receive service at each server. After service completion at the Jth server, the customers return to the first queue for a new cycle of service.*

We denote the number of customers initially residing at queue j by n_j. We assume that there are J customers circulating through the network and that initially there is one customer in each queue, that is, $n_j = 1$ for $1 \leq j \leq J$. Figure 1.7 shows the initial state of the tandem network, customers are represented by the symbol '•'.

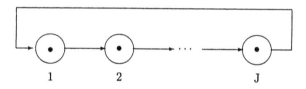

Figure 1.7: The closed tandem queueing system at initial state $n_j = 1$ for $1 \leq j \leq J$.

Let $\sigma_j(k)$ denote the kth service time at queue j and let $x_j(k)$ be the time of the kth service completion at node j, then the time evolution of the system can be described by a J-dimensional vector $x(k) = (x_1(k), \ldots, x_J(k))$ following the homogeneous equation

$$x(k+1) = A(k) \otimes x(k), \qquad (1.23)$$

where the matrix $A(k)$ looks like

$$A(k-1) = \begin{pmatrix} \sigma_1(k) & \varepsilon & \cdots & \varepsilon & \sigma_1(k) \\ \sigma_2(k) & \sigma_2(k) & \varepsilon & \cdots & \varepsilon \\ & & \ddots & & \vdots \\ \cdots & \varepsilon & \sigma_{J-1}(k) & \sigma_{J-1}(k) & \varepsilon \\ \cdots & \cdots & \varepsilon & \sigma_J(k) & \sigma_J(k) \end{pmatrix} \qquad (1.24)$$

for $k \geq 1$. Observe that $A(k)$ is irreducible and that equation (1.23) is noticeably the standard autonomous equation.

Example 1.5.2 *We now consider the open variant of the tandem network in Example 1.5.1. Let queue 0 represent an external arrival stream of customers. Each customer who arrives at the system has to pass through queues 1 to J and then leaves the system. We assume that the system starts empty. Denoting the number of customers initially present at queue j by n_j, we assume $n_j = 0$ for $1 \leq j \leq J$. Figure 1.8 shows the initial state of the tandem network.*

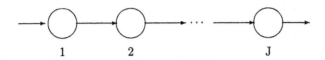

Figure 1.8: The open tandem queueing system at initial state $n_j = 0$ for $1 \leq j \leq J$.

Again, we let $x_j(k)$ denote the time of the kth service completion at station j. In particular, we let $x_0(k)$ denote the kth arrival epoch at the system. The time evolution of the system can then be described by a $(J + 1)$-dimensional vector $x(k) = (x_0(k), \ldots, x_J(k))$ following the homogeneous equation

$$x(k + 1) = A(k) \otimes x(k), \tag{1.25}$$

where the matrix $A(k - 1)$ looks like

$$\begin{pmatrix} \sigma_0(k) & \varepsilon & \varepsilon & \cdots & \varepsilon \\ \sigma_0(k) \otimes \sigma_1(k) & \sigma_1(k) & \varepsilon & \cdots & \varepsilon \\ \sigma_0(k) \otimes \sigma_1(k) \otimes \sigma_2(k) & \sigma_1(k) \otimes \sigma_2(k) & \sigma_2(k) & \cdots & \varepsilon \\ & & \vdots & & \\ \sigma_0(k) \otimes \cdots \otimes \sigma_J(k) & \sigma_1(k) \otimes \cdots \otimes \sigma_J(k) & \sigma_2(k) \otimes \cdots \otimes \sigma_J(k) & \cdots & \sigma_J(k) \end{pmatrix} \tag{1.26}$$

for $k \geq 1$.

Alternatively, we could describe the system via a J dimensional vector $\hat{x}(k) = (\hat{x}_1(k), \ldots, \hat{x}_J(k))$ following the inhomogeneous equation

$$\hat{x}(k + 1) = \hat{A}(k) \otimes \hat{x}(k) \oplus B(k) \otimes \tau(k + 1), \tag{1.27}$$

where the matrix $\hat{A}(k)$ looks like (1.26), except for the first column and the first row which are missing, that is, $(\hat{A}(k))_{ij} = (A(k))_{i+1\,j+1}$ for $1 \leq i, j \leq J$; the *vector $B(k)$ is given by*

$$B(k) = \begin{pmatrix} \sigma_1(k+1) \\ \sigma_1(k+1) \otimes \sigma_2(k+1) \\ \vdots \\ \sigma_1(k+1) \otimes \sigma_2(k+1) \otimes \cdots \otimes \sigma_J(k+1) \end{pmatrix}$$

for $k \geq 0$; and

$$\tau(k) = \sum_{i=1}^{k} \sigma_0(i)$$

denotes the kth arrival time. Notice that matrices $A(k)$ and $\hat{A}(k)$ are reducible and that (1.27) is the standard non-autonomous equation. Notice that $B_j(0)$, for $1 \leq j \leq J$, denotes the time it takes the first customer from entering the system until departing from station j, c.f. Lemma 1.4.3.

The transformation from the homogenuous equation (1.25) to the in-homogenuous equation (1.27) is the inverse transformation to the one described in Section 1.4.3.

Example 1.5.3 *(Example 1.5.2 revisited) We consider the system as described in the above example. However, in contrast to Example 1.5.2, we let $x_j(k)$ denote the time of the kth beginning of service at station j, with $1 \leq j \leq J$. The standard non-autonomous equation now reads*

$$x(k+1) = A(k) \otimes x(k) \oplus B(k) \otimes \tau(k+1), \tag{1.28}$$

with $A(k)$ given by

$$(A(k))_{ij} = \begin{cases} \varepsilon & \text{for } i < j, \\ \sigma_j(k) \otimes \bigotimes_{h=j}^{i-1} \sigma_h(k+1) & \text{for } i \geq j, \end{cases}$$

for $1 \leq i, j \leq J$, where we set $\sigma_j(0) = 0$, and

$$B(k) = \begin{pmatrix} 0 \\ \sigma_1(k+1) \\ \sigma_1(k+1) \otimes \sigma_2(k+1) \\ \vdots \\ \sigma_1(k+1) \otimes \sigma_2(k+1) \otimes \cdots \otimes \sigma_{J-1}(k+1) \end{pmatrix}$$

for $k \geq 0$. An element $B_j(0)$ denotes the time it takes the first customer from entering the system until reaching station j, c.f. Lemma 1.4.3. Notice that $A(k)$ is reducible.

Example 1.5.1 and Example 1.5.2 model sequences of departure times from the queues via a max-plus recurrence relation and a model for beginning of service times is given in Example 1.5.3. We now turn to another important application of max-plus linear models: waiting times.

Example 1.5.4 *Consider the open tandem network described in Example 1.5.3. Let $W_j(k)$ be the time the kth customer arriving at the network spends in the system until the beginning of her/his service at station j. Then, the vector of waiting times $W(k) = (W_1(k), \ldots, W_J(k))$ follows the recurrence relation*

$$W(k+1) = A(k) \otimes C(\sigma_0(k+1)) \otimes W(k) \oplus B(k), \quad k \geq 0,$$

with $W(0) = (0, \ldots, 0)$ and $C(r)$ a matrix with diagonal entries $-r$ and all other entries equal to ε, see Section 1.4.4.

Taking $J = 1$, the above recurrence relation for the waiting times reads

$$\begin{aligned} W(k+1) &= \sigma_1(k) \otimes (-\sigma_0(k+1)) \otimes W(k) \oplus 0 \\ &= \max(\sigma_1(k) - \sigma_0(k+1) + W(k), 0), \quad k \geq 0, \end{aligned}$$

with $\sigma_1(0) = 0$, which is Lindley's equation for the actual waiting time in a $G/G/1$ queue.

If we had let $x(k)$ describe departure times at the stations, c.f. Example 1.5.2, then $W(k)$ would yield the vector of sojourn times of the kth customer. In other words, $W_j(k)$ would model the time the kth customer arriving at the network spends in the system until leaving station j.

In the above examples the positions which are equal to ε are fixed and the randomness is generated by letting the entries different from ε be random variables. The next example is of a different kind. Here, the matrix as a whole is random, that is, the values of the elements are completely random in the sense that an element can with positive probability be equal to ε or finite.

Example 1.5.5 *(Baccelli & Hong, [7]) Consider a cyclic tandem queueing network consisting of a single server and a multi server, each with deterministic service time. Service times at the single-server station equal σ, whereas service times at the multi-server station equal σ'. Three customers circulate in the network. Initially, one customer is in service at station 1, the single server, one customer is in service at station 2, the multi-server, and the third customer is just about to enter station 2. The time evolution of this network is described by a max-plus linear sequence $x(k) = (x_1(k), \ldots, x_4(k))$, where $x_1(k)$ is the kth beginning of service at the single-server station and $x_2(k)$ is the kth departure epoch at the single-server station; $x_3(k)$ is the kth beginning of service at the multi-server station and $x_4(k)$ is the kth departure epoch from the multi-server station. The system then follows*

$$x(k+1) = D_2 \otimes x(k),$$

where

$$D_2 = \begin{pmatrix} \sigma & \varepsilon & \sigma' & \varepsilon \\ \sigma & \varepsilon & \varepsilon & \varepsilon \\ \varepsilon & e & \varepsilon & e \\ \varepsilon & \varepsilon & \sigma' & \varepsilon \end{pmatrix},$$

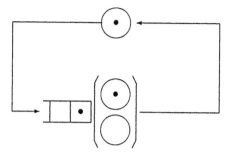

Figure 1.9: The initial state of the multi-server system (three customers).

with $x(0) = (0, 0, 0, 0)$. For a detailed discussion of the above model, see Section B in the Appendix. Figure 1.9 shows the initial state of this system.

Consider the cyclic tandem network again, but one of the servers of the multi-server station has broken down. The system is thus a tandem network with two single server stations. Initially one customer is in service at station 1, one customer is in service at station 2, and the third customer is waiting at station 2 for service. This system follows

$$x(k + 1) = D_1 \otimes x(k),$$

where

$$D_1 = \begin{pmatrix} \sigma & \varepsilon & \sigma' & \varepsilon \\ \sigma & \varepsilon & \varepsilon & \varepsilon \\ \varepsilon & e & \sigma' & \varepsilon \\ \varepsilon & \varepsilon & \sigma' & \varepsilon \end{pmatrix},$$

with $x(0) = (0, 0, 0, 0)$, see Section B in the Appendix for details. Figure 1.10 shows the initial state of the system with breakdown.

Assume that whenever a customer enters station 2, the second server of the multi server station breaks down with probability θ. Let $A_\theta(k)$ have distribution

$$P(A_\theta(k) = D_1) = \theta$$

and

$$P(A_\theta(k) = D_2) = 1 - \theta,$$

then

$$x_\theta(k + 1) = A_\theta(k) \otimes x_\theta(k)$$

describes the time evolution of the system with breakdowns. That the above recurrence relation indeed models the sample path dynamic of the system with breakdowns is not obvious and a proof can be found in [7]. See also Section 1.5.3.3.

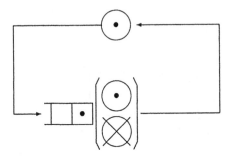

Figure 1.10: The initial state of the multi-server system with breakdown (three customers).

1.5.3 Sample Path Dynamics

This section provides the analysis of the sample path dynamic of a queueing network. Section 1.5.3.1 introduces a recursive formula describing the departure times from queues. From this sample path recurrence relation, we derive in Section 1.5.3.2 a max-plus linear model of the departure times in a queueing network. The relation between max-plus models for departure times and for begin of service times are discussed in Section 1.5.3.3. Finally, in Section 1.5.3.4, we study max-plus linear queueing networks for which the structure of the max-plus model is time independent.

1.5.3.1 The General Sample Path Recursion

Consider a queueing network satisfying condition ($\mathbf{A}$). Let B_j denote the buffer size of node j and S_j the number of service places, respectively, i.e., node j has $P_j \overset{\text{def}}{=} B_j + S_j$ places, where we adopt the convention $\infty + n = \infty = n + \infty$ for $n \in \mathbb{N}$. The number of items initially present at node j is n_j, with $n_j \leq P_j$. We denote the kth service time at node j by $\sigma_j(k)$ and the kth departure epoch at j by $x_j(k)$. In particular, for $k \leq \min(n_j, S_j)$, $\sigma_j(k)$ is the residual service time of an item initially in service at j. For technical reasons, we set $x_j(k) = \varepsilon$ for $k < 0$.

At node j initially n_j items are present. Therefore, the item that is the first to arrive at j only finds a service place if the $(\max(1 + n_j - S_j, 0))^{th}$ departure from j has taken place. Indeed, if $n_j - S_j < 0$, then the first arriving item finds a service place upon arrival. If $n_j - S_j \geq 0$, the arriving item has to wait for a service place. From FCFS follows that this place only becomes available if the $(1 + n_j - S_j)^{th}$ departure has taken place. More general, the mth item arriving at j cannot be served before the $(\max(m + n_j - S_j, 0))^{th}$ departure has taken place. We now let $d(j, k)$ denote the arrival number of the kth item departing from j, where we set $d(j, k) = 0$ if the kth item departing from j is initially present at j. Then, the kth item departing from j can only be served at j if departure

$$c(j,k) = d(j,k) + n_j - S_j \qquad (1.29)$$

has taken place. If the kth item departing from j initially resides at a service place at j (where we assume $n_j > S_j$), we set $c(j,k) = 0$, and if this item was in position m in the initial queue at j, we set $c(j,k) = m - S_j$. We call $c(j,k)$ the *service index*. For example, if j is an infinite server, i.e., $S_j = \infty$, then $c(j,k) = -\infty$ for all k such that the kth item departing from j was not initially present at j; which means that all items find a service place upon arrival. If the network is open, we set $c(0,k) = k - 1$ for all k, in words: the kth interarrival time is initiated by the $(k-1)^{st}$ arrival.

We now consider the arrival process at j more closely. Let the kth item departing from j be constituted out of the items which triggered the $(a_i(j,k))^{th}$ departure from $i \in \mathcal{A}(j,k)$. If the item was initially present at j, set $a_j(j,k) = 0$ and $\mathcal{A}(j,k) = \{j\}$. Then, the item that constitutes the kth departure from j arrives at j at time

$$
\begin{aligned}
\alpha_j(k) &= \max(x_i(a_i(j,k)) \; : \; i \in \mathcal{A}(j,k)) \\
&= \bigoplus_{i \in \mathcal{A}(j,k)} x_i(a_i(j,k)) \qquad (1.30)
\end{aligned}
$$

and we call $a_i(j,k)$ the *arrival index*. If the network is open, we set $a_i(0,k) = \varepsilon$ for all i and all k, which is tantamount to assuming that the source does not have to wait for arrivals.

FCFS queueing discipline implies that the service of the kth item departing from j starts at

$$
\begin{aligned}
\beta_j(k) &= \max(\,\alpha_j(k), x_j(c(j,k))\,) \\
&= \alpha_j(k) \oplus x_j(c(j,k)) \\
&\overset{(1.30)}{=} \bigoplus_{i \in \mathcal{A}(j,k)} x_i(a_i(j,k)) \oplus x_j(c(j,k))\,. \qquad (1.31)
\end{aligned}
$$

Let the item triggering the kth departure from j receive the $(s(j,k))^{th}$ service time at j. We call $s(j,k)$ the *service-time index*. For example, if j is a single-server node, then the FCFS queueing discipline implies $s(j,k) = k$. If the network is open, we assume $s(0,k) = k$, that is, the kth arrival occurs upon completion of the kth interarrival time. Utilising (1.31), the service completion time of the kth item departing from j is given by

$$
\begin{aligned}
\gamma_j(k) &= \beta_j(k) + \sigma_j(s(j,k)) \\
&= \beta_j(k) \otimes \sigma_j(s(j,k)) \\
&\overset{(1.31)}{=} \left(\bigoplus_{i \in \mathcal{A}(j,k)} x_i(a_i(j,k)) \oplus x_j(c(j,k)) \right) \otimes \sigma_j(s(j,k))\,. \qquad (1.32)
\end{aligned}
$$

In order to determine the departure epochs at j, we have to study the resequencing and the blocking mechanism.

First, we consider the resequencing mechanism. If j is a resequencing node, then finished items can only leave the node when all items that arrived at the node before have been served completely. Let $\rho_j(k)$ be the time when the kth item departing from j is ready to leave the node. Furthermore, let $a(j,k)$ be the arrival number of the item that triggers the kth departure at j. For $k \leq n_j$, we associated numbers $1 \ldots n_j$ to the n_j items initially present, so that $a(j,k)$ is defined for all $k \geq 1$. The index $a(j,k)$ counts the arrivals *after* possible join operations. The set of all items arriving prior to the kth item departing from j is given by

$$\{k' \, : \, a(j,k') < a(j,k)\}$$

and the kth item departing from node j is ready to leave node j at

$$\rho_j(k) \;=\; \gamma_j(k) \oplus \bigoplus_{k' \in \{l \, : \, a(j,l) < a(j,k)\}} \gamma_j(k')\,.$$

We now set

$$\mathcal{R}(j,k) \;=\; \{k' \, : \, a(j,k') \leq a(j,k)\}$$

and call $\mathcal{R}(j,k)$ the *resequencing domain*. Note that $k \in \mathcal{R}(j,k)$. If j is no resequencing node, then $\mathcal{R}(j,k) = \{k\}$. The resequencing mechanism can then be expressed through

$$\rho_j(k) \;=\; \bigoplus_{k' \in \mathcal{R}(j,k)} \gamma_j(k')$$

$$\stackrel{(1.32)}{=} \bigoplus_{k' \in \mathcal{R}(j,k)} \left(\bigoplus_{i \in \mathcal{A}(j,k')} x_i(a_i(j,k')) \oplus x_j(c(j,k')) \right) \otimes \sigma_j(s(j,k')). \quad (1.33)$$

We now turn to the blocking mechanism. At node j' there are initially $n_{j'}$ items present. Therefore, the first $P_{j'} - n_{j'}$ items arriving at j' certainly find a place at node j'. However, in general, the mth arriving item finds only a place at j' if the $(m - (P_{j'} - n_{j'}))^{th}$ departure from j' has taken place. Let the kth departure from j trigger the $(d_{j'}(j,k))^{th}$ arrival at j'. Then, the kth departure from j can only take place if the $(d_{j'}(j,k) - (P_{j'} - n_{j'}))^{th}$ departure from j' has taken place. We call

$$b_{j'}(j,k) \;=\; d_{j'}(j,k) - (P_{j'} - n_{j'})$$

the *blocking index*. If the right-hand side of the above equation is smaller than zero, then the kth departing item from j will never be blocked. For example, if j' has infinitely many places, then $b_{j'}(j,k) = -\infty$ for all j and k. Therefore, the kth departure epoch at j satisfies

$$x_j(k) \leq \max\Big(\rho_j(k),\, \max\big(x_i(\,b_i(j,k)\,) \, : \, i \in \mathcal{B}(j,k)\big)\Big)$$

$$= \rho_j(k) \oplus \bigoplus_{i \in \mathcal{B}(j,k)} x_i(b_i(j,k))\,. \qquad (1.34)$$

If the network is open, we set $b_i(0, k) = \varepsilon$ for all i and all k, which is tantamount to assuming that the source cannot be blocked.

Combining (1.34) with (1.32) yields the kth departure time from j; and we obtain the following general recurrence relation for the departure times in a queueing network:

$$x_j(k) = \bigoplus_{i \in B(j,k)} x_i(b_i(j, k))$$

$$\oplus \bigoplus_{k' \in \mathcal{R}(j,k)} \left(\bigoplus_{i \in A(j,k')} x_i(a_i(j, k')) \oplus x_j(c(j, k')) \right) \otimes \sigma_j(s(j, k')), \quad (1.35)$$

for $j \leq J$. We call the above recurrence relation the *general sample-path formula* (GSPF). It will provide the basis of our further analysis.

Definition 1.5.1 *We say that the GSPF is of order $M(k)$ if the right-hand side of (1.35) contains at most the values $x(k), x(k - 1), \ldots, x(k - M(k))$. Furthermore, we say that the GSPF is of finite order if*

$$M = \sup\{M(k) : k \in \mathbb{N}\} < \infty \,.$$

The next lemma provides structural conditions for finiteness of the order of the GSPF.

Lemma 1.5.1 *Consider a queueing network satisfying condition* (**A**). *If the network admits no routing and no internal overtaking, and if all resequencing queues have only finitely many service places, then the associated GSPF is of finite order. Moreover, for all $j \leq J$, we obtain for the arrival index*

$$a_i(j, k) = (k - n_j) 1_{k > n_j}, \quad \text{for } i \in A(j) \,,$$

for the service index

$$c(j, k) = (k - S_j) 1_{S_j < k} \,,$$

for the service-time index

$$s(j, k) = k \,,$$

for the resequencing domain

$$\mathcal{R}(j, k) = \{k, \ldots, k - S_j + 1\} \cap \mathbb{N}$$

and for the blocking index

$$b_i(j, k) = k - (P_i - n_i), \quad \text{for } i \in B(j) \,.$$

If the network is open, we obtain for all k and all $i \leq J$

$$c(0, k) = k - 1, \quad s(0, k) = k \quad \text{and} \quad a_i(0, k) = \varepsilon = b_i(0, k) \,.$$

Proof: Internal overtake-freeness implies that items leave the nodes in the same order as they arrive at them. In particular, the first n_j departures from j are triggered by the n_j items initially present at j, which implies $a_i(j,k) = 0$ for $k \leq n_j$. Consider the case $k > n_j$. Under FCFS items are served in the order of their arrival. Therefore, the kth item departing from j triggered the $(k - n_j)^{th}$ arrival at j. Under 'no routing' each departure from the nodes $i \in \mathcal{A}(j)$ causes an arrival at j. Therefore, the $(k - n_j)^{th}$ arrival at j corresponds to the $(k - n_j)^{th}$ departure from $i \in \mathcal{A}(j)$, which gives

$$a_i(j,k) = (k - n_j)1_{k > n_j} .$$

FCFS implies that the items are served in the order of their arrival which is, by internal overtake freeness, the order in which the items depart. Therefore, the first S_j items departing from j can immediately receive a service place, which gives

$$c(j,k) = 0 , \quad \text{for } k \leq S_j .$$

If $n_j > S_j$, then for $n_j \geq k > S_j$, the kth item departing from j is initially at a waiting place $k - S_j$ at j. The definition of $c(j,k)$ therefore implies

$$c(j,k) = k - S_j , \quad \text{for } S_j < k \leq n_j .$$

For $k > \max(n_j, S_j)$, the kth item departing from j constitutes the $(k - n_j)^{th}$ arrival at j, that is, $d(j,k) = k - n_j$ in the defining relation (1.29). This yields

$$c(j,k) = k - S_j \quad \text{for } n_j < k .$$

Combining these results yields

$$c(j,k) = (k - S_j)1_{S_j < k} .$$

Internal overtake-freeness together with FCFS implies that the kth item departing from j is also the item that initiated the kth service time at j, that is, $s(j,k) = k$.

We now turn to the resequencing domain. The items leave the node in the same order as they arrive at it. For $k \leq S_j$, the kth item departing from j can only leave the node if the items with departure number $k' \leq k$ are completely serviced, that is,

$$\mathcal{R}(j,k) = \{k'|k' \leq k\} , \quad k \leq S_j .$$

Consider the item that triggers the $(S_j + 1)^{st}$ departure from j. This item could only be serviced at j because a free service place was available. We assumed that there is no internal overtaking, so this departure was triggered by the first departure from j. Hence, the $(S_j + 1)^{st}$ item can only be delayed by the items that arrived before and are still in service, that is, by the items with departure numbers $2, \ldots, S_j$. This gives

$$\mathcal{R}(j, S_j + 1) = \{2 \leq k' \leq S_j + 1\} .$$

Induction with respect to k completes the proof.

Eventually, we deal with the blocking index. The kth item departing from j might possibly split up and require a free place at each node $i \in \mathcal{B}(j)$. 'No routing' implies that the kth departure from j constitutes the kth arrival at $i \in \mathcal{B}(j)$. Therefore, $d_i(j, k)$ in the definition of $b_i(j, k)$ equals k.

From the particular form of the indices follows that the GSPF is of finite order, which completes the proof. $\square$

Example 1.5.6 (Example 1.5.1 cont.) *Consider the closed tandem queueing system, that is, all nodes have infinite capacity $P_j = \infty$ and initially one item resides at each node, that is, $n_j = 1$ for $j \leq J$. This network satisfies condition (A) and Lemma 1.5.1 implies $a_i(k, j) = c(j, k) = k-1$, for $k \geq 2$, and $a_i(k, j) = c(j, k) = 0$, for $k = 1$. Furthermore, $\mathcal{R}(j, k) = \{k\}$, $s(j, k) = k$ and $b_i(j, k) = -\infty$ for all j and k. In particular, for nodes j, with $2 \leq j \leq J$, we obtain $\mathcal{A}(j) = \{j - 1\}$, whereas for node 1 we have $\mathcal{A}(1) = \{J\}$. The GSPF now reads*

$$x_j(k)=x_{j-1}(k-1) \oplus x_j(k-1) \otimes \sigma_j(k), \qquad (1.36)$$

for $j \leq J$ and $k \in \mathbb{N}$, where we let $x_0(k) = x_J(k)$. We can write (1.36) in vectorial notation. In order to do so, we set

$$(A(k-1))_{ij} = \begin{cases} \sigma_j(k) \text{ for } i = j \text{ or } i \in \mathcal{A}(j), \\ \varepsilon \qquad else. \end{cases}$$

This yields the matrix given in (1.24). Consequently, (1.36) reads

$$x(k+1) = A(k) \otimes x(k),$$

for $k \in \mathbb{N}$.

Example 1.5.7 *Consider a $GI/G/2/\infty$ resequencing queue, that is, $S_1 = 2$ and $P_1 = \infty$. Let the system be initially empty. This system satisfies condition (A) and Lemma 1.5.1 implies $c(1, k) = k - 2$, for $k \geq 3$, and $c(1, k) = 0$, for $k = 1, 2$. Furthermore, $s(1, k) = a_i(1, k) = k$, $b_i(1, k) = -\infty$ for all j and k. The resequencing domain is given by $\mathcal{R}(j, k) = \{k, k - 1\}$ for $k \geq 2$. Then the GSPF reads*

$$x_1(k) = \Big((x_0(k-1) \oplus x_1(k-3)) \otimes \sigma_1(k-1) \Big)$$
$$\oplus \Big((x_0(k) \oplus x_1(k-2)) \otimes \sigma_1(k) \Big),$$

for $j \leq J$.

Example 1.5.8 (Example 1.5.2 cont.) *Consider the open tandem queueing system again, that is, all nodes have infinite capacity $P_j = \infty$ and the system starts empty, that is, $n_j = 0$ for $j \leq J$. This network satisfies condition (A) and Lemma 1.5.1 implies $a_i(k, j) = k$ and $c(j, k) = k - 1$, for $k \geq 1$. Furthermore, $\mathcal{R}(j, k) = \{k\}$, $s(j, k) = k$ and $b_i(j, k) = -\infty$ for all j and k. In particular, for*

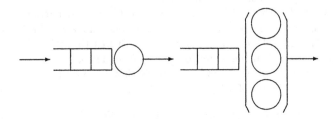

Figure 1.11:   The initial state of an open tandem system with a multi-server station.

nodes j, with $1 \leq j \leq J$, we obtain $\mathcal{A}(j) = \{j - 1\}$, where node 0 represents the source. The GSPF now reads

$$x_j(k) = x_{j-1}(k) \oplus x_j(k - 1) \otimes \sigma_j(k), \qquad (1.37)$$

for $j \leq J$ and $k \in \mathbb{N}$. Note that $x(k)$ occurs on both sides of the above recurrence relation. This is different from the situation in Example 1.5.6, where in the corresponding recurrence relation (1.36) $x(k)$ occurs only on the right-hand side. However, in the following section we provide the means for transforming (1.37) into a vectorial form.

Example 1.5.9 *Consider the following open queueing system. Let queue 0 represent an external arrival stream of customers. Each customer who arrives at the system has to pass through station 1 and 2, where station 1 is a single-server station with unlimited buffer space and station 2 is multi-server station with 3 identical servers and unlimited buffer space. We assume that the system starts empty, i.e., $n_1 = 0 = n_2$. Figure 1.11 shows the initial state of the network.*

Provided that the service times at station 2 are deterministic, this network satisfies condition $(\mathbf{A})$ and Lemma 1.5.1 implies $a_i(k, j) = k$, for $k \geq 1$. Furthermore, $\mathcal{R}(j, k) = \{k\}$, $s(j, k) = k$ and $b_i(j, k) = -\infty$ for all j and k. Furthermore, $c(1, k) = k - 1$ and $c(2, k) = k - 3$. In particular, for nodes j, with $1 \leq j \leq J$, we obtain $\mathcal{A}(j) = \{j - 1\}$, where node 0 represents the source. The GSPF now reads

$$\begin{aligned}
x_0(k) &= x_0(k - 1) \otimes \sigma_0(k), \\
x_1(k) &= \big(x_0(k) \oplus x_1(k - 1)\big) \otimes \sigma_1(k), \\
x_2(k) &= \big(x_1(k) \oplus x_2(k - 3)\big) \otimes \sigma_2,
\end{aligned} \qquad (1.38)$$

for $k \in \mathbb{N}$, where σ_2 denotes the service time at station 2. Note that, like in the previous example, $x(k)$ occurs on both sides of the above recurrence relation.

In the subsequent section we will show how a GSPF of finite order can be algebraically simplified by means of max-plus algebra.

1.5.3.2 The Standard Max-Plus Linear Model

In this section we transform (1.35) into a standard max-plus linear model. In particular, we will show how a GSPF of finite order can be transformed into a first-order GSPF.

In what follows we assume that the GSPF is of finite order. We now define $J \times J$ dimensional matrices $A_m(k)$, where $0 \le m \le M$, with

$$(A_m(k))_{ji} = \begin{cases} \sigma_j(s(j, k')) & \text{if } a_i(j, k') = k - m, \text{ for } i \in \mathcal{A}(j, k') \vee k' \in \mathcal{R}(j, k), \\ & \text{or, if } c(j, k') = k - m, \text{ for } k' \in \mathcal{R}(j, k), \\ 0 & \text{if } b_i(j, k) = k - m, \text{ for } i \in \mathcal{B}(j, k), \\ \varepsilon & \text{else}, \end{cases}$$

(1.39)

cf. equation (1.6) which is the Petri net counterpart of the above definition. Then, recurrence relation (1.35) reads

$$x(k) = \bigoplus_{m=0}^{M} A_m(k) \otimes x(k - m).$$

(1.40)

In what follows, we will transform (1.40) into a recurrence relation of type $x(k + 1) = A(k) \otimes x(k)$, where we follow the line of argument in Section 1.4.2.

If $A_0(k)$ is a lower triangular matrix, then a finite number p exists, such that

$$A_0^*(k) = \bigoplus_{i=0}^{p} A_0^i(k),$$

where $A_0^i(k)$ denotes the ith power of $A_0(k)$, see (1.5) for a definition. We now turn to the algebraic manipulation of (1.40). Set

$$b(k) = \bigoplus_{m=1}^{M} A_m(k) \otimes x(k - m),$$

then (1.40) reduces to

$$x(k) = A_0(k) \otimes x(k) \oplus b(k).$$

(1.41)

For fixed k, the above equation can be written $x = A \otimes x \oplus b$. It is well known that $x = A^* \otimes b$ solves this equation, see Theorem 3.17 in [10]. Therefore, (1.9) can be written

$$x(k) = A_0^*(k) \otimes b(k),$$

or, more explicitly,

$$x(k) = A_0^*(k) \otimes \bigoplus_{m=1}^{M} A_m(k) \otimes x(k - m).$$

(1.42)

The difference between (1.40) and (1.42) is that the latter contains no 0th-order recurrence relation, that is, $x(k)$ occurs only on the left-hand side of the equation.

As a next step we transform (1.42) into a first-order recurrence relation. In order to do so, we set

$$\tilde{x}(k) = (\, x(k), x(k-1), \ldots, x(k-M+1)\,)^T$$

and

$$\tilde{A}(k-1) = \begin{pmatrix} A_0^*(k) \otimes A_1(k) & A_0^*(k) \otimes A_2(k) & \cdots & \cdots & A_0^*(k) \otimes A_M(k) \\ E & \mathcal{E} & \cdots & \cdots & \mathcal{E} \\ \mathcal{E} & E & \ddots & & \mathcal{E} \\ \vdots & & \ddots & & \vdots \\ \mathcal{E} & \mathcal{E} & \cdots & E & \mathcal{E} \end{pmatrix}.$$

Then, (1.40) (and therefore (1.35)) can be written

$$\tilde{x}(k) = \tilde{A}(k-1) \otimes \tilde{x}(k-1)\,,$$

or, in standard form,

$$\tilde{x}(k+1) = \tilde{A}(k) \otimes \tilde{x}(k)\,. \tag{1.43}$$

The above recurrence relation is the standard max-plus linear representation of the departure times in a queueing network.

Definition 1.5.2 *We call a queueing network max-plus linear if the departure times from the queues admit a representation like in (1.43).*

Example 1.5.10 (Example 1.5.6 cont.) *For the closed tandem network, we obtain $M = 1$ and $A_0(k) = \mathcal{E}$. Hence, $A_0^*(k) = E$ and the standard max-plus linear model reads $x(k+1) = A_1(k) \otimes x(k)$, where $A_1(k)$ is the same as the matrix $A(k)$ as defined in (1.24).*

Example 1.5.11 (Example 1.5.8 cont.) *The open tandem queueing system is of order $M = 1$ and we obtain*

$$A_0(k) = \begin{pmatrix} \varepsilon & \varepsilon & \cdots & & \varepsilon \\ \sigma_1(k) & \varepsilon & \cdots & & \varepsilon \\ & & \ddots & & \\ \cdots & \varepsilon & \sigma_{J-1}(k) & \varepsilon & \varepsilon \\ \cdots & & \varepsilon & \sigma_J(k) & \varepsilon \end{pmatrix}$$

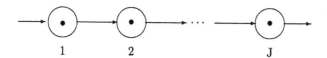

Figure 1.12: The open tandem queueing system at initial state $n_j = 1$ for $1 \leq j \leq J$.

and

$$A_1(k) = \begin{pmatrix} \sigma_0(k) & \varepsilon & \cdots & & & \varepsilon \\ \varepsilon & \sigma_1(k) & \varepsilon & \cdots & & \varepsilon \\ & & & \ddots & & \\ & & \cdots & \varepsilon & \sigma_{J-1}(k) & \varepsilon \\ & & \cdots & \varepsilon & \varepsilon & \sigma_J(k) \end{pmatrix}.$$

For this particular example, p, cf. (1.8), turns out to be $J + 1$, which gives

$$A_0^*(k) = \bigoplus_{i=0}^{J+1} A_0^i(k).$$

Let $A(k)$ be defined as in (1.26), then it is easily checked that $A(k-1) = \bigoplus_{i=0}^{J+1} A_0^i(k) \otimes A_1(k)$.

Example 1.5.12 *We now consider the open tandem queueing system in Example 1.5.2 but with initial population $n_j = 1$ for $1 \leq j \leq J$. Figure 1.12 shows the initial state of the tandem network, where customers are represented by the symbol '•'.*
This yields $A_0(k) = \mathcal{E}$, which implies $A_0^*(k) = E$ and

$$A_1(k) = \begin{pmatrix} \sigma_0(k) & \varepsilon & \cdots & & & \varepsilon \\ \sigma_0(k) & \sigma_1(k) & \varepsilon & & \cdots & \varepsilon \\ & & & \ddots & & \\ & \cdots & \varepsilon & \sigma_{J-2}(k) & \sigma_{J-1}(k) & \varepsilon \\ & & \cdots & \varepsilon & \sigma_{J-1}(k) & \sigma_J(k) \end{pmatrix} \quad (1.44)$$

for $k \geq 1$. Therefore, we obtain $x(k+1) = A(k) \otimes x(k)$ as our max-plus model, with $A(k) = A_1(k+1)$. Comparing (1.44) with (1.26) illustrates the sensitivity of the max-plus model with respect to the initial population.

Example 1.5.13 (Example 1.5.9 cont.) *For the open tandem system with multi-server station we obtain*

$$A_0 = \begin{pmatrix} \varepsilon & \varepsilon & \varepsilon \\ \sigma_1(k) & \varepsilon & \varepsilon \\ \varepsilon & \sigma_2 & \varepsilon \end{pmatrix} ,$$

$$A_1 = \begin{pmatrix} \sigma_0(k) & \varepsilon & \varepsilon \\ \varepsilon & \sigma_1(k) & \varepsilon \\ \varepsilon & \varepsilon & \varepsilon \end{pmatrix} ,$$

$A_2 = \mathcal{E}$, *and*

$$A_3 = \begin{pmatrix} \varepsilon & \varepsilon & \varepsilon \\ \varepsilon & \varepsilon & \varepsilon \\ \varepsilon & \varepsilon & \sigma_2 \end{pmatrix} .$$

We compute $A_0^*(k)$ as follows

$$A_0^*(k) = \begin{pmatrix} e & \varepsilon & \varepsilon \\ \sigma_1(k) & e & \varepsilon \\ \sigma_1(k) \otimes \sigma_2 & \sigma_2 & e \end{pmatrix} .$$

We now set

$$\tilde{x}(k) = (x(k), x(k-1), x(k-2))^T$$

and

$$\tilde{A}(k-1) = \begin{pmatrix} A_0^*(k) \otimes A_1(k) & \mathcal{E} & A_0^*(k) \otimes A_3(k) \\ E & \mathcal{E} & \mathcal{E} \\ \mathcal{E} & E & \mathcal{E} \end{pmatrix}$$

$$= \begin{pmatrix} \sigma_0(k) & \varepsilon & \varepsilon & & \varepsilon & \varepsilon & \varepsilon \\ \sigma_0(k) \otimes \sigma_1(k) & \sigma_1(k) & \varepsilon & \mathcal{E} & \varepsilon & \varepsilon & \varepsilon \\ \sigma_0(k) \otimes \sigma_1(k) \otimes \sigma_2 & \sigma_1(k) \otimes \sigma_2 & \varepsilon & & \varepsilon & \varepsilon & \sigma_2 \\ \\ & E & & \mathcal{E} & & \mathcal{E} & \mathcal{E} \\ \\ & \mathcal{E} & & E & & \mathcal{E} & \mathcal{E} \end{pmatrix}$$

and obtain

$$\tilde{x}(k+1) = \tilde{A}(k) \otimes \tilde{x}(k)$$

as max-plus model for the system, which recovers recurrence relation (1.38).

Example 1.5.14 *We consider the system in Example 1.5.9 again, but with initially one customer in service at station 1 and 3 customers in service at station 2. Figure 1.13 shows the initial state of the tandem network, where customers are represented by the symbol '•'.*
Computing like in the previous example, this system follows

$$\hat{x}(k+1) = \hat{A}(k) \otimes \hat{x}(k) ,$$

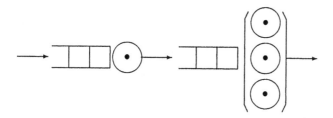

Figure 1.13: The initial state of the open tandem system with a multi-server station and 4 initial customers.

with

$$\hat{A}(k-1) = \begin{pmatrix} \sigma_0(k) & \varepsilon & \varepsilon & & \varepsilon & \varepsilon & \varepsilon \\ \sigma_1(k) & \sigma_1(k) & \varepsilon & \mathcal{E} & \varepsilon & \varepsilon & \varepsilon \\ \varepsilon & \varepsilon & \varepsilon & & \varepsilon & \sigma_2 & \sigma_2 \\ & & E & & \mathcal{E} & & \mathcal{E} \\ & & \mathcal{E} & & E & & \mathcal{E} \end{pmatrix}$$

and obtain

$$\tilde{x}(k+1) \;=\; \tilde{A}(k) \otimes \tilde{x}(k)$$

as max-plus model for the system. The above recurrence relations reads in explicit form

$$x_0(k) = x_0(k-1) \otimes \sigma_0(k)\,,$$
$$x_1(k) = \Big(x_0(k-1) \oplus x_1(k-1)\Big) \otimes \sigma_1(k)\,,$$
$$x_2(k) = \Big(x_1(k-3) \oplus x_2(k-3)\Big) \otimes \sigma_2\,,$$

for $k \in \mathbb{N}$. Comparing $\hat{A}(k)$ with $\tilde{A}(k)$ in Example 1.5.9 (respectively, the above set of equations and (1.38)), illustrates the influence of the initial population on the max-plus model.

We obtain the following characterization of the max-plus linearity of a queueing network.

Theorem 1.5.1 *Consider a queueing network satisfying* (**A**). *The queueing network is max-plus linear if and only if the associated GSPF is of finite order.*

Proof: The GSPF can be transformed into matrix form of fixed dimension if and only if the GSPF is of finite order. □

The following example shows that Markovian routing implies that the GSPF is not of finite order and thereby rules out max-plus linearity.

Example 1.5.15 *Consider a queueing network with three nodes. Upon leaving node 1 items move independently of everything else with probability $p > 0$ to node 2 and with probability $1-p > 0$ to node 3. For the sake of simplicity, assume that initially no item is present at node 2. The kth departure from node 2 depends on the $(k+m)^{th}$ departure from node 1 if exactly m items have been directed to node 2, that is, $a_1(2,k) = k+m$; and the probability that $a_1(2,k) = k+m$ is equal to $p^k(1-p)^m$. Hence, all numbers m have positive probability, which implies that with positive probability $a_1(2,k) = k+m$ for all $m \in \mathbb{N}$. Therefore, the GSPF is not of finite order.*

1.5.3.3   Departure Times vs. Beginning of Service Times

We developed max-plus linear models for departure times from queues. Following the same line of argument one can derive a similar recurrence relation for the beginning of service times at queues. Using such a beginning of service time based approach, a queueing network would be called max-plus linear if the beginning of service times at the queues admitted a representation like in (1.43), cf. Definition 1.5.2. The difference between a max-plus linear recurrence relation for departure times and beginning of service times is illustrated by Example 1.5.2 together with Example 1.5.3.

Apart from modeling beginning of service times rather than departure times one might include both, beginning of service times and departure times, into the state-vector; see Example 1.5.5. In the following we provide our standard max-plus model for the system in Example 1.5.5. Recall that $x_2(k)$ denotes the kth departure time from the single-server station and that $x_4(k)$ denotes the kth departure time from the multi-server station. Applying Lemma 1.5.1, the GSPF for the system with no breakdown reads:

$$x_2(k+1) = (x_2(k) \oplus x_4(k)) \otimes \sigma$$
$$x_4(k+1) = (x_2(k-1) \oplus x_4(k-1)) \otimes \sigma',$$

for $k \geq 0$, where we set

$$x_2(0) = x_4(0) = 0 \quad \text{and} \quad x_2(-1) = x_4(-1) = \varepsilon. \tag{1.45}$$

The GSPF for the system with breakdown reads:

$$x_2(k+1) = (x_2(k) \oplus x_4(k)) \otimes \sigma$$
$$x_4(k+1) = (x_2(k-1) \oplus x_4(k)) \otimes \sigma',$$

for $k \geq 0$, and (1.45). Both GSPF's are of order $M = 2$. In order to obtain the standard max-plus linear model, we therefore have to enlarge the state-space. This leads for the system with no breakdown to the following standard max-plus linear model:

$$\begin{pmatrix} x_2(k+1) \\ x_4(k+1) \\ x_2(k) \\ x_4(k) \end{pmatrix} = \begin{pmatrix} \sigma & \sigma & \varepsilon & \varepsilon \\ \varepsilon & \varepsilon & \sigma' & \sigma' \\ e & \varepsilon & \varepsilon & \varepsilon \\ \varepsilon & e & \varepsilon & \varepsilon \end{pmatrix} \otimes \begin{pmatrix} x_2(k) \\ x_4(k) \\ x_2(k-1) \\ x_4(k-1) \end{pmatrix},$$

for $k \geq 0$. The standard model for the system with breakdown reads:

$$
\begin{pmatrix} x_2(k+1) \\ x_4(k+1) \\ x_2(k) \\ x_4(k) \end{pmatrix} = \begin{pmatrix} \sigma & \sigma & \varepsilon & \varepsilon \\ \varepsilon & \sigma' & \sigma' & \varepsilon \\ e & \varepsilon & \varepsilon & \varepsilon \\ \varepsilon & e & \varepsilon & \varepsilon \end{pmatrix} \otimes \begin{pmatrix} x_2(k) \\ x_4(k) \\ x_2(k-1) \\ x_4(k-1) \end{pmatrix},
$$

for $k \geq 0$. Note that the qualitative aspects of the model have not been altered: the matrix for the system with no breakdown is still irreducible, whereas the matrix for the system with breakdown is reducible.

The main difference between the models in Example 1.5.5 and the one above is that the above state-vectors comprise k^{th} and $(k+1)^{st}$ departure times, whereas in the original models the state-vectors only contained k^{th} departure and beginning of service times. However, to model a randomly occurring breakdown, we require a model whose state space only contains time variables referring to the same transition, which is achieved by the original model. Hence, the standard max-plus linear model is not always appropriate and providing a good model remains an art.

1.5.3.4 Models with Fixed Support

For stability or ergodicity results the mere existence of a max-plus linear model is not sufficient. For this type of analysis one requires a certain structural insensitivity of the transition matrix $\tilde{A}(k)$ of the standard max-plus model, namely, that $\tilde{A}(k)$ has fixed support; see Definition 1.4.1.

As we will explain in Chapter 2, if a queueing network is max-plus linear, then Kingman's subadditive ergodic theorem applies, and we obtain the ergodicity of the maximal growth rate $\max(x_1(k), \ldots, x_J(k))/k$. If $\tilde{A}(k)$ has fixed support, then the ergodicity of the maximal growth rate implies that of the individual growth rates $x_j(k)/k$ (which are related to the inverse throughput of a station in a queueing networks). With respect to applications, ergodicity of the individual growth rate is of key importance. Unfortunately, the fixed-support condition imposes strong restrictions on the class of queueing networks for which ergodicity and stability results can be obtained. More specifically, a max-plus linear representation of departure times via a matrix with fixed support has the following interpretation: *The kth beginning of service at j is always triggered by the $(k-1)^{st}$ departure(s) of the same set of nodes, that is, these nodes do not vary over time.*

In what follows we will give a necessary and sufficient condition for a queueing system to be max-plus linear with fixed support.

Theorem 1.5.2 *Consider a queueing network satisfying* (**A**). *The network is max-plus linear with fixed support if and only if the network admits no routing, no internal overtaking and all resequencing nodes have only finitely many service places.*

Proof: If the network admits no routing, no internal overtaking and if all resequencing nodes have finitely many service places, then Lemma 1.5.1 implies that the corresponding GSPF is of finite order. Hence, Theorem 1.5.1 applies and max-plus linearity follows. The position of the service times as well as that of the zeros in the matrices in recurrence relation (1.39) only depends on the arrival, service and blocking indices and the resequencing domain. The special form of these indicators, as stated in Lemma 1.5.1, implies that the resulting matrix has fixed support.

Now suppose that the queueing network is max-plus linear with fixed support. Then the interactions between departure epochs are time invariant, which rules out routing, internal overtaking and a resequencing node with infinitely many service places. □

Example 1.5.16 *Consider a GI/G/s/∞ system satisfying* (**A**) *with* $s > 1$. *If the system is a resequencing queue, then it is internal overtake-free and a max-plus linear model exists, see Example 1.5.7. On the other hand, if the system does not operate with resequencing, then, in general, this system is not max-plus linear because it admits internal overtaking. However, if the service times are deterministic, then internal overtaking is ruled out and a max-plus linear model exists.*

In the following section we will give a simple characterization of networks with fixed support.

1.5.4 Invariant Queueing Networks

Let $\mathcal{K}$ be the countable set of items moving through the network, that is, we count the items present in the network. The items initially present in the network can be easily counted. Furthermore, if, during the operation of the network, items are generated via a fork mechanism, we count them as new ones, as we do for items arriving from the outside.

In what follows we describe the path of an item through the network. There are two kinds of new items: those created by the external source, and those that result when an existing item splits up into (sub) items. In the latter case, the original item ceases to exist. On the other hand, items also vanish if they leave the network or if they are consumed by a join mechanism in order to generate a new (super) item. The route of item $k \in \mathcal{K}$ is given by

$$w(k) = (w(k,1), \ldots, w(k, S(k))),$$

where $S(k) \in \mathbb{N} \cup \{\infty\}$ is called the length of $w(k)$. The elements $w(k,n) \in \{1, \ldots, J\}$ are called *stages* of route $w(k)$. If an item k that is created by a fork operation is immediately afterwards consumed by an join operation, we set $w(k) = (0)$ and take $S(k) = 0$ as the length of the route of k.

All nodes out of the set $J(k) = \{w(k,n) : n \leq S(k)\}$ are visited by item k. More precisely, the first visit of k at node $j \in J(k)$ is represented by

$$v_j(k,1) = (k,m) \quad \text{with} \quad m = \inf\{n \in \mathbb{N} : w(k,n) = j\}.$$

In a similar way, we find the pair $v_j(k, 2)$ which represents k's second visit at j. In total, k visits j $v_j(k)$ times and we set $v_j(k, m) = -1$ for all $m > v_j(k)$. Furthermore, we set $v_j(k, m) = -1$ for all $m \geq 1$ and $j \notin J(k)$.

We denote the number of the arrival triggered by k at node $w(k, n)$ by $A(k, n)$. In the same way, we let $D(k, n)$ denote the number of the departure triggered by item k at node $w(k, n)$. Then, k's first visit to node j triggers the $(A(v_j(k, 1)))^{th}$ arrival at j and the $(D(v_j(k, 1)))^{th}$ departure from j. Consider two items: k and k', and a node j with $j \in J(k) \cap J(k')$. We say that the distance between k and k' upon their first arrival at j is given by

$$|A(v_j(k, 1)) - A(v_j(k', 1))|.$$

Let $A(-1) = \infty$, then

$$d_j(k, k'; m) = |A(v_j(k, m)) - A(v_j(k', m))|$$

is the distance between k and k' upon their m^{th} visit at node j. If k visits node j at least m times but k' does not, then $d_j(k, k'; m) = \infty$. On the other hand, if k and k' both visit j less than m times, then $d_j(k, k'; m) = 0$. For $0 < d_j(k, k'; m) < \infty$, there are exactly $d_j(k, k'; m)$ arrivals between the arrival of k and k' on their m^{th} visit at j.

We can now easily detect whether two items overtake each other at a certain node j: for k, k' with $j \in J(k) \cap J(k')$ we set

$$f_j(k, k'; m) = \left| A(v_j(k, m)) - D(v_j(k, m)) - \left(A(v_j(k', m)) - D(v_j(k', m)) \right) \right|,$$

and otherwise zero. Then, $f_j(k, k'; m)$ is greater than zero if k and k' do overtake each other during their m^{th} visit. We now call

$$\delta_j(k, k'; m) = d_j(k, k'; m) + \infty 1_{f_j(k, k'; m) > 0}$$

the *distance between k and k' at j*. The distance equals ∞ if either only one of the items visits j m times or if the items overtake each other during their m^{th} visit. If the distances between all items visiting a node j are finite, then the items arrive at the node in exactly the same order in which they departed from it.

Definition 1.5.3 *We call a queueing network invariant if for all $k, k' \in \mathcal{K}$, $m \in \mathbb{N}$ and for all $j \leq J$*

$$\delta(k, k') = \delta_j(k, k'; m).$$

The invariance of a queueing network is tantamount to viewing the distances between items as constant. For example, if item k and k' both visit node j and k' is three places ahead of k, that is, there are two items between k' and k, then there are exactly two items between k and k' at every node these two items visit. On the other hand, if k visits a node k' does not visit, then they have no common node on their route. This gives rise to the following:

Theorem 1.5.3 *Provided the queueing network satisfies* (**A**), *then the queueing network is invariant if and only if it admits no routing, no internal overtaking and all resequencing nodes have only finitely many service places.*

Combining the above theorem with Theorem 1.5.2, we can summarize our analysis as follows.

Corollary 1.5.1 *A queueing network satisfying* (**A**) *is max-plus linear with fixed support if and only if it is invariant.*

Remark 1.5.3 *Consider the network shown in Figure 1.5. Formally, this network is invariant and satisfies condition* (**A**). *Hence, Corollary 1.5.1 implies that the network is max-plus linear. However, this reasoning is not correct! To see this, recall that we assumed for our analysis that the networks contain no isolated fork or join operations, see Remark 1.5.2. Since the network in Figure 1.5 contains an isolated join operation, Corollary 1.5.1 does not apply to this network. However, we may consider the equivalent network, as shown in Figure 1.6, that falls into our framework. This network is not invariant and applying Corollary 1.5.1 we (correctly) conclude that the networks in Figure 1.6 and Figure 1.5, respectively, are not max-plus linear.*

1.5.5   Condition (A) Revisited

This section provides a detailed discussion of assumption (**A**). Section 1.5.5.1 discusses the assumption that the routing is state independent. Section 1.5.5.2 discusses queueing disciplines other than FCFS. Blocking schemes other than blocking-after-service are addressed in Section 1.5.5.3. Finally, Section 1.5.5.4 treats batch processing.

1.5.5.1   State-Dependent Dynamics

In max-plus linear models we have no information about the physical state of the system in terms of queue lengths. Therefore, any dependence of the service times on the physical state cannot be covered by a max-plus linear model. For example, in many situations items are divided into classes. These classes determine the route and/or the service time distribution of items along their route. Due to lack of information about the actual queue-length vector at a node, we cannot determine the class of the item being served, that is, classes may not influence the service time or the routing decisions. Hence, a queueing network is only max-plus linear if there is only one class of items present. For the same reasons, state-dependent queueing disciplines, like processor sharing, cannot be incorporated into a max-plus linear model.

For the sake of completeness we remark that in some cases class-dependent service times can be incorporated into a max-plus linear model. For example, in a GI/G/1/∞ system with two customer classes where F_1 is the service time distribution of class 1 customers and F_2 is the service time distribution of class 2 customers, and where an arriving customer is of class 1 with probability p, we can consider a new service time distribution G which is the mixture of F_1 and F_2 with weights p and $(1 - p)$, respectively. Then the resulting single-class model mimics the dynamic of the queue with two classes and is max-plus linear. However, apart from such model isomorphisms, multi-class queueing systems are not max-plus linear.

Another example of such a model isomorphism is the round robin routing discipline: A node j sends items to nodes $i_1, \ldots, i_n$; the first item is sent to node i_1, the second to node i_2 and so on; once n items have left the node, the cycle starts again. As Krivulin shows in [78], a node with 'round robin' routing discipline can be modeled by a max-plus linear model if this particular node is replaced by a subnetwork of n nodes.

1.5.5.2 Queueing Disciplines

State-dependent queueing disciplines like processor sharing are ruled out by max-plus linearity as explained in Section 1.5.5.1. This extends to queueing disciplines that require information on the physical state of the system, like the last come, first served rule.

For the sake of completeness, we remark that Baccelli et al. discuss in Section 1.2.3 of [10] a production network where three types of parts, called p_1 to p_3, are produced on three machines. It is assumed that the sequencing of part types on the machines is known and fixed. Put another way, the machines do not operate according to FCFS but process the parts according to a fixed sequence. Consider machine M_i which is visited by, say, parts of type p_1 and p_2. If the sequencing of parts at machine M_i is (p_1, p_2), then machine M_i synchronizes the p_1 and p_2 arrival stream in such a way that it always first produces on a p_1 part and then on a p_2 part. Hence, the kth beginning of service on a p_2 part equals the maximum of the $(k-1)^{st}$ departure time of a p_1 part and the kth arrival time of a p_2 part. This system is max-plus linear even though the machines do not operate according to FCFS and there are several types of customers. However, it should be clear from the model that the fact that this system is max-plus linear stems from the particular combination of priority service discipline and classes of customers (parts).

1.5.5.3 Blocking Schemes

We have already considered blocking after service of manufacturing type. Another frequently used blocking scheme is blocking before service, that is, an item is only processed if a place is available at the next station. Under blocking before service, the basic recurrence relation (1.35) reads

$$x_j(k) = \bigoplus_{i \in \mathcal{B}(j,k)} x_i(b_i(j,k)) \otimes \sigma_j(s(j,k)) \oplus \bigoplus_{i \in \mathcal{A}(j,k)} x_i(a_i(j,k)) \otimes \sigma_j(s(j,k))$$
$$\oplus x_j(c(j,k)) \otimes \sigma_j(s(j,k)), \tag{1.46}$$

for $j \leq J$ and $k \in \mathbb{N}$. The standard (max+)-model follows just as easily. We remark that blocking schemes considered here can be extended by including transportation time between nodes, see [14].

Another extension of the blocking mechanism is the so-called *general blocking mechanism*: items that have been successfully processed are put in an output buffer; they leave the output buffer when they find a free place at the next node. This scheme allows the server to process items even though the node is blocked. See [54] for more details.

For max-plus linear systems with fixed support, variable origins are ruled out, that is, each arrival originates from the same node or, if j is a join node, from the same set of nodes. Therefore, it is not necessary to assume a particular blocking discipline like FBFU for systems without variable origins.

1.5.5.4 Batching

Consider a GI/G/1/∞ system with batch arrivals. For the sake of simplicity, assume that the batch size equals two. Hence, the first batch is constituted out of the first two arriving items. More generally, the kth batch is constituted out of the items that triggered the $(2k)^{th}$ and $(2k-1)^{st}$ arrival at the queue. This implies that the arrival index $a_0(k,j)$ equals $2k$ and is therefore not bounded. In other words, the order of the GSPF is not bounded for the above system. Therefore, no standard max-plus linear model exists.

1.5.6 Beyond Fixed Support: Patterns

In this section we explain how new max-plus linear models can be obtained from existing ones through a kind of stochastic mixing.

So far, we considered the initial population and the physical layout of the queueing network as given. Consider, for the sake of simplicity, the GSPF of a queueing network with fixed support as given in (1.36). The GSPF depends via the arrival and blocking index on the initial population $(n_1, \ldots, n_J)$. There is no mathematical reason why n_j should not depend on k. For example, Baccelli and Hong [7] consider a window flow control model where the initial population is non-unique. In particular, they consider two versions of the system, one started with initial population $n^1 = (n_1^1, \ldots, n_J^1)$ and the other with $n^2 = (n_1^2, \ldots, n_J^2)$. The idea behind this is that the version with n^1 is the window flow system under normal load whereas the n^2 version represents the window flow control under reduced load, that is, with fewer items circulating through the system. Both versions of the system are max-plus linear with fixed support, that is, there exists $\tilde{A}^1(k)$ and $\tilde{A}^2(k)$, so that $\tilde{x}^1(k+1) = \tilde{A}^1(k) \otimes \tilde{x}^1(k)$ represents the time evolution under n^1 and $\tilde{x}^2(k+1) = \tilde{A}^2(k) \otimes \tilde{x}^2(k)$ that under n^2. Now assume that after the kth departure epoch the system runs under normal load with probability p and under reduced load with probability $1-p$. Define $\tilde{A}(k)$ so that $P(\tilde{A}(k) = \tilde{A}^1(k)) = p$ and $P(\tilde{A}(k) = \tilde{A}^2(k)) = 1-p$, then $\tilde{x}(k+1) = \tilde{A}(k) \otimes \tilde{x}(k)$ models the window flow control scheme with stochastic change of load. In particular, $\tilde{A}(k)$ fails to have fixed support, which stems from the fact that the support of $\tilde{A}^1(k)$ and $\tilde{A}^2(k)$ doesn't coincide.

Another example of this kind is the multi-server queue with a variable number of servers modeling breakdowns of servers. See Example 1.5.5. Observe that the multi-server queue with breakdowns fails to have fixed support.

Consider a sequence $\{\tilde{A}(k) : k \in \mathbb{N}\}$, like in the above example, and assume for the sake of simplicity that $\tilde{A}(k)$ has only finitely many outcomes. Put another way, $\{\tilde{A}(k) : k \in \mathbb{N}\}$ is a stochastic mixture of finitely many deterministic max-plus linear systems. We say that $\{\tilde{A}(k) : k \in \mathbb{N}\}$ admits a pattern if $N \geq 1$ exists such that with positive probability $\hat{A} = \tilde{A}(k + N) \otimes \cdots \otimes \tilde{A}(k + 1)$, where $\hat{A}$ is an irreducible matrix of cyclicity one and its eigenspace is of dimension one. As we will explain in Section 2.5, if $\{\tilde{A}(k) : k \in \mathbb{N}\}$ admits a pattern, then $\tilde{x}(k)$ converges in total variation towards a unique stationary regime. Moreover, it can be shown that $\tilde{x}(k)$ couples in almost surely finite time with the stationary version. In other words, the concept of a pattern plays an important role for the stability theory of systems that fail to have fixed support.

This approach extends the class of systems that can be analyzed via max-plus stability theory. However, mixing max-plus linear systems in the above way is not straightforward and for a particular system we have to prove that the way in which we combine the elementary systems reflects the dynamic of the compound system. See Section B in the Appendix where the correctness of the multi-server model with breakdowns is shown. Furthermore, the existence of a pattern requires that a finite product of possible outcomes of the transition dynamic of the system results in a matrix which satisfies certain conditions. Unfortunately, this property is analytic and cannot be expressed in terms of the model alone.

We conclude with the remark that, even though the fixed support condition can be relaxed, we still need a GSPF that is of finite order to obtain a max-plus linear model. In other words, we are still limited to systems satisfying condition (**A**), that is, the discussion in Section 1.5.5 remains valid.

1.6 Bounds and Metrics

Our study is devoted to max-plus linear systems. Specifically, we are interested in the asymptotic growth rate of a max-plus linear system. A prerequisite for this type of analysis is that we provide bounds and metrics, respectively, on $\mathbb{R}_{max}$. Section 1.6.1 discusses bounds, which serve as substitutes for norms, for semirings. In Section 1.6.2, we turn to the particular case of the max-plus semiring. In Section 1.6.3, we illustrate how $\mathbb{R}_{max}$ can be made a metric space.

1.6.1 Real-Valued Upper Bounds for Semirings

To set the stage, we state the definition of an upper bound on a semiring.

Definition 1.6.1 *Let R be a non-empty set. Any mapping $\| \cdot \| : R \to [0, \infty)$ is called an upper bound on set R, or, an upper bound on R for short. We write $\| \cdot \|_R$ for such an upper bound when we want to indicate the set on which the upper bound is defined.*

Let $\mathcal{R} = (R, \oplus, \otimes, e, \varepsilon)$ be a semiring. If $|| \cdot ||$ is an upper bound on the set R such that for any $r, s \in R$ it holds that

$$||r \oplus s|| \leq ||r|| + ||s|| \quad and \quad ||r \otimes s|| \leq ||r|| + ||s|| ,$$

then $|| \cdot ||$ is called an upper bound on semiring $\mathcal{R}$, or, an upper bound on $\mathcal{R}$ for short.

On $\mathbb{R}_{\max}$ we introduce the following upper bound

$$||r||_\oplus = \begin{cases} |r| & \text{for } r \in (-\infty, \infty), \\ 0 & \text{otherwise.} \end{cases}$$

That $|| \cdot ||_\oplus$ is indeed an upper bound on $\mathbb{R}_{\max}$ follows easily from the fact that for any $x, y \in \mathbb{R}_{\max}$ it holds

$$x \odot y \leq ||x \odot y||_\oplus \leq ||x||_\oplus + ||y||_\oplus , \quad \odot = \oplus, \otimes . \tag{1.47}$$

The upper bound $|| \cdot ||_\oplus$ is extended to matrices in the obvious way: for $A \in \mathbb{R}_{\max}^{I \times J}$ let

$$||A||_\oplus = \max\{||A_{ij}||_\oplus : 1 \leq i \leq I, \, 1 \leq j \leq J\}$$

$$= \bigoplus_{i=1}^{I} \bigoplus_{j=1}^{J} ||A_{ij}||_\oplus .$$

Note that $||\mathcal{E}||_\oplus = ||E||_\oplus = 0$ and $|| \cdot ||_\oplus$ thus fails to be a norm on $\mathbb{R}_{\max}^{I \times J}$.

Let A be a random element in $\mathbb{R}_{\max}^{I \times J}$ defined on a probability space $(\Omega, \mathcal{A}, P)$. We call A *integrable* if

$$\mathbb{E}[\,||A||_\oplus\,] < \infty .$$

Hence, A is integrable if

$$\mathbb{E}[\,1_{A_{ij}>\varepsilon}|A_{ij}|\,] < \infty , \quad 1 \leq i \leq I, 1 \leq j \leq J .$$

In words, integrability of a matrix is defined through integrability of its non-ε elements. If A is integrable, then the expected value of A is given by the matrix $\mathbb{E}[A]$ with

$$(\mathbb{E}[A])_{ij} = \begin{cases} \mathbb{E}[\,1_{A_{ij}>\varepsilon}A_{ij}\,] & \text{for } P(A_{ij} \neq \varepsilon) > 0, \\ \varepsilon & \text{for } P(A_{ij} = \varepsilon) = 1, \end{cases}$$

for $1 \leq i \leq I, 1 \leq j \leq J$. Note that if $x \in \mathbb{R}^J$ and $A \in \mathbb{R}_{\max}^{I \times J}$ are integrable and A is a.s. regular, then $A \otimes x$ is integrable.

We show that $|| \cdot ||_\oplus$ is indeed an upper bound.

Lemma 1.6.1 *For $A \in \mathbb{R}_{\max}^{I \times K}$ and $B \in \mathbb{R}_{\max}^{K \times J}$,*

$$|| A \otimes B ||_\oplus \leq || A ||_\oplus + || B ||_\oplus .$$

Furthermore, for $A \in \mathbb{R}_{\max}^{I \times J}$ and $B \in \mathbb{R}_{\max}^{I \times J}$,

$$|| A \oplus B ||_\oplus \leq || A ||_\oplus + || B ||_\oplus .$$

Proof: Observe that $A_{ij} \leq \| A \|_{\oplus}$. Making use of the fact that $\oplus$ is idempotent, we calculate as follows:

$$\| A \otimes B \|_{\oplus} = \bigoplus_{i=1}^{I} \bigoplus_{j=1}^{J} \| (A \otimes B)_{ij} \|_{\oplus}$$

$$= \bigoplus_{i=1}^{I} \bigoplus_{j=1}^{J} \left\| \bigoplus_{k=1}^{K} A_{ik} \otimes B_{kj} \right\|_{\oplus}$$

$$\overset{(1.47)}{\leq} \bigoplus_{i=1}^{I} \bigoplus_{j=1}^{J} \left\| \bigoplus_{k=1}^{K} \| A_{ik} \|_{\oplus} \otimes \| B_{kj} \|_{\oplus} \right\|_{\oplus}$$

$$\leq \bigoplus_{i=1}^{I} \bigoplus_{j=1}^{J} \left\| \bigoplus_{k=1}^{K} \| A \|_{\oplus} \otimes \| B \|_{\oplus} \right\|_{\oplus}$$

$$= \| A \|_{\oplus} \otimes \| B \|_{\oplus}$$

$$= \| A \|_{\oplus} + \| B \|_{\oplus} .$$

The proof of the second part of the lemma follows from the same line of argument and is therefore omitted. $\square$

An immediate consequence of the definition of $\| \cdot \|_{\oplus}$ is that the $\| \cdot \|_{\oplus}$-value of a matrix A is always bounded by the sum of all possible sub-matrices of A. The following corollary gives a precise statement.

Corollary 1.6.1 *For $A \in \mathbb{R}_{\max}^{I \times J}$, $B \in \mathbb{R}_{\max}^{K \times L}$ and $C = (A, B) \in \mathbb{R}_{\max}^{I \times J} \times \mathbb{R}_{\max}^{K \times L}$ it holds that*

$$\| C \|_{\oplus} \leq \| A \|_{\oplus} + \| B \|_{\oplus} .$$

1.6.2 General Upper Bounds over the Max-Plus Semiring

In the previous section, we required that an upper-bound maps the elements of a given semiring on $[0, \infty)$. We now extend the usual order relation on $\mathbb{R}$ to $\mathbb{R}_{\max}$ by setting $\varepsilon \leq x$ for all $x \in \mathbb{R}_{\max}$. Recalling that ε is the zero-element of $\mathbb{R}_{\max}$, the natural extension of Definition 1.6.1 to the max-plus semiring is as follows.

Definition 1.6.2 *A mapping $\| \cdot \| : \mathbb{R}_{\max} \to \mathbb{R}_{\max}$ is called a max-plus upper bound if*

- *for any $r \in \mathbb{R}_{\max}$ it holds $\| r \| \geq \varepsilon$,*

- *for any $r, s \in \mathbb{R}_{\max}$ is holds*

$$\| r \oplus s \| \leq \| r \| \otimes \| s \| \quad and \quad \| r \otimes s \| \leq \| r \| \otimes \| s \| .$$

We introduce on $\mathbb{R}_{\max}^{I \times J}$ the following max-plus upper bounds:

$$\| A \|_{\min} \overset{\text{def}}{=} \min_{1 \leq i \leq I} \min_{1 \leq j \leq J} A_{ij}$$

and

$$||A||_{\max} \stackrel{\text{def}}{=} \max_{1 \le i \le I} \max_{1 \le j \le J} A_{ij} = \bigoplus_{i=1}^{I} \bigoplus_{j=1}^{J} A_{ij} .$$

A direct consequence of the above definitions is that for any $A \in \mathbb{R}_{\max}^{I \times J}$

$$||A||_{\min} \le ||A||_{\max} \le ||A||_{\oplus} .$$

The main difference between $||A||_{\min}, ||A||_{\max}$ and $||A||_{\oplus}$ is that $||A||_{\min}, ||A||_{\max}$ can have negative values whereas the definition of $|| \cdot ||_{\oplus}$ implies that for any A it holds that $||A||_{\oplus} \ge 0$. More precisely, if $A_{ij} \in [-\infty, 0)$ for all elements (i, j), then $||A||_{\max} < ||A||_{\oplus}$. Hence, $||A||_{\min}, ||A||_{\max}$ and $||A||_{\oplus}$ are max-plus upper bounds but only $||A||_{\oplus}$ is an upper bound. For example, let

$$A = \begin{pmatrix} -1 & \varepsilon \\ -4 & -3 \end{pmatrix} ,$$

then

$$||A||_{\max} = -1 < 4 = ||A||_{\oplus} .$$

On the other hand, if all finite elements of A are greater than or equal to 0 and if A has at least one finite element, then $||A||_{\max} = ||A||_{\oplus}$. For example, let

$$A = \begin{pmatrix} 1 & \varepsilon \\ 4 & 3 \end{pmatrix} ,$$

then

$$||A||_{\max} = 4 = ||A||_{\oplus} .$$

Lemma 1.6.2 *For $A \in \mathbb{R}_{\max}^{I \times K}$ and $B \in \mathbb{R}_{\max}^{K \times J}$,*

$$|| A \otimes B ||_{\max} \le || A ||_{\max} \otimes || B ||_{\max} ,$$

and

$$|| A \otimes B ||_{\min} \ge || A ||_{\min} \otimes || B ||_{\min} .$$

Furthermore, for $A \in \mathbb{R}_{\max}^{I \times J}$ and $B \in \mathbb{R}_{\max}^{I \times J}$,

$$|| A \oplus B ||_{\max} \le || A ||_{\max} \oplus || B ||_{\max}$$

and

$$|| A \oplus B ||_{\min} \ge || A ||_{\min} \oplus || B ||_{\min} .$$

Proof: Observe that $A_{ij} \leq \|A\|_{\max}$. Calculation yields

$$\|A \otimes B\|_{\max} = \bigoplus_{i=1}^{I} \bigoplus_{j=1}^{J} (A \otimes B)_{ij}$$

$$= \bigoplus_{i=1}^{I} \bigoplus_{j=1}^{J} \bigoplus_{k=1}^{K} A_{ik} \otimes B_{kj}$$

$$\leq \bigoplus_{j=1}^{J} \bigoplus_{k=1}^{K} \|A\|_{\max} \otimes B_{kj}$$

$$= \|A\|_{\max} \otimes \bigoplus_{j=1}^{J} \bigoplus_{k=1}^{K} B_{kj}$$

$$\leq \|A\|_{\max} \otimes \|B\|_{\max}.$$

For $\|\cdot\|_{\min}$ we elaborate on the fact that $A_{ij} \geq \|A\|_{\min}$ and calculate as follows

$$\|A \otimes B\|_{\min} = \min_{1 \leq i \leq I} \min_{1 \leq j \leq J} (A \otimes B)_{ij}$$

$$= \min_{1 \leq i \leq I} \min_{1 \leq j \leq J} \bigoplus_{k=1}^{K} A_{ik} \otimes B_{kj}$$

$$\geq \min_{1 \leq j \leq J} \bigoplus_{k=1}^{K} \|A\|_{\min} \otimes B_{kj}$$

$$= \|A\|_{\min} \otimes \min_{1 \leq j \leq J} \bigoplus_{k=1}^{K} B_{kj}$$

$$\geq \|A\|_{\min} \otimes \|B\|_{\min}.$$

The proof of the second part of the lemma follows from the same line of argument and is therefore omitted. $\square$

For $A, B \in \mathbb{R}_{\max}^{I \times J}$, let $A - B$ denote the component-wise difference, that is, $(A - B)_{ij} = A_{ij} - B_{ij}$, where we set $(A - B)_{ij} = \epsilon$ if both A_{ij} and B_{ij} are equal to ϵ. Recall that the positions of finite elements of a matrix $A \in \mathbb{R}_{\epsilon}^{I \times J}$ is given by set of edges of the communication graph of A, denoted by $\mathcal{D}(A)$. More precisely, A_{ij} is finite if $(j, i) \in \mathcal{D}(A)$.

Lemma 1.6.3 *Let $A, B \in \mathbb{R}_{\max}^{I \times J}$ be regular and let x, y be J dimensional vectors with finite entries. If $\mathcal{D}(A) = \mathcal{D}(B)$, then it holds that*

$$\|A \otimes x - B \otimes y\|_{\max} \leq \|A - B\|_{\max} + \|x - y\|_{\max}.$$

Proof: Let $j^A(i)$ be such that

$$(A \otimes x)_i = A_{ij^A(i)} + x_{j^A(i)} = \bigoplus_{j=1}^{J} A_{ij} \otimes x_j.$$

and $j^B(i)$ be such that

$$(B \otimes y)_i = B_{ij^B(i)} + y_{j^B(i)} = \bigoplus_{j=1}^{J} B_{ij} \otimes y_j.$$

Regularity of A and B implies that $A_{ij^A(i)}$ and $B_{ij^B(i)}$ are finite. Moreover, the fact that the positions of finite entries of A and B coincide implies that $A_{ij^B(i)}$ and $B_{ij^A(i)}$ are finite as well. This yields

$$B_{ij^A(i)} + y_{j^A(i)} \leq B_{ij^B(i)} + y_{j^B(i)}.$$

Hence,

$$A_{ij^A(i)} + x_{j^A(i)} - (B_{ij^B(i)} + y_{j^B(i)}) \leq A_{ij^A(i)} + x_{j^A(i)} - (B_{ij^A(i)} + y_{j^A(i)}).$$

Note that for any i

$$(A \otimes x)_i - (B \otimes y)_i \leq A_{ij^A(i)} - B_{ij^A(i)} + x_{j^A(i)} - y_{j^A(i)}$$
$$\leq ||A - B||_{\max} + ||x - y||_{\max}.$$

Taking the maximum with respect to i on the left-hand side of the above formula proves the claim. $\square$

1.6.3 The Max-Plus Semiring as a Metric Space

If we want to equip $\mathbb{R}_{\max}$ with a metric, then a natural first choice would be $d(A, B) = ||A - B||_\oplus$. Such a definition is to no avail, since $\mathbb{R}_{\max}$ is a semiring, and we cannot give meaning to the '-' operation. However, we may equip $\mathbb{R}_{\max}$ with a metric through embedding $\mathbb{R}_{\max}$ into $[0, \infty)$. We will illustrate this in Section 1.6.3.1.

Elaborating on the projective space, a metric can be introduced that is most helpful in studying max-plus linear recurrence relations. This approach is presented in Section 1.6.3.2.

1.6.3.1 Exponential Lifting

$\mathbb{R}_{\max}$ can be embedded into $[0, \infty)$ in the following way. We map $x \in \mathbb{R}_{\max}$, where x is different from ε, onto e^x and for $x = \varepsilon$, we set $e^x = e^{-\infty} = 0$. With the help of the mapping e^x we are able to introduce the following metric on $\mathbb{R}_{\max}^{I \times J}$

$$d(A, B) \stackrel{\text{def}}{=} \max\left(e^{\max(A_{ij}, B_{ij})} - e^{\min(A_{ij}, B_{ij})} : 1 \leq i \leq I, 1 \leq j \leq J\right)$$
$$= \max\left(\left|e^{A_{ij}} - e^{B_{ij}}\right| : 1 \leq i \leq I, 1 \leq j \leq J\right).$$

For $A \in \mathbb{R}_{\max}^{I \times J}$, let $e^A \in \mathbb{R}^{I \times J}$ be given through $(e^A)_{ij} = e^{A_{ij}}$, for $1 \leq i \leq I$, $1 \leq j \leq J$. With this notation, we obtain

$$||e^A||_\oplus = d(A, \mathcal{E}). \tag{1.48}$$

Consider the metric space $(\mathbb{R}_{\max}^{I \times J}, d(\cdot, \cdot))$. A sequence $\{A(k)\}$ of matrices in $\mathbb{R}_{\max}^{I \times J}$ converges to a matrix A in $\mathbb{R}_{\max}^{I \times J}$, in symbols $A(k) \xrightarrow{d} A$, if and only if for any element (i, j) it holds that

$$\lim_{k \to \infty} d(A_{ij}(k), A_{ij}) = 0 \,.$$

Hence, if $A_{ij} \in \mathbb{R}$, then

$$\lim_{k \to \infty} A_{ij}(k) = A_{ij}$$

and if $A_{ij} = \varepsilon$, then $A_{ij}(k)$ tends to $-\infty$ for k towards ∞.

Example 1.6.1 *The mapping $\|\cdot\|_{\max} : \mathbb{R}_{\max}^{I \times J} \to \mathbb{R}_{\max}$ is continuous with respect to the topology induced by the metric $d(\cdot, \cdot)$. To see this consider $A(k) \in \mathbb{R}_{\max}^{I \times J}$, for $k \in \mathbb{N}$, with*

$$A(k) \xrightarrow{d} A$$

for some matrix $A \in \mathbb{R}_{\max}^{I \times J}$. Continuity of the maximum operation then yields that

$$\lim_{k \to \infty} \|A(k)\|_{\max} = \|A\|_{\max},$$

which implies continuity of $\|\cdot\|_{\max}$. More specifically, recall that $\mathcal{D}(A)$ is the set of edges of the communication graph of A indicating the positions of the finite elements of A. Hence, $A(k) \xrightarrow{d} A$ implies

$$\forall (j, i) \in \mathcal{D}(A) : \quad \lim_{k \to \infty} A_{ij}(k) = A_{ij} \in \mathbb{R},$$

whereas for $(j, i) \notin \mathcal{D}(A)$ we have

$$\lim_{k \to \infty} A_{ij}(k) = -\infty.$$

Provided that $\mathcal{D}(A)$ contains at least one element, continuity of the maximum operation yields that

$$\lim_{k \to \infty} \|A(k)\|_{\max} = \lim_{k \to \infty} \bigoplus_{(j,i) \in \mathcal{D}(A)} A_{ij}(k) = \|A\|_{\max}.$$

In case $\mathcal{D}(A) = \emptyset$, we obtain again by continuity of the maximum operation that

$$\lim_{k \to \infty} \|A(k)\|_{\max} = -\infty = \|A\|_{\oplus}.$$

Apart from the $\|\cdot\|_{\max}$ upper bound, what kind of mappings are continuous with respect to the metric $d(\cdot, \cdot)$ on $\mathbb{R}_{\max}$? In the following we give a partial answer by showing that any continuous mapping from $\mathbb{R}$ to $\mathbb{R}$ that satisfies a certain technical condition can be continuously extended to $\mathbb{R}_{\max}$. Let g be a continuous real-valued mapping defined on $\mathbb{R}$. We extend g to a mapping $\tilde{g} : \mathbb{R}_{\max} \to \mathbb{R}_{\max}$ by setting

$$\tilde{g}(x) = g(x), \quad x \in \mathbb{R}, \tag{1.49}$$

and

$$\tilde{g}(\varepsilon) = \lim_{x \to -\infty} g(x), \tag{1.50}$$

provided that the limit exists. In particular, we set $\tilde{g}(x) = \varepsilon$ if

$$\limsup_{k \to \infty} g(x(k)) = \liminf_{k \to \infty} g(x(k)) = -\infty.$$

The following lemma shows that $\hat{g}$ is again a continuous mapping.

Lemma 1.6.4 *Let $g : \mathbb{R} \to \mathbb{R}$ be continuous and let $\tilde{g}$ be defined as in (1.49) and (1.50). Then $\tilde{g}$ is continuous with respect to the topology induced by the metric $d(\cdot; \cdot)$.*

Proof: Let $x(k) \in \mathbb{R}_{\max}$ be a sequence such that $x(k) \overset{d}{\to} x$ for $x \in \mathbb{R}_{\max}$. If $x \neq \varepsilon$, then

$$\tilde{g}(x(k)) \overset{d}{\to} \tilde{g}(x) = \lim_{k \to \infty} d(\tilde{g}(x(k)), \tilde{g}(x))$$
$$= \lim_{k \to \infty} \left(e^{\max(\tilde{g}(x(k)), \tilde{g}(x))} - e^{\min(\tilde{g}(x(k)), \tilde{g}(x))} \right)$$
$$= \lim_{k \to \infty} \left(e^{\max(g(x(k)), g(x))} - e^{\min(g(x(k)), g(x))} \right).$$

Since, $g(\cdot)$, e^x, max and min are continuous as mappings on $\mathbb{R}$, the above equality implies that $\tilde{g}(x(k)) \overset{d}{\to} \tilde{g}(x)$, which shows the continuity of $\tilde{g}(\cdot)$ on $\mathbb{R}$.

Now let $x = \varepsilon$ and assume that $x(k) \in \mathbb{R}$ for $k \in \mathbb{N}$. Convergence of $x(k)$ towards ε implies that $x(k)$ tends to $-\infty$ for k towards ∞. By (1.50), it follows

$$\lim_{k \to \infty} \tilde{g}(x(k)) = \lim_{k \to \infty} g(x(k))$$
$$= \tilde{g}(\varepsilon)$$
$$= \tilde{g}\left(\lim_{k \to \infty} x(k) \right),$$

which yields continuity of $\tilde{g}(\cdot)$ in ε. $\square$

We conclude this section with some thoughts on the definition of the upper bound $|| \cdot ||_{\oplus}$. In the light of the above analysis, one might be tempted to introduce the bound

$$|||A||| = ||e^A||_{\oplus} \overset{(1.48)}{=} d(A, \mathcal{E}).$$

Unfortunately, $||| \cdot |||$ is not an upper bound on $\mathbb{R}_{\max}$. To see this, consider the matrix

$$A = \begin{pmatrix} 1 & 1 \\ 1 & 1 \end{pmatrix}.$$

For this particular matrix one obtains

$$||A||_{\oplus} = 1 \quad \text{and} \quad |||A||| = e^1,$$

and

$$A \otimes A = \begin{pmatrix} 2 & 2 \\ 2 & 2 \end{pmatrix}$$

implies that

$$\|A \otimes A\|_{\oplus} = 2 = \|A\|_{\oplus} + \|A\|_{\oplus} ,$$

$$\|\|A \otimes A\|\| = e^2 > 2e^1 = \|\|A\|\| + \|\|A\|\| .$$

Hence, $\|\|\cdot\|\|$ fails to be an upper bound on $\mathbb{R}_{\max}$ (that $\|\cdot\|_{\oplus}$ is indeed an upper bound on $\mathbb{R}_{\max}$ has been shown in Lemma 1.6.1).

1.6.3.2 A Metric on the Projective Space

On $\mathbb{PR}^J$ we define the *projective norm* by

$$\|\overline{X}\|_{\mathbb{P}} \stackrel{\text{def}}{=} \|X\|_{\max} - \|X\|_{\min} , \quad X \in \overline{X} .$$

It is easy to check that $\|\overline{X}\|_{\mathbb{P}}$ does not depend on the representative X. Furthermore, $\|\overline{X}\|_{\mathbb{P}} \geq 0$ for any $\overline{X} \in \mathbb{PR}^J$ and

$$\|\overline{X}\|_{\mathbb{P}} = 0 \quad \text{if and only if} \quad \overline{X} = \overline{0} ,$$

that is, $\|\overline{X}\|_{\mathbb{P}} = 0$ if and only if for any $X \in \overline{X}$ it holds that all components are equal. For $\mu \in \mathbb{R}$, let $\mu \cdot X$ be defined as the component-wise conventional multiplication of X by μ. Thus $\mu \cdot \overline{X} = \overline{\mu X}$, which implies

$$\|\mu \cdot \overline{X}\|_{\mathbb{P}} = |\mu| \cdot \|\overline{X}\|_{\mathbb{P}} , \quad \mu \in \mathbb{R}, \, \overline{X} \in \mathbb{PR}^J .$$

In the same vein, for $X, Y \in \mathbb{R}^J$, let $X + Y$ be defined as the component-wise conventional addition of X and Y, then $\|\cdot\|_{\mathbb{P}}$ satisfies the triangular inequality. To see this, let $\overline{X}, \overline{Y} \in \mathbb{PR}^J$, then, for any $X \in \overline{X}$ and $Y \in \overline{Y}$,

$$\begin{aligned}
\|\overline{X} + \overline{Y}\|_{\mathbb{P}} &= \max_i (X_i + Y_i) - \min_i (X_i + Y_i) \\
&\leq \max_i (\max_j (X_j) + Y_i) - \min_i (\min_j (X_j) + Y_i) \\
&= \max_j X_j - \min_j X_j + \max_i (Y_i) - \min_i (Y_i) \\
&= \|\overline{X}\|_{\mathbb{P}} + \|\overline{Y}\|_{\mathbb{P}} .
\end{aligned}$$

Hence, $\|\cdot\|_{\mathbb{P}}$ is indeed a norm on $\mathbb{PR}^J$. We extend the definition of $\|\cdot\|_{\mathbb{P}}$ to $\mathbb{PR}^J_{\max}$ by adopting the convention that $x - \varepsilon = \infty$ for $x \neq \varepsilon$ and $\varepsilon - \varepsilon = \varepsilon + \infty = 0$. However, $\|\cdot\|_{\mathbb{P}}$ fails to be a norm on $\mathbb{PR}^J_{\max}$: for any $X \in \mathbb{R}^J_{\max}$ with at least one finite element and at least one element equal to ε it holds that $\|\overline{X}\|_{\mathbb{P}} = \infty$, whereas a norm is by definition a mapping onto $\mathbb{R}$.

On $\mathbb{PR}^J$, we define $X - Y$ as the component-wise conventional difference of X and Y. With this definition, we obtain a metric $d_{\mathbb{P}}(\cdot, \cdot)$ on $\mathbb{PR}^J$ in the natural way: for $X \in \overline{X}$ and $Y \in \overline{Y}$ set

$$d_{\mathbb{P}}(\overline{X}, \overline{Y}) = \|\overline{X} - \overline{Y}\|_{\mathbb{P}} ,$$

or, more explicitly,

$$
\begin{aligned}
d_{\mathbb{P}}(\overline{X}, \overline{Y}) &= \|X - Y\|_{\max} - \|Y - X\|_{\min} \\
&= \max_{j}(X_j - Y_j) - \min_{j}(X_j - Y_j) \\
&= \max_{j}(Y_j - X_j) - \min_{j}(Y_j - X_j),
\end{aligned}
$$

where for the last equality we have used that $\max_j(X_j - Y_j) = -\min_j(Y_j - X_j)$. The metric $d_{\mathbb{P}}(\cdot, \cdot)$ is called *projective metric*. We extend the definition of $d_{\mathbb{P}}(\cdot, \cdot)$ to $\mathbb{PR}_{\max}^J$ by adopting the convention that $\varepsilon - x = \varepsilon$, for $x \neq \varepsilon$. Note that $d_{\mathbb{P}}(\cdot, \cdot)$ fails to be a metric on $\mathbb{PR}_{\max}^J$. To see this, let $\overline{Y}$ be such that for $Y \in \overline{Y}$ it holds that all components of Y are equal to ε. Then, for any $\overline{X} \in \mathbb{PR}^J$, it follows that $d_{\mathbb{P}}(\overline{X}, \overline{Y}) = 0$.

Chapter 2

Ergodic Theory

Ergodic theory for stochastic max-plus linear systems studies the asymptotic behavior of the sequence

$$x(k+1) = A(k) \otimes x(k), \quad k \geq 0,$$

where $\{A(k)\}$ is a sequence of regular matrices in $\mathbb{R}_{\max}^{J \times J}$ and $x(0) = x_0 \in \mathbb{R}_{\max}^J$. One distinguishes between two types of asymptotic results:

(**Type I**) *first-order* limits

$$\lim_{k \to \infty} \frac{x(k)}{k},$$

(**Type II**) *second-order* limits of type

$$(a) \quad \lim_{k \to \infty} \left(x_i(k) - x_j(k) \right) \quad \text{and} \quad (b) \quad \lim_{k \to \infty} \left(x_j(k+1) - x_j(k) \right).$$

A first-order limit of departure times is an inverse throughput in a queuing network. For example, the throughput of the tandem queuing network in Example 1.5.2 can be obtained from

$$\lim_{k \to \infty} \frac{k}{x_J(k)},$$

provided that the limit exists.

Second-order limits are related to steady-state waiting times and cycle times. Consider the closed tandem network in Example 1.5.1. There are J customers circulating through the system. Thus, the kth and the $(k+J)^{th}$ departure from queue j refers to the same (physical) customer and the cycle time of this customer equals

$$x_j(k+J) - x_j(k).$$

Hence, the existence of the second-order limit $x_j(k+1) - x_j(k)$ implies limit results on steady-state cycle times of customers. For more examples of the modeling of performance characteristics of queuing systems via first-order and second-order expressions we refer to [10, 77, 84].

The chapter is organized as follows. Section 2.1 and Section 2.2 are devoted to limits of type I. Section 2.1 presents background material from the theory of deterministic max-plus systems. In Section 2.2 we present Kingman's celebrated subadditive ergodic theorem. We will show that max-plus recurrence relations constitute in a quite natural way subadditive sequences and we will apply the subadditive ergodic theorem in order to obtain a first ergodic theorem for max-plus linear systems. Limits of type IIa will be addressed in Section 2.3, where the stability theorem for waiting times in max-plus linear networks is addressed. In Section 2.4, limits of type I and type IIa will be discussed. This section is devoted to the study of max-plus linear systems $\{x(k)\}$ such that the relative difference between the components of $x(k)$ constitutes a Harris recurrent Markov chain. Section 2.5 and Section 2.6 are devoted to limits of type IIb and type I. In Section 2.5, we study ergodic theorems in the so called projective space. In Section 2.6, we show how the type I limit can be represented as a second-order limit.

2.1   Deterministic Limit Theory (Type I)

This section provides results from the theory of deterministic max-plus linear systems that will be needed for ergodic theory of max-plus linear stochastic systems. This monograph is devoted to stochastic systems and we state the results presented in this section without proof. To begin with, we state the celebrated cyclicity theorem for deterministic matrices, which is of key importance for our analysis.

Let $A \in \mathbb{R}_{\max}^{J \times J}$, if $x \in \mathbb{R}_{\max}^{J}$ with at least one finite element and $\lambda \in \mathbb{R}_{\max}$ satisfy

$$\lambda \otimes x = A \otimes x,$$

then we call λ an *eigenvalue of A* and x an *eigenvector associated with* λ. Note that the set of all eigenvectors associated with an eigenvalue is a vector space. We denote the set of eigenvectors of A by $V(A)$. The following theorem states a key result from the theory of deterministic max-plus linear systems, namely, that any irreducible square matrix in the max-plus semiring possesses a unique eigenvalue. Recall that $x^{\otimes n}$ denotes the nth power of $x \in \mathbb{R}_{\max}$, see equation (1.5).

Theorem 2.1.1 *(Cohen et al. [33, 34] and Heidergott et al. [65]) For any irreducible matrix $A \in \mathbb{R}_{\max}^{J \times J}$, uniquely defined integers $c(A)$, $\sigma(A)$ and a uniquely defined real number $\lambda = \lambda(A)$ exist such that for all $n \geq c(A)$:*

$$A^{n+\sigma(A)} = \lambda^{\otimes \sigma(A)} \otimes A^n.$$

In the above equation, $\lambda(A)$ is the eigenvalue of A; the number $c(A)$ is called the coupling time of A and $\sigma(A)$ is called the cyclicity of A.

Moreover, for any finite initial vector $x(0)$ the sequence $x(k+1) = A \otimes x(k)$, $k \geq 0$, satisfies

$$\lim_{k \to \infty} \frac{x_j(k)}{k} = \lambda, \quad 1 \leq j \leq J.$$

The above theorem can be seen as the max-plus analog of the Perron-Frobenius theorem in conventional linear algebra and it is for this reason that it is sometimes referred to as 'max-plus Perron-Frobenius theorem.' We illustrate the above definition with a numerical example.

Example 2.1.1 *Matrix*

$$A = \begin{pmatrix} 1 & \varepsilon & 2 & \varepsilon \\ 1 & \varepsilon & \varepsilon & \varepsilon \\ \varepsilon & e & \varepsilon & e \\ \varepsilon & \varepsilon & 2 & \varepsilon \end{pmatrix}$$

has eigenvalue $\lambda(A) = 1$ and coupling time $c(A) = 4$. The critical graph of A consists of the circuits $(1,1)$ and $((1,2),(2,3),(3,1))$, and A is thus of cyclicity $\sigma(A) = 1$. In accordance with Theorem 2.1.1, $A^{n+1} = 1 \otimes A^n$, for $n \geq 4$ and

$$\lim_{k \to \infty} \frac{(A^k \otimes x_0)_j}{k} = 1, \quad 1 \leq j \leq 4,$$

for any finite initial condition x_0. For matrix

$$B = \begin{pmatrix} 1 & \varepsilon & 2 & \varepsilon \\ 1 & \varepsilon & \varepsilon & \varepsilon \\ \varepsilon & e & 2 & e \\ \varepsilon & \varepsilon & 2 & \varepsilon \end{pmatrix}$$

we obtain $\lambda(B) = 2$, coupling time $c(B) = 4$. The critical graph of B consists of the selfloop $(3,3)$, which implies that $\sigma(B) = 1$. Theorem 2.1.1 yields $B^{n+1} = 2 \otimes B^n$, for $n \geq 4$ and

$$\lim_{k \to \infty} \frac{(B^k \otimes x_0)_j}{k} = 2, \quad 1 \leq j \leq 4,$$

for any finite initial condition x_0. Matrix

$$C = \begin{pmatrix} \varepsilon & \varepsilon & 7 & \varepsilon \\ 3 & \varepsilon & \varepsilon & \varepsilon \\ \varepsilon & e & \varepsilon & e \\ \varepsilon & \varepsilon & 7 & \varepsilon \end{pmatrix}$$

has eigenvalue $\lambda(C) = 3.5$, coupling time $c(C) = 4$. The critical graph of C consists of the circuit $((3,4),(4,3))$, which implies that $\sigma(C) = 2$. Theorem 2.1.1 yields $C^{n+2} = 3.5^{\otimes 2} \otimes C^n = 7 \otimes C^n$, for $n \geq 4$ and

$$\lim_{k \to \infty} \frac{(C^k \otimes x_0)_j}{k} = 3.5, \quad 1 \leq j \leq 4,$$

for any finite initial condition x_0.

Let $A \in \mathbb{R}_{\max}^{J \times J}$ and recall that the communication graph of A is denoted by $\mathcal{G}(A)$. For each circuit $\xi = ((i = i_1, i_2), (i_2, i_3), \ldots, (i_n, i_{n+1} = i))$, with arcs (i_m, i_{m+1}) in $\mathcal{G}(A)$ for $1 \leq m \leq n$, we define the average weight of ξ by

$$w(\xi) = \frac{1}{n} \bigotimes_{m=1}^{n} A_{i_{m+1} i_m} = \frac{1}{n} \sum_{m=1}^{n} A_{i_{m+1} i_m} \, .$$

Let $\mathcal{C}(A)$ denote the set of all circuits in $\mathcal{G}(A)$. One of the main results of deterministic max-plus theory is that for any irreducible square matrix A its eigenvalue can be obtained from

$$\lambda = \max_{\xi \in \mathcal{C}(A)} w(\xi) \, .$$

In words, the eigenvalue is equal to the maximal average circuit weight in $\mathcal{G}(A)$.

A circuit ξ in $\mathcal{G}(A)$ is called *critical* if its average weight is maximal, that is, if $w(\xi) = \lambda$. The critical graph of A, denoted by $\mathcal{G}^c(A)$, is the graph consisting of those nodes and arcs that belong to a critical circuit in $\mathcal{G}(A)$. Eigenvectors of A are characterized through the critical graph. However, before we are able to present the precise statement we have to introduce the necessary concepts from graph theory.

Let (E, V) denote a graph with set of nodes E and edges V. A graph is called *strongly connected* if for any two different nodes $i \in E$ and $j \in E$ there exists a path from i to j. For $i, j \in E$, we say that $i \mathcal{R} j$ if either $i = j$ or there exists a path from i to j and from j to i. We split (E, V) up into equivalence classes $(E_1, V_1), \ldots, (E_q, V_q)$ with respect to the relation $\mathcal{R}$. Any equivalence class (E_i, V_i), $1 \leq i \leq q$, constitutes a strongly connected graph. Moreover, (E_i, V_i) is maximal in the sense that we cannot add a node from (E, V) to (E_i, V_i) such that the resulting graph would still be strongly connected. For this reason we call $(E_1, V_1), \ldots, (E_q, V_q)$ *maximal strongly connected subgraphs* (m.s.c.s.) of (E, V). Note that this definition implies that an isolated node or a node with just incoming or outgoing arcs constitutes a m.s.c.s. with an empty arc set. We define the reduced graph, denoted by $(\tilde{E}, \tilde{V})$, by $\tilde{E} = \{1, \ldots, q\}$ and $(i, j) \in \tilde{V}$ if there exists $(k, l) \in V$ with $k \in E_i$ and $l \in E_j$. The *cyclicity* of a strongly connected graph is the greatest common divisor of the lengths of all circuits, whereas the cyclicity of a graph is the least common multiple of the cyclicities of the maximal strongly connected sub-graphs. As shown in [10], the cyclicity of a square matrix A (that is, $\sigma(A)$ in Theorem 2.1.1) is given by the cyclicity of the critical graph of A. A class of matrices that is of importance in applications are irreducible square matrices whose critical graph has a single m.s.c.s. of cyclicity one. Following [65], we call such matrices *primitive*. In the literature, primitive matrices are also referred to as *scs1-cyc1* matrices. For example, matrices A and B in Example 2.1.1 are primitive whereas matrix C in Example 2.1.1 is not.

Example 2.1.2 *We revisit the open tandem queuing system with initially one customer present at each server. The max-plus model for this system is given in*

Example 1.5.12. Suppose that the service times are deterministic, that is, $\sigma_j = \sigma_j(k)$ for $k \in \mathbb{N}$ and $0 \le j \le J$. The communication graph of $A = A_1(k)$ consists of the circuit $((0,1), (1,2), \ldots, (J,0))$ and the recycling loops $(0,0)$, $(1,1)$ to (J,J). Set

$$L = \{j : \sigma_j = \max\{\sigma_i : 0 \le i \le J\}\}.$$

We distinguish between three cases.

- *If $1 = |L|$, then the critical graph of A consists of the node $j \in L$ and the arc (j,j). The critical graph has thus a single m.s.c.s. of cyclicity one, A is therefore primitive.*

- *If $1 < |L| < J$, then the critical graph of A consists of the nodes $j \in L$ and the arcs (j,j), $j \in L$. The critical graph has thus $|L|$ m.s.c.s. each of which has cyclicity one and A fails to be primitive.*

- *If $|L| = J$, then the critical graph and the communication graph coincide and A. The critical graph has a single m.s.c.s. of cyclicity one, and A is primitive.*

Let $A \in \mathbb{R}_{\max}^{J \times J}$ be irreducible. Denote by A_λ the normalized matrix, that is, the matrix which is obtained by subtracting (in conventional algebra) the eigenvalue of A from all components, in formula: $(A_\lambda)_{ij} = A_{ij} - \lambda$, for $1 \le i, j \le J$. The eigenvalue of a normalized matrix is e. For a normalized matrix of dimension $J \times J$ we set

$$A^+ \overset{\text{def}}{=} \bigoplus_{k \ge 1} (A_\lambda)^k. \tag{2.1}$$

It can be shown that $A^+ = A_\lambda \oplus (A_\lambda)^2 \oplus \cdots \oplus (A_\lambda)^J$. See, for example, Lemma 2.2 in [65]. The eigenspaces of A and A_λ are equal. To see this, let $\mathbf{e}$ denote the vector with all components equal to e; for $x \in V(A)$, it then holds that

$$\lambda \otimes x = A \otimes x \Leftrightarrow x = A \otimes x - \lambda \otimes \mathbf{e} \Leftrightarrow e \otimes x = A_\lambda \otimes x.$$

The following theorem is an adaptation of Theorem 3.101 in [10] which characterizes the eigenspace of A_λ. We write $A_{\cdot i}$ to indicate the ith column of A.

Theorem 2.1.2 *(Baccelli et al. [10]) Let A be irreducible and let A^+ be defined as in (2.1).*

(i) If i belongs to the critical graph, then $A_{\cdot i}^+$ is an eigenvector of A.

(ii) For i, j belonging to the critical graph, there exists $a \in \mathbb{R}$ such that

$$a \otimes A_{\cdot i}^+ = A_{\cdot j}^+$$

if and only if i, j belong to the same m.s.c.s.

(iii)   Every eigenvector of A can be written as a linear combination of critical columns, that is, for every $x \in V(A)$ it holds that

$$x = \bigoplus_{i \in G^c(A)} a_i \otimes A^+_{\cdot i},$$

where $G^c(A)$ denotes the set of nodes belonging to the critical graph, and $a_i \in \mathbb{R}^J_{\max}$ such that

$$\bigoplus_{i \in G^c(A)} a_i \neq \varepsilon.$$

Example 2.1.3 *Consider the matrix*

$$A = \begin{pmatrix} 0 & -2 \\ 1 & 0 \end{pmatrix}.$$

A is irreducible with eigenvalue 0 and the critical graph of A consists of the nodes $\{1, 2\}$ and recycling loops $(1, 1)$ and $(2, 2)$. The critical graph has thus two m.s.c.s., namely, the recycling loops $(1, 1)$ and $(2, 2)$, and $\sigma(A) = 1$. For A it holds that

$$A = A^n = A_\lambda = A^+, \quad n \in \mathbb{N}.$$

Theorem 2.1.2 yields the following representation of the eigenspace of A: A vector $x \in \mathbb{R}^2_{\max}$ belongs to $V(A)$ if and only if numbers $a_1, a_2 \in \mathbb{R}_{\max}$ exist with $a_1 \oplus a_2 \neq \varepsilon$ (in words: at least one of two numbers is finite) such that

$$\begin{pmatrix} x_1 \\ x_2 \end{pmatrix} = a_1 \otimes \begin{pmatrix} 0 \\ 1 \end{pmatrix} \oplus a_2 \otimes \begin{pmatrix} -2 \\ 0 \end{pmatrix},$$

see (1.3) for the definition of scalar multiplication of vectors.

Let $A \in \mathbb{R}^{J \times J}_{\max}$ be irreducible with cyclicity one. Recall that we call $v, w \in \mathbb{R}^J_{\max}$ linear dependent if an $\alpha \in \mathbb{R}$ exists such that $v = \alpha \otimes w$. We say that *the eigenvector of A is unique* if any two eigenvectors of A are linear dependent, or, equivalently, if there exists $v \in \mathbb{R}^J$ such that

$$V(A) = \{\alpha \otimes v : \alpha \in \mathbb{R}\}.$$

This can conveniently be expressed by saying that the eigenspace of A reduces to a single point in $\mathbb{R}^J_{\max}$.

An important consequence of Theorem 2.1.2 is that eigenvectors of primitive matrices are unique. Primitive matrices enjoy the additional properties that, for sufficiently large k, $A^k \otimes x$ becomes an eigenvector of A for any finite vector x. These properties of primitive matrices will be of use in Section 2.5 and Section 2.6. The precise statement is as follows.

Corollary 2.1.1 *If $A \in \mathbb{R}^{J \times J}_{\max}$ is a primitive matrix, then the eigenvector of A is unique.*

Let $x(k+1) = A \otimes x(k)$, for $k \geq 0$, and let $x(0)$ be a finite vector. Then, it holds that $x(k) \in V(A)$ for $k \geq c(A)$. Specifically, it holds that

$$x(k+1) = \lambda \otimes x(k), \quad k \geq c(A),$$

where λ denotes the eigenvalue of A, and consequently, for $k \geq c(A)$, it holds that $||x(k)||_{\mathbb{P}} = a$ for some finite constant a.

Proof: Because A is primitive, the critical graph has only one m.s.c.s. Thus, by Theorem 2.1.2 (ii), there exists i_0 in the critical graph such that

$$A_{\cdot i}^+ = \alpha_i \otimes A_{\cdot i_0}, \quad i \in G^c(A).$$

Hence, by Theorem 2.1.2 (iii), any eigenvector v of A can be written

$$v = \bigoplus_{i \in G^c(A)} a_i \otimes A_{\cdot i}^+$$

$$= \bigoplus_{i \in G^c(A)} a_i \otimes \left(\alpha_i \otimes A_{\cdot i_0}^+ \right)$$

$$= \left(\bigoplus_{i \in G^c(A)} a_i \otimes \alpha_i \right) \otimes A_{\cdot i_0}^+$$

$$= \gamma \otimes A_{\cdot i_0}^+,$$

where

$$\gamma = \bigoplus_{i \in G^c(A)} a_i \otimes \alpha_i \in \mathbb{R}_{\max},$$

which establishes uniqueness of the eigenvector.

We now turn to the proof of the second part of the corollary. Since A is primitive, $\sigma(A)$ in Theorem 2.1.1 is equal to one. This yields for $k \geq c(A)$: $A^{k+1} = \lambda \otimes A^k$ for any $k \geq c(A)$. Multiplying both sides of the above equation with the initial vector x_0 concludes the proof. $\square$

Eigenvalues and eigenvectors of matrices over the max-plus semi-ring can be computed in an iterative way. A classical reference is [73]. For more methods for computing max-plus eigenvalues and eigenvectors we refer to [10, 65]. A recent alternative method based on policy iteration is given in [32], see also [65] for a detailed discussion. A general approach for computing cycle times (gives eigenvalues only) for so-called min-max-plus systems (an extension of max-plus linear systems) is established in [57, 56, 49]. Algorithms for computing eigenvalues and eigenvectors of both max-plus and min-max-plus systems can be found in [98, 101]. In particular, the algorithm given in [98] yields an upper bound for the cyclicity of a matrix in the max-plus semiring. Computing the eigenvalue of a matrix A can be achieved in polynomial time. In contrast to this, computing the coupling time is NP-hard (in the number of circuits of the critical graph), see [25]. Feasible upper bounds for the coupling time can be found in [60] and [25].

2.2   Subadditive Ergodic Theory (Type I)

Subadditive ergodic theory is based on Kingman's subadditive ergodic theorem [74, 75] and its application to generalized products of random matrices. We start with an elementary result which appears as an exercise in [91]. A sequence $a = \{a_n : n \in \mathbb{N}\}$ of real numbers is called subadditive if

$$a_{m+n} \leq a_n + a_m , \quad \text{for } n, m \geq 1 .$$

If a is subadditive, then a_n/n has a limit as $n \to \infty$, which may be $-\infty$. To see this, note that for given m, any n can be written as $n = k_n m + l_n$, where $l_n < m$ and k_n is a multiplier that depends on n. The subadditivity of a implies

$$a_n = a_{k_n m + l_n} \leq k_n a_m + a_{l_n} .$$

Dividing both sides by n yields

$$\frac{a_n}{n} = \frac{k_n}{n} a_m + \frac{1}{n} a_{l_n} .$$

Noticing that $k_n/n \leq 1/m$ and $k_n/n \to 1/m$, we have

$$\limsup_n \frac{a_n}{n} \leq \frac{a_m}{m} .$$

Since m is arbitrary, we may take the infimum w.r.t. m over the right-hand side and get

$$\limsup_n \frac{a_n}{n} \leq \liminf_m \frac{a_m}{m} .$$

Therefore, the limit a_n/n exists (and is equal to $\liminf_n a_n/n$).

Kingman's [75] result is formulated in terms of *subadditive processes*. These are double indexed processes $X = \{X_{mn} : m, n \in \mathbb{N}\}$ satisfying the following conditions:

(S1) If $i < j < k$, then $X_{ik} \leq X_{ij} + X_{jk}$ a.s.

(S2) For $m \geq 0$, the joint distributions of the process $\{X_{m+1n+1} : m < n\}$ are the same as those of $\{X_{mn} : m < n\}$.

(S3) The expected value $g_n = \mathbb{E}[X_{0n}]$ exists and satisfies $g_n \geq -cn$ for some finite constant $c > 0$ and all $n \geq 1$.

A consequence of **(S1)**, **(S3)** and the elementary result given above is that

$$\lambda = \lim_{n \to \infty} \frac{g_n}{n}$$

exists and is finite. We can now state Kingman's subadditive ergodic theorem: if X is a subadditive process (that is, **(S1)**, **(S2)** and **(S3)** hold), then the limit

$$\xi = \lim_{n \to \infty} \frac{X_{0n}}{n}$$

exists almost surely, and $\mathbb{E}[\xi] = \lambda$. Condition **(S2)**, on the shift $\{X_{mn}\} \rightarrow \{X_{m+1n+1}\}$, is a stationarity condition. If all events defined in terms of X that are invariant under this shift have probability zero or one, then X is *ergodic*. In this case, as discussed in Kingman [75], the limiting random variable ξ is almost surely constant and equal to λ. Note that the limit also holds when expected values are considered.

We now turn to homogeneous equations, that is, to max-plus linear systems whose dynamic can be described via

$$x(k+1) = A(k) \otimes x(k),$$

for $k \geq 0$, with $x(0) = x_0$ given. In particular, we write

$$x(n+1, x_0) = \bigotimes_{k=0}^{n} A(k) \otimes x_0, \quad n \geq 0, \tag{2.2}$$

to indicate the initial value of the sequence. Recall that $\mathbf{e}$ denotes the vector with all components equal to e. We set

$$x_{nm} = \bigotimes_{k=n}^{m-1} A(k) \otimes \mathbf{e}$$

From this we recover $x(k+1, \mathbf{e})$ through $x_{0k+1} = x(k+1, \mathbf{e})$.

Lemma 2.2.1 *Let $\{A(k)\}$ be a stationary sequence of a.s. regular and integrable matrices in $\mathbb{R}_{\max}^{J \times J}$. Then $\{-||x_{nm}||_{\min} : m > n \geq 0\}$ and $\{||x_{nm}||_{\max} : m > n \geq 0\}$ are subadditive ergodic processes.*

Proof: For $x, y \in \mathbb{R}_{\max}^{J}$, let $x \leq y$ denote the component-wise order. Note that $x \leq y$ implies $||x||_{\max} \leq ||y||_{\max}$; in particular, $x \leq ||x||_{\max} \otimes \mathbf{e}$, where $\mathbf{e}$ denotes the vector whose components are equal to e (we refer to (1.3) for a definition of the $\otimes$-product of a scalar and a vector). Furthermore, for any $A \in \mathbb{R}_{\max}^{J \times J}$ it holds that $x \leq y$ implies $A \otimes x \leq A \otimes y$. Combining these statements it follows for $x \in \mathbb{R}_{\max}^{J}$ and $A \in \mathbb{R}_{\max}^{J \times J}$:

$$||A \otimes x||_{\max} \leq ||A \otimes (||x||_{\max} \otimes \mathbf{e})||_{\max}. \tag{2.3}$$

In the same vein, for $x \in \mathbb{R}_{\max}^{J}$ and $A \in \mathbb{R}_{\max}^{J \times J}$:

$$||A \otimes x||_{\min} \geq ||A \otimes (||x||_{\min} \otimes \mathbf{e})||_{\min}. \tag{2.4}$$

We now show the subadditive property of $||x_{nm}||_{\max}$. For $0 \leq n < p < m$, we obtain

$$\|x_{nm}\|_{\max} = \left\|\bigotimes_{i=n}^{m-1} A(i) \otimes \mathbf{e}\right\|_{\max}$$

$$= \left\|\bigotimes_{i=p}^{m-1} A(i) \otimes x_{np}\right\|_{\max}$$

$$\overset{(2.3)}{\leq} \left\|\bigotimes_{i=p}^{m-1} A(i) \otimes \left(\|x_{np}\|_{\max} \otimes \mathbf{e}\right)\right\|_{\max}$$

$$= \left\|\|x_{np}\|_{\max} \otimes \left(\bigotimes_{i=p}^{m-1} A(i) \otimes \mathbf{e}\right)\right\|_{\max}$$

$$= \|x_{np}\|_{\max} + \left\|\bigotimes_{i=p}^{m-1} A(i) \otimes \mathbf{e}\right\|_{\max}$$

$$= \|x_{np}\|_{\max} + \|x_{pm}\|_{\max},$$

which establishes (S1) for $\|x_{nm}\|_{\max}$. The proof that (S1) holds for $-\|x_{nm}\|_{\min}$ as well follows the same line of argument: for $0 \leq n < p < m$,

$$\|x_{nm}\|_{\min} = \left\|\bigotimes_{i=n}^{m-1} A(i) \otimes \mathbf{e}\right\|_{\min}$$

$$= \left\|\bigotimes_{i=p}^{m-1} A(i) \otimes x_{np}\right\|_{\min}$$

$$\overset{(2.4)}{\geq} \left\|\bigotimes_{i=p}^{m-1} A(i) \otimes \left(\|x_{np}\|_{\min} \otimes \mathbf{e}\right)\right\|_{\min}$$

$$= \left\|\|x_{np}\|_{\min} \otimes \left(\bigotimes_{i=p}^{m-1} A(i) \otimes \mathbf{e}\right)\right\|_{\min}$$

$$= \|x_{np}\|_{\min} + \left\|\bigotimes_{i=p}^{m-1} A(i) \otimes \mathbf{e}\right\|_{\min}$$

$$= \|x_{np}\|_{\min} + \|x_{pm}\|_{\min},$$

which establishes (S1) for $-\|x_{nm}\|_{\min}$.

The stationarity condition (S2) follows immediately from the stationarity of $\{A(k)\}$.

We now turn to condition (**S3**). We have assumed that each row of $A(k)$ contains at least one non-ε element, which implies $x(k, \mathbf{e}) \in \mathbb{R}^J$ for any k. We may now prove by induction that $x(k, \mathbf{e})$ is absolutely integrable where we use the fact that (i) $|\min(a, b)|, |\max(a, b)| \leq |a| + |b|$, (ii) $A(k)$ is integrable, and that (iii) the initial condition $\mathbf{e}$ of $x(k, \mathbf{e})$ is integrable. From

$$\mathbb{E}[\,||x_{0k}||_{\max}\,] = \mathbb{E}[\,||x(k, \mathbf{e})||_{\max}\,] \tag{2.5}$$

it follows that x_{0k} is integrable for any k. Let $|||A|||$ denote the smallest non-ε element of A (note that (i) and (ii) above imply that $\mathbb{E}[\,|||A(k)|||\,]$ is finite). With this definition it is immediate that

$$\sum_{j=0}^{k-1} \mathbb{E}[\,|||A(j)|||\,] \leq \mathbb{E}[\,||x(k, \mathbf{e})||_{\max}\,]. \tag{2.6}$$

Stationarity of $\{A(k)\}$ implies that $\mathbb{E}[\,|||A(k)|||\,] = c$ for any k. Integrability of $A(k)$ together with the fact that there are at least J finite elements in $A(k)$ yields $c > -\infty$. We obtain from (2.6):

$$-k\,|c| \leq \mathbb{E}[\,||x(k, \mathbf{e})||_{\max}\,]$$
$$\overset{(2.5)}{=} \mathbb{E}[\,||x_{0k}||_{\max}\,],$$

which establishes (**S3**) for $\{||x_{nm}||_{\max} : m \geq 1; m > n \geq 0\}$.

We now turn to $\{-||x_{nm}||_{\min} : m \geq 1; m > n \geq 0\}$. Following the above line of argument it holds that, for $k \in \mathbb{N}$,

$$\mathbb{E}[\,||x_{0k}||_{\min}\,] = \mathbb{E}[\,||x(k, \mathbf{e})||_{\min}\,] < \infty$$

and

$$\sum_{j=0}^{k-1} \mathbb{E}[\,||A(j)||_{\max}\,] \geq \mathbb{E}[\,||x_{0k}||_{\min}\,].$$

Hence,

$$\sum_{j=0}^{k-1} -\mathbb{E}[\,||A(j)||_{\max}\,] \leq \mathbb{E}[\,-||x_{0k}||_{\min}\,],$$

for $k \in \mathbb{N}$, and for $\tilde{c} = \mathbb{E}[\,||A(1)||_{\max}\,]$, we obtain

$$-|\tilde{c}|\,k \leq \mathbb{E}[\,-||x_{0k}||_{\min}\,],$$

which concludes the proof of the lemma. $\square$

The above lemma provides the means of applying Kingman's subadditive ergodic theorem to $||x(k)||_{\min}$ and $||x(k)||_{\max}$, respectively. The precise statement is given in the following theorem.

Theorem 2.2.1 *Let $\{A(k)\}$ be a stationary sequence of a.s. regular, integrable square matrices. Then, finite constants λ^{top} and λ^{bot} exist, so that for all (non-random) finite initial conditions x_0:*

$$\lambda^{\text{bot}} \stackrel{\text{def}}{=} \lim_{k \to \infty} \frac{||x(k)||_{\min}}{k} \leq \lambda^{\text{top}} \stackrel{\text{def}}{=} \lim_{k \to \infty} \frac{||x(k)||_{\max}}{k} \qquad a.s.$$

and

$$\lim_{k \to \infty} \frac{1}{k} \mathbb{E}[||x(k)||_{\min}] = \lambda^{\text{bot}} \leq \lim_{k \to \infty} \frac{1}{k} \mathbb{E}[||x(k)||_{\max}] = \lambda^{\text{top}}.$$

The above limits also hold for random initial conditions provided that the initial condition is a.s. finite and integrable.

Proof: Lemma 2.2.1 applies and subadditivity of $||x(k, \mathbf{e})||_{\min}$ and $||x(k, \mathbf{e})||_{\max}$, respectively, follows. Therefore, Kingman's subadditive ergodic theorem applies and the proof with respect to the limit of $||x(k, \mathbf{e})||_{\min}$ as k tends to ∞ and the limit of $||x(k, \mathbf{e})||_{\max}$ as k tends to ∞ follows.

It remains to be shown that the limit exists for any finite initial condition. To see this note that for any finite initial condition y it holds that:

$$\begin{aligned}
||y||_{\min} + ||x(k, \mathbf{e})||_{\max} &= ||x(k, ||y||_{\min} \otimes \mathbf{e})||_{\max} \\
&\leq ||x(k, y)||_{\max} \\
&\leq ||x(k, ||y||_{\max} \otimes \mathbf{e})||_{\max} \\
&= ||y||_{\max} + ||x(k, \mathbf{e})||_{\max}
\end{aligned}$$

(for a proof use the fact that $x \leq y$ implies $A \otimes x \leq A \otimes y$). Thus,

$$||y||_{\min} + ||x(k, \mathbf{e})||_{\max} \leq ||x(k, y)||_{\max} \leq ||y||_{\max} + ||x(k, \mathbf{e})||_{\max}$$

and, by similar arguments,

$$||y||_{\min} + ||x(k, \mathbf{e})||_{\min} \leq ||x(k, y)||_{\min} \leq ||x(k, \mathbf{e})||_{\min} + ||y||_{\max}.$$

Therefore, for $k > 0$,

$$\frac{1}{k}||y||_{\min} + \frac{1}{k}||x(k, \mathbf{e})||_{\max} \leq \frac{1}{k}||x(k, y)||_{\max} \leq \frac{1}{k}||x(k, \mathbf{e})||_{\max} + \frac{1}{k}||y||_{\max} \tag{2.7}$$

and

$$\frac{1}{k}||y||_{\min} + \frac{1}{k}||x(k, \mathbf{e})||_{\min} \leq \frac{1}{k}||x(k, y)||_{\min} \leq \frac{1}{k}||x(k, \mathbf{e})||_{\min} + \frac{1}{k}||y||_{\max}. \tag{2.8}$$

Letting k tend to infinity, it follows from (2.7) that the limits of $||x(k, \mathbf{e})||_{\max}/k$ and $||x(k, y)||_{\max}/k$ coincide. In the same vein, (2.8) implies that the limits of $||x(k, \mathbf{e})||_{\min}/k$ and $||x(k, y)||_{\min}/k$ coincide. If, in addition, x_0 is integrable, we first prove by induction that $x(k, x_0)$ is integrable for any $k > 0$. Then, we take expected values in (2.7) and (2.8). Using the fact that, by Kingman's subadditive ergodic theorem, the limits of $\mathbb{E}[||x(k, \mathbf{e})||_{\max}]/k$ and $\mathbb{E}[||x(k, \mathbf{e})||_{\min}]/k$ as k tends to ∞ exist, the proof follows from letting k tend to ∞. $\square$

The constant λ^{top} is called the *top* or *maximal Lyapunov exponent* of $\{A(k)\}$ and λ^{bot} is called the *bottom Lyapunov exponent*.

Remark 2.2.1 *Irreducibility is a sufficient condition for $A(k)$ to be a.s. regular, see Remark 1.4.1. Therefore, in the literature, Theorem 2.2.1 is often stated with irreducibility as a condition.*

Remark 2.2.2 *Note that integrablity of $\{A(k)\}$ is a necessary condition for applying Kingman's subadditive ergodic theorem in the proof of the path-wise statement in Theorem 2.2.1.*

Remark 2.2.3 *Provided that (i) any finite element of $A(k)$ is positive, (ii) $A(k)$ is a.s. regular, and (iii) the initial state x_0 is positive, the statement in Theorem 2.2.1 holds for $\|\cdot\|_\oplus$ as well. This stems from the fact that under conditions (i) to (iii) it holds that $\|A(k)\|_{\max} = \|A(k)\|_\oplus$. In particular, following the line of argument in the proof of Lemma 2.2.1, one can show that under the conditions of the lemma the sequence $\|x_{nm}\|_\oplus$ constitutes a subadditive process.*

2.2.1 The Irreducible Case

In this section, we consider stationary sequences $\{A(k)\}$ of integrable and irreducible matrices in $\mathbb{R}_{\max}^{J\times J}$ with the additional property that all finite elements are non-negative and that all diagonal elements are non-negative. We consider $x(k+1) = A(k) \otimes x(k)$, $k \geq 0$, and recall that $x(k)$ may model an autonomous system (for example, a closed queuing network). See Section 1.4.3. Indeed, $A(k)$ for the closed tandem queuing system in Example 1.5.1 is irreducible. As we will show in the following theorem, the setting of this section implies that $\lambda^{\text{top}} = \lambda^{\text{bot}}$, which in particular implies convergence of $x_i(k)/k$, $1 \leq i \leq J$. The condition that all finite elements of $A(k)$ are non-negative is not very restrictive when working with queuing networks. Here the non-ε elements of $A(k)$ represent sums of service times at the stations, which are by definition non-negative. In contrast, the assumption that all diagonal elements are non-negative (and thus different from ε) is indeed a restriction as illustrated by Example 1.5.5. The following theorem goes back to Cohen [35] and Baccelli et al. [10].

Theorem 2.2.2 *Let $\{A(k)\}$ be a stationary sequence of integrable and irreducible matrices in $\mathbb{R}_{\max}^{J\times J}$ such that all finite elements are non-negative and all diagonal elements are different from ε. Then, a finite constant λ exists, so that for any non-random finite initial condition x_0:*

$$\lim_{k\to\infty} \frac{x_j(k)}{k} = \lim_{k\to\infty} \frac{\|x(k)\|_{\min}}{k} = \lim_{k\to\infty} \frac{\|x(k)\|_{\max}}{k} = \lambda \quad a.s. \tag{2.9}$$

and

$$\lim_{k\to\infty} \frac{1}{k}\mathbb{E}[x_j(k)] = \lim_{k\to\infty} \frac{1}{k}\mathbb{E}[\|x(k)\|_{\min}] = \lim_{k\to\infty} \frac{1}{k}\mathbb{E}[\|x(k)\|_{\max}] = \lambda\,,$$

for $1 \leq j \leq J$. The above limits also hold true for random initial conditions provided that the initial condition is a.s. finite and integrable.

Proof: The existence of the limits (except that for $x_j(k)/k$) is guaranteed by Theorem 2.2.1 and in order to prove the theorem we have to show that the component-wise limits (that is, the limit of $x_j(k)/k$ as k tends to ∞, for $1 \leq j \leq J$) equal the limits of $|| \cdot ||_{\min}$ and $|| \cdot ||_{\max}$.

Irreducibility of $A(k)$ implies that $A(k)$ has fixed support and the communication graph of $A(k)$ is thus non-random. We have assumed that all elements different from ε are non-negative and all diagonal elements are non-negative. Hence, Lemma 1.4.1 applies and

$$G(k) = \bigotimes_{j=k-J}^{k-1} A(j), \quad k \geq J,$$

has all elements larger than or equal to zero for all k. This implies for any component j

$$x_j(k, \mathbf{e}) = \bigoplus_{i=1}^{J} (G(k))_{ji} \otimes x_i(k - J, \mathbf{e})$$

$$\geq \bigoplus_{i=1}^{J} 0 \otimes x_i(k - J, \mathbf{e})$$

$$= ||x(k - J, \mathbf{e})||_{\max},$$

for $k \geq J$, which yields

$$||x(k, \mathbf{e})||_{\min} \geq ||x(k - J, \mathbf{e})||_{\max}. \tag{2.10}$$

By (2.10),

$$\frac{1}{k}||x(k, \mathbf{e})||_{\min} \geq \frac{1}{k}||x(k - J, \mathbf{e})||_{\max},$$

which implies

$$\lambda^{\text{bot}} = \lim_{k \to \infty} \frac{||x(k, \mathbf{e})||_{\min}}{k} \geq \lim_{k \to \infty} \frac{||x(k, \mathbf{e})||_{\max}}{k} = \lambda^{\text{top}} \quad \text{a.s.}$$

By Theorem 2.2.1, it holds that $\lambda^{\text{bot}} \leq \lambda^{\text{top}}$ and we have thus shown $\lambda^{\text{bot}} = \lambda^{\text{top}}$. In other words, setting $\lambda \stackrel{\text{def}}{=} \lambda^{\text{bot}} = \lambda^{\text{top}}$ we have shown

$$\lim_{k \to \infty} \frac{||x(k, \mathbf{e})||_{\min}}{k} = \lim_{k \to \infty} \frac{||x(k, \mathbf{e})||_{\max}}{k} = \lambda \quad \text{a.s.} \tag{2.11}$$

and from

$$||x(k, \mathbf{e})||_{\max} \geq x_j(k, \mathbf{e}) \geq ||x(k, \mathbf{e})||_{\min}, \quad 1 \leq j \leq J,$$

follows:

$$\lim_{k \to \infty} \frac{x_j(k, \mathbf{e})}{k} = \lambda \quad \text{a.s.} \tag{2.12}$$

for $1 \leq j \leq J$.

Like for the proof of Theorem 2.2.1, we show that the limits in (2.11) and (2.12) are independent of the initial condition. This concludes the proof of the first part of the theorem.

We now turn to the proof of the second part of the theorem. Let λ, as defined in the first part of Theorem 2.2.2, exist. Then Theorem 2.2.1 yields,

$$\lambda = \lim_{k \to \infty} \frac{1}{k} \mathbb{E}[\|x(k)\|_{\max}] = \lim_{k \to \infty} \frac{1}{k} \mathbb{E}[\|x(k)\|_{\min}]$$

and

$$\frac{1}{k} \mathbb{E}[\|x(k)\|_{\min}] \leq \frac{1}{k} \mathbb{E}[x_j(k)] \leq \frac{1}{k} \mathbb{E}[\|x(k)\|_{\max}]$$

implies

$$\lambda = \lim_{k \to \infty} \frac{1}{k} \mathbb{E}[x_j(k)],$$

for $1 \leq j \leq J$. $\square$

The constant λ, as defined in (2.9) in Theorem 2.2.2, is called *max-plus Lyapunov exponent* of the sequence of random matrices $\{A(k)\}$. There is no ambiguity in denoting the Lyapunov exponent of $\{A(k)\}$ and the eigenvalue of a matrix A by the same symbol, since for $A(k) = A$, for all k, the Lyapunov exponent of $\{A(k)\}$ is just the eigenvalue of A.

Remark 2.2.4 *Depending on the sequence $\{A(k)\}$, it is sometimes possible to replace an element of x_0 that is equal to ε by a finite element without changing the value of $x(k)$, for $k \geq 1$. In these cases, Theorem 2.2.2 applies even though not all elements of x_0 are finite.*

Remark 2.2.5 *We say that $A, B \in \mathbb{R}_{\max}^{J \times J}$ have the same structure if any element (ij) is either finite in A and B, or, is equal to ε (that is, the arc sets of communication graph of A and B coincide). The irreducibility condition in the above theorem can be replaced by the following weaker condition. There exists a.s. a sequence $\{m_n\}$ with $\lim_{n \to \infty} m_n = \infty$, such that $A(k + m_n)$, $1 \leq k \leq J$, have the same structure and are irreducible.*

Remark 2.2.6 *If the initial condition x_0 is positive, then the statement in Theorem 2.2.2 holds for $\| \cdot \|_{\oplus}$ as well. See Remark 2.2.3 for details.*

Computing exactly, or approximating the Lyapunov exponent of products of matrices over the max-plus semiring is a long standing problem [35, 96, 93, 10, 36, 46, 11, 50, 21, 8, 7, 42]. Only for special cases exact formulae are known. Upper and lower bounds can be found in [14, 18, 53, 28, 29]. In [12] approaches are described which use parallel simulation to estimate the ratio $x_j(k)/k$ for large k. When it comes to discrete event systems, Lyapunov exponents measure the cycle time, i.e., the average time between two events. A classical reference on Lyapunov exponents of products of random matrices is [24] and a more recent one, dedicated to non-negative matrices, is [66].

Consider the system in Example 1.5.1. If we assume that (i) the service times $\sigma_j(k)$ are i.i.d. with finite mean for each j and (ii) the sequences $\{\sigma_j(k)\}$ $(1 \leq j \leq J)$ are mutually independent, then Theorem 2.2.2 applies (indeed, $\{A(k)\}$ is an i.i.d. sequence of irreducible matrices with fixed support).

Comparing the conditions in Theorem 2.2.2 with those in Theorem 2.2.1, Theorem 2.2.2 imposes the additional conditions that (i) the matrices are irreducible (and have thus fixed support), (ii) all elements different from ε are non-negative and that (iii) all diagonal elements are non-negative. However, conditions (i)-(iii) are only needed to establish the pathwise statement in Theorem 2.2.2. Hence, the second part of Theorem 2.2.2 is valid under weaker conditions. The exact statement is as follows:

Corollary 2.2.1 *Let $\{A(k)\}$ be a stationary sequence of a.s. regular and integrable matrices in $\mathbb{R}_{\max}^{J \times J}$. If*

- $\lambda^{\text{bot}} \geq \lambda^{\text{top}}$, *and*

- *the initial condition is integrable,*

then

$$\lim_{k \to \infty} \frac{1}{k} \mathbb{E}[x_j(k)] = \lambda \,,$$

for all components $1 \leq j \leq J$ of $x(k)$.

Proof: By assumption,

$$\lim_{k \to \infty} \frac{\|x(k)\|_{\min}}{k} = \lim_{k \to \infty} \frac{\|x(k)\|_{\max}}{k} = \lambda \,,$$

with $\lambda = \lambda^{\text{bot}} = \lambda^{\text{top}}$, and Theorem 2.2.1 yields

$$\lim_{k \to \infty} \frac{\|x(k)\|_{\min}}{k} = \lim_{k \to \infty} \frac{1}{k} \mathbb{E}[\|x(k)\|_{\min}] = \lambda \,,$$

$$\lim_{k \to \infty} \frac{\|x(k)\|_{\max}}{k} = \lim_{k \to \infty} \frac{1}{k} \mathbb{E}[\|x(k)\|_{\max}] = \lambda \,.$$

For any $k \in \mathbb{N}$ and $1 \leq j \leq J$,

$$\frac{1}{k} \mathbb{E}[\|x(k)\|_{\min}] \leq \frac{1}{k} \mathbb{E}[x_j(k)] \leq \frac{1}{k} \mathbb{E}[\|x(k)\|_{\max}]$$

and taking limits yields

$$\lambda = \lim_{k \to \infty} \frac{1}{k} \mathbb{E}[x_j(k)] \,,$$

which concludes the proof. $\square$

2.2.2 The Reducible Case

The setup is as in the previous section except that we now suppose that $A(k)$ has fixed support and drop the assumption that it is irreducible. An example of a model that has fixed support but fails to be irreducible is the open tandem queuing system in Example 1.5.2. We study the homogeneous equation

$$x(k+1) = A(k) \otimes x(k), \quad k \geq 0.$$

Notice that this setup comprises inhomogeneous equations, such as the standard autonomous equation as well, see Section 1.4.3 for details.

To deal with reducible matrices $A(k)$, we decompose $A(k)$ into its 'irreducible' components. The ergodic theorem, to be proved presently, then states that the Lyapunov exponent of the overall matrix is given by the maximal top Lyapunov exponent of its irreducible components. However, before we are able to present the ergodic theorem and give the proof, we need to introduce some concepts from graph theory. For the basic definitions we refer to Section 2.1.

Let $\{A(k)\}$ be a sequence of matrices in $\mathbb{R}_{\max}^{J \times J}$ with fixed support. If we replace any element of $A(k)$ that is different from ε by e, then the resulting communication graph of $A(k)$, denoted by $\mathcal{G}_e(A)$, is independent of k (and thus non-random). Let $\mathcal{G}_e^r(A)$ denote the reduced graph of $\mathcal{G}_e(A)$. We denote by $[i] \stackrel{\text{def}}{=} \{j \in \{1, \dots, J\} : i\mathcal{R}j\}$ the set of nodes of the m.s.c.s. that contains i. The set of all nodes j such that there exists a path from j to i in $\mathcal{G}_e(A)$ is denoted by $\pi^+(i)$. Furthermore, we set $\pi^*(i) = \{i\} \cup \pi^+(i)$; and we define predecessor sets

$$[\leq i] = \bigcup_{j \in \pi^*(i)} [j]$$

and $[< i] = [\leq i] \setminus [i]$. We denote by $\lambda_{[i]}^{\text{top}}$ the top Lyapunov exponent associated with the matrix obtained by restricting $A(k)$ to the nodes in $[i]$. In case i is an isolated node or node with only incoming or outgoing arcs, we set $\lambda_{[i]}^{\text{top}} = \varepsilon$. The following theorem goes back to [6].

Theorem 2.2.3 *Let $\{A(k)\}$ be a stationary sequence of integrable matrices in $\mathbb{R}_{\max}^{J \times J}$ with fixed support such that with probability one all finite elements are nonnegative and the diagonal elements are different from ε. For any (non-random) finite initial value x_0 it holds true that*

$$\lim_{k \to \infty} \frac{x_j(k)}{k} = \lambda_j \quad a.s. \, ,$$

with

$$\lambda_j = \bigoplus_{i \in \pi^*(j)} \lambda_{[i]}^{\text{top}} \, ,$$

and

$$\lim_{k \to \infty} \frac{1}{k} \mathbb{E}[x_j(k)] = \lambda_j \, ,$$

for $1 \leq j \leq J$. The above limits also hold for random initial conditions provided that the initial condition is a.s. finite and integrable.

Proof: Under the conditions of the theorem, it a.s. holds, for any k, that $||x(k)||_{\max} = ||x(k)||_\oplus$, see Remark 2.2.3. In the following proof we will only work with upper bounds on the growth rate $||x(k)||_{\max}/k$ and thus adopt the notation $||\cdot||_\oplus$ for the maximal element of a vector/matrix.

Let $A_{[i]\,[i]}(k)$ denote the matrix that is obtained from $A(k)$ by restricting $A(k)$ to the nodes in $[i]$ and write $x_{[i]}(k)$ for $x(k)$ restricted to the nodes in $[i]$. To understand the difficulty that arises when proving the theorem, it is worth noting that in general

$$\lim_{k\to\infty} \frac{1}{k}||x_{[i]}(k)||_\oplus \neq \lambda_{[i]}^{\text{top}} \text{ a.s.}$$

This stems from the fact that $\lambda_{[i]}^{\text{top}}$ is the top Lyapunov exponent of the matrix restricted to the nodes in $[i]$, whereas $x_{[i]}(k)$ is also influenced by nodes others than those in $[i]$ namely those in $[\leq i] \setminus [i]$.

We now turn to the proof. In the same way as we have defined $A_{[i]\,[i]}(k)$ and $x_{[i]}(k)$, we write $A_{[\leq i]\,[\leq i]}(k)$ for the restriction of $A(k)$ to the nodes in $[\leq i]$ and $x_{[\leq i]}(k)$ for $x(k)$ restricted to the nodes in $[\leq i]$. By Theorem 2.2.1, the maximal Lyapunov exponent of $A_{[\leq i]\,[\leq i]}(k)$, given by $\lambda_{[\leq i]}^{\text{top}}$, exists (indeed, Theorem 2.2.1 applies to reducible matrices). Note that

$$\frac{1}{k}||x_{[i]}(k)||_\oplus \leq \frac{1}{k}||x_{[\leq i]}(k)||_\oplus$$

and thus

$$\limsup_{k\to\infty} \frac{1}{k}||x_{[i]}(k)||_\oplus \leq \limsup_{k\to\infty} \frac{1}{k}||x_{[\leq i]}(k)||_\oplus$$
$$= \lambda_{[\leq i]}^{\text{top}}. \tag{2.13}$$

Fixed support of $A(k)$ implies that $\mathcal{G}_e(A)$ is non-random. Node i can be reached from any node $h \in \pi^*(i)$ and since $A(k)$ is of dimension $J \times J$ such a path is at most of length J. We have assumed that the diagonal elements of $A(k)$ are all different from ε. Hence, if there is a path of length l from h to i, then there is for any $p \geq l$ a path of length p from h to i (just add sufficiently many loops of length one at h). Any finite element of $A(k)$ is positive and paths have therefore positive weights. We thus obtain for any $j \in [i]$

$$x_j(k) \geq \bigoplus_{h\in\pi^*(i)} x_h(k-J)$$
$$= ||x_{[\leq i]}(k-J)||_\oplus, \tag{2.14}$$

for $k \geq J$. Therefore,

$$||x_{[i]}(k)||_\oplus \geq ||x_{[\leq i]}(k-J)||_\oplus,$$

for $k \geq J$, which implies that

$$\liminf_{k\to\infty} \frac{1}{k}||x_{[i]}(k)||_\oplus \geq \liminf_{k\to\infty} \frac{1}{k}||x_{[\leq i]}(k)||_\oplus$$
$$= \lambda_{[\leq i]}^{\text{top}} \quad \text{a.s.}$$

Together with (2.13) we obtain

$$\lim_{k \to \infty} \frac{1}{k} \|x_{[i]}(k)\|_\oplus = \lambda^{top}_{[\leq i]} \quad \text{a.s.} \tag{2.15}$$

By (2.14), it holds a.s. for any $j \in [i]$ that

$$\frac{1}{k}\|x_{[i]}(k)\|_\oplus \geq \frac{1}{k}x_j(k) \geq \frac{1}{k}\|x_{[\leq i]}(k-J)\|_\oplus , \tag{2.16}$$

and by (2.15) it follows that

$$\lim_{k \to \infty} \frac{1}{k}\|x_{[i]}(k)\|_\oplus = \lim_{k \to \infty} \frac{1}{k}\|x_{[\leq i]}(k-J)\|_\oplus = \lambda^{top}_{[\leq i]} , \tag{2.17}$$

which yields

$$\lim_{k \to \infty} \frac{1}{k}x_j(k) = \lambda^{top}_{[\leq i]} \quad \text{a.s.} , \quad j \in [i] .$$

In the integrable case, (2.16) implies

$$\frac{1}{k}\mathbb{E}[\|x_{[\leq i]}(k)\|_\oplus] \geq \frac{1}{k}\mathbb{E}[x_j(k)] \geq \frac{1}{k}\mathbb{E}[\|x_{[\leq i]}(k-J)\|_\oplus] .$$

By Theorem 2.2.1, the expected values on the right-hand side and on the left-hand side in the above inequality converge to $\lambda^{top}_{[\leq i]}$ as k tends to ∞. Hence,

$$\lim_{k \to \infty} \frac{1}{k}\mathbb{E}[x_j(k)] = \lambda^{top}_{[\leq i]} , \quad j \in [i] .$$

It remains to be shown that

$$\lambda^{top}_{[\leq i]} = \bigoplus_{j \in \pi^*(i)} \lambda_{[j]} . \tag{2.18}$$

The reduced graph $\mathcal{G}^r_e(A)$ is acyclic and we obtain

$$x_{[i]}(k+1) = A_{[i][i]}(k) \otimes x_{[i]}(k) \oplus s(i, k+1) , \tag{2.19}$$

where

$$s(i, k+1) \overset{\text{def}}{=} A_{[i][<i]}(k) \otimes x_{[<i]}(k)$$

and $A_{[i][<i]}(k)$ is defined in the obvious way. By definition,

$$\|s(i, k+1)\|_\oplus \leq \|A_{[i][<i]}(k)\|_\oplus \otimes \|x_{[<i]}(k)\|_\oplus$$
$$\leq \|A(k)\|_\oplus \otimes \|x_{[<i]}(k)\|_\oplus . \tag{2.20}$$

Note that

$$\lim_{k \to \infty} \frac{1}{k}\|A(k)\|_\oplus = 0 \quad \text{a.s.} \tag{2.21}$$

Indeed, integrability of $\{A(k)\}$ together with stationarity and ergodicity implies that

$$\lim_{k \to \infty} \frac{1}{k}\sum_{n=1}^{k} \|A(k)\|_\oplus = \mathbb{E}[\|A(1)\|_\oplus] < \infty \quad \text{a.s.}$$

(integrability of $||A(1)||_\oplus$ is guaranteed by integrability of $A(1)$), which gives

$$\mathbb{E}[||A(1)||_\oplus] = \lim_{k \to \infty} \frac{1}{k} \sum_{n=1}^{k} ||A(n)||_\oplus$$

$$= \lim_{k \to \infty} \frac{k-1}{k} \frac{1}{k-1} \sum_{n=1}^{k-1} ||A(n)||_\oplus + \lim_{k \to \infty} \frac{1}{k} ||A(k)||_\oplus$$

$$= \mathbb{E}[||A(1)||_\oplus] + \lim_{k \to \infty} \frac{1}{k} ||A(k)||_\oplus \quad \text{a.s.}$$

and thus establishes (2.21).

We obtain from (2.20) together with (2.21)

$$\limsup_{k \to \infty} \frac{1}{k} ||s(i, k+1)||_\oplus \leq \lambda^{top}_{[<i]} \quad \text{a.s.}$$

At the same time, following the line of argument that has lead to (2.14), we obtain

$$||s(i, k+1)||_\oplus \geq ||x_{[<i]}(k-J)||_\oplus \quad \text{a.s.},$$

which implies

$$\liminf_{k \to \infty} \frac{1}{k} ||s(i, k+1)||_\oplus \geq \lambda^{top}_{[<i]} \quad \text{a.s.}$$

and thus

$$\lim_{k \to \infty} \frac{1}{k} ||s(i, k+1)||_\oplus = \lambda^{top}_{[<i]} \quad \text{a.s.}$$

It is clear from the definition of $s(i, k)$ that

$$||x_{[\leq i]}(k)||_\oplus \geq ||s(i, k)||_\oplus,$$

so that

$$\frac{1}{k} ||x_{[\leq i]}(k)||_\oplus \geq \frac{1}{k} ||s(i, k)||_\oplus,$$

which in turn implies

$$\lambda^{top}_{[\leq i]} \geq \lambda^{top}_{[<i]}. \tag{2.22}$$

Now suppose that $\lambda^{top}_{[\leq i]} > \lambda^{top}_{[<i]}$. The existence of the individual limits implies that for sufficiently large $K \in \mathbb{N}$ it holds that

$$A_{[i]\,[i]}(k) \otimes x_{[i]}(k) \geq s(i, k+1), \quad k \geq K.$$

Accordingly, equation (2.19) reads

$$x_{[i]}(k+1) = A_{[i]\,[i]}(k) \otimes x_{[i]}(k) \geq s(i, k+1), \quad k \geq K,$$

which, by Theorem 2.2.1, yields

$$\lim_{k \to \infty} \frac{1}{k} ||x_{[i]}(k)||_\oplus = \lambda^{top}_{[i]} \quad \text{a.s.}$$

and, by (2.17), this implies

$$\lambda_{[\leq i]}^{top} = \lambda_{[i]}^{top}.$$

We have thus shown that

$$\lambda_{[\leq i]}^{top} > \lambda_{[<i]}^{top} \quad \Rightarrow \quad \lambda_{[\leq i]}^{top} = \lambda_{[i]}^{top}. \tag{2.23}$$

Combining (2.22) and (2.23) we reach at:

$$\lambda_{[\leq i]}^{top} = \lambda_{[<i]}^{top} \oplus \lambda_{[i]}^{top}.$$

Any node $i \in \mathcal{G}_e(A)$ belongs to a m.s.c.s. that is represented in $\mathcal{G}_e^r(A)$ by the single node $[i]$. Let $\pi([i])$ denote the set of direct predecessors of $[i]$ in $\mathcal{G}_e^r(A)$ and set $\pi([i]) = \emptyset$ if there is no predecessor. Each element of $\pi([i])$ represents a m.s.c.s. in $\mathcal{G}_e(A)$ and we denote by $\tau(i)$ the set of nodes in $\mathcal{G}_e(A)$ that belong to the m.s.c.s. corresponding to the elements of $\pi([i])$. If $\pi([i]) = \emptyset$, we set $\tau(i) = \emptyset$. Then

$$\lambda_{[<i]}^{top} = \bigoplus_{j \in \tau(i)} \lambda_{[\leq j]}^{top}$$

and inserting this into the above equation yields

$$\lambda_{[\leq i]}^{top} = \lambda_{[i]}^{top} \oplus \bigoplus_{j \in \tau(i)} \lambda_{[\leq j]}^{top}.$$

We now repeat the argument until applying τ yields no more nodes. In particular, going from $\tau(i)$ to $\{\tau(j) : j \in \tau(i)\}$ and so forth, we will eventually cover the set $\pi^*(i)$. This concludes the proof of (2.18). □

Remark 2.2.7 *Suppose that the conditions in Theorem 2.2.3 are satisfied. Continuity of the operators* max *and* min *yields that it holds with probability one that*

$$\lambda^{bot} = \min(\lambda_j : 1 \leq j \leq J)$$

and

$$\lambda^{top} = \max(\lambda_j : 1 \leq j \leq J).$$

The vector $\vec{\lambda} = (\lambda_1, \lambda_2, \ldots, \lambda_J)$, with λ_j defined in Theorem 2.2.3, is called the *Lyapunov vector* of $\{A(k)\}$. In the light of Theorem 2.2.2 we can state that irreducibility of $\{A(k)\}$ is a sufficient condition for the components of $\vec{\lambda}$ to be equal.

Recalling that $\lim_{k \to \infty} x_j(k)/k$ is the (asymptotic) speed with which transition j operates, the above theorem matches our intuition that the (asymptotic) speed with which the system operates is determined by the slowest component of the system. In terms of queuing networks, the throughput of a system is determined by the smallest throughput of one of its components. Moreover, if the queuing network is irreducible in the max-plus sense, then the throughput is the same at any station.

The key conditions on $A(k)$ are that any element of $A(k)$ is either equal to ε or non-negative, that the elements on the diagonal are non-negative and that it has fixed support. As we have already explained, the condition that any element different from ε has only non-negative values is a natural condition for queuing systems, and all examples presented in this monograph enjoy this property. The fixed support condition is satisfied by the queuing systems in Example 1.5.1 and Example 1.5.2. An example of a system that fails to have fixed support is given in Example 1.5.5. Such a system cannot be analyzed via the subadditive ergodic theory developed so far.

2.2.3 Variations and Extensions

One of the marvels of max-plus theory is that the existence of the top and bottom Lyapunov exponent follows so easily from Kingman's subadditive ergodic theorem. See the proof of Theorem 2.2.1. However, the conditions in Theorem 2.2.1 are too weak to guarantee that the top and bottom Lyapunov exponents are equal, or, in other words, that the individual growth rates (that is, $\lim_{k\to\infty} x_i(k)/k$, $1 \leq j \leq J$) have the same limit. In this section, we discuss approaches to establish equality of the top and bottom Lyapunov exponent without imposing conditions on the elements of $A(k)$.

2.2.3.1 The 'Up-Crossing' Property

In order to show that the individual growth rates coincide we had to impose the assumption that (i) any non-ε element of $A(k)$ is non-negative, that (ii) all diagonal elements are non-negative, and that (iii) $A(k)$ has fixed support. The 'non-negativity' condition on the finite elements causes no restriction for queuing systems. Therefore, we focus in this section on a relaxation of the 'fixed support' and the 'diagonal' condition.

Inspecting the proof of Theorem 2.2.2 one sees that what is actually needed is the following 'up-crossing' property: a subsequence $\{x(k_n)\}$ and a constant M exist, such that for any $n \geq 1$

$$\|x(k_n + M)\|_{\min} \geq a_n + b_n \|x(k_n)\|_{\max} \quad \text{a.s.},$$

with

$$\lim_{n\to\infty} \frac{a_n}{n} = 0 \quad \text{and} \quad \lim_{n\to\infty} b_n = 1,$$

see (2.10) on page 72 in the proof of Theorem 2.2.2, where $a_n = 0$ and $b_n = 1$ for all n. Indeed, Vincent uses in [102] this type of condition to show that the top and bottom Lyapunov exponent coincide. Provided that finite elements of $A(k)$ are positive, the diagonal condition together with fixed support are sufficient for the above 'up-crossing' property to hold, see Lemma 1.4.1.

2.2.3.2 The 'Memory Loss' Property

In this section we present an alternative approach to finding sufficient conditions for $\lambda^{\text{top}} = \lambda^{\text{bot}}$. This approach goes back to [48, 84] and applies to sequences with countable state-space.

The key observation for this approach is the following. Let $A \in \mathbb{R}_{\max}^{J \times J}$ be such that any two columns of A are linear dependent. Then, a finite number a exists such that

$$||A \otimes x||_{\max} - ||A \otimes x||_{\min} = a, \quad x \in \mathbb{R}^J \qquad (2.24)$$

(for a proof use the argument put forward in the proof of Corollary 2.1.1). A matrix with the property that any two columns are linear dependent is said to be of *rank 1*. While the notation of rank 1 is undisputed, there are several notions of rank in the literature, see [37] and [103].

Definition 2.2.1 *A sequence $\{A(k)\}$ of square matrices is said to have memory loss property (MLP) if there exists an N such that $A(N-1) \otimes A(N-2) \otimes \cdots \otimes A(0)$ with positive probability has only mutually linear dependent columns, i.e., is of rank 1.*

Let A be a matrix with mutually linear dependent columns and assume that $\{A(k)\}$ has MLP with respect to A and N, that is, assume that a finite number N exists such that $P(A(N-1) \otimes A(N-2) \otimes \cdots \otimes A(0) = A) > 0$ and A is of rank 1. Let

$$T_0 = \inf\{k \geq N - 1 : A(k) \otimes A(k-1) \otimes \cdots \otimes A(k-N+1) = A\}$$

denote the first time a partial product of the series of matrix generates A. This gives

$$x(T_0) = A \otimes \bigotimes_{k=0}^{T_0-N} A(k) \otimes x_0,$$

where we set the product to E for $T_0 = N - 1$ and we assume that $x_0 \in \mathbb{R}^J$. By (2.24) a finite number a exists such that

$$||x(T_0)||_{\max} - ||x(T_0)||_{\min} = a,$$

for any finite initial value x_0. For $n \geq 0$, introduce the time of the $(n+1)^{st}$ occurrence of the event that a partial product of $\{A(k)\}$ generates A by

$$T_{n+1} = \inf\{k \geq N + T_n : A(k) \otimes A(k-1) \otimes \cdots \otimes A(k-N+1) = A\} \quad (2.25)$$

and we obtain

$$||x(T_k)||_{\max} - ||x(T_k)||_{\min} = a, \quad k \geq 0. \qquad (2.26)$$

If $\{A(k)\}$ is stationary and ergodic, then $\lim_{n \to \infty} T_n = \infty$ and $T_n < \infty$ with probability one; for details see Section E.3 in the Appendix. Specifically, by equation (2.26),

$$\lim_{k \to \infty} \frac{1}{T_k} ||x(T_k)||_{\max} - \frac{1}{T_k} ||x(T_k)||_{\min} = 0 \quad \text{a.s.} \qquad (2.27)$$

If, in addition, $\{A(k)\}$ is a sequence of a.s. regular and integrable matrices, Theorem 2.2.1 yields

$$\lim_{k\to\infty}\frac{1}{T_k}||x(T_k)||_{\max} = \lambda^{\text{top}} \quad\text{and}\quad \lim_{k\to\infty}\frac{1}{T_k}||x(T_k)||_{\min} = \lambda^{\text{bot}}$$

with probability one, and equality of λ^{top} and λ^{bot} follows from (2.27). We summarize our analysis in the following theorem:

Theorem 2.2.4 *Let $\{A(k)\}$ be a stationary and ergodic sequence of integrable and a.s. regular matrices in $\mathbb{R}_{\max}^{J\times J}$. If $\{A(k)\}$ has MLP, then a finite constant λ exists such that, for any (non-random) finite initial conditions x_0:*

$$\lim_{k\to\infty}\frac{x_j(k)}{k} = \lim_{k\to\infty}\frac{||x(k)||_{\min}}{k} = \lim_{k\to\infty}\frac{||x(k)||_{\max}}{k} = \lambda \quad a.s.$$

and

$$\lim_{k\to\infty}\frac{1}{k}\mathbb{E}\big[x_j(k)\big] = \lim_{k\to\infty}\frac{1}{k}\mathbb{E}\big[||x(k)||_{\min}\big] = \lim_{k\to\infty}\frac{1}{k}\mathbb{E}\big[||x(k)||_{\max}\big] = \lambda,$$

for $1 \le j \le J$. The above limits also hold for random initial conditions provided that the initial condition is a.s. finite and integrable.

It is worth noting that, in contrast to Theorem 2.2.3, the Lyapunov exponent is unique, or, in other words, the components of the Lyapunov vector are equal. In view of Theorem 2.2.2 the above theorem can be phrased as follows: Theorem 2.2.2 remains valid in the presence of reducible matrices if MLP is satisfied.

MLP is a technical condition and typically impossible to verify directly. A sufficient condition for $\{A(k)\}$ to have MLP is the following:

(C) There exists a primitive matrix C and $N \in \mathbb{N}$ such that

$$P\Big(A(N-1) \otimes A(N-2) \otimes \cdots \otimes A(0) = C \Big) > 0.$$

The following lemma illustrates the close relationship between primitive matrices and matrices of rank 1.

Lemma 2.2.2 *If A is primitive with coupling time c, then A^c has only finite entries and is of rank 1. Moreover, for any matrix A that has only finite entries it holds that A is of rank 1 if and only if the projective image of A is a single point in the projective space.*

Proof: We first prove the second part of the lemma. '$\Rightarrow$': Let $A \in \mathbb{R}_{\max}^{I\times J}$ be such that all elements are finite and that it is of rank 1. Denote the jth column of A by $A_{\cdot j}$. Since A is of rank 1, there exits finite numbers α_j, with $2 \le j \le J$, such that $A_{\cdot 1} = \alpha_j \otimes A_{\cdot j}$ for $2 \le j \le J$. Hence, for $x \in \mathbb{R}^J$ it holds that

$$A \otimes x = \bigotimes_{j=1}^{J}\alpha_j \otimes x_j \otimes A_{\cdot 1}, \qquad (2.28)$$

with $\alpha_1 = 0$. Let $\gamma_x \overset{\text{def}}{=} \bigotimes_{j=1}^{J} \alpha_j \otimes x_j$. Let $y \in \mathbb{R}^J$, with $x \neq y$. By (2.28), $A \otimes x = \gamma_x \otimes A_{.1}$ and $A \otimes y = \gamma_y \otimes A_{.1}$, which implies that $A \otimes x$ and $A \otimes y$ are linear dependent. Hence, the projective image of A contains only the single point $\overline{A_{.1}}$.

'$\Leftarrow$': We give a proof by contradiction. Suppose that A is not of rank 1, then there exist at least two columns $A_{.j}$ and $A_{.i}$ of A such that $A_{.j}$ and $A_{.i}$ are linear independent. Then $x^i, x^j \in \mathbb{R}^J$ can be chosen such that $A \otimes x^i = \beta^i \otimes A_{.i}$ and $A \otimes x^j = \beta^j \otimes A_{.j}$ for finite constants β^i, β^j. Since $A_{.j}$ and $A_{.i}$ are linear independent, the projective image of A contains at least the two distinct points $\overline{A_{.i}}$ and $\overline{A_{.j}}$.

We now turn to the proof of the first part of the lemma. For $1 \leq j \leq J$, let $\mathbf{e}_j$ be the vector with ε entries except for element j which is equal to e. Hence, $A^c \otimes \mathbf{e}_j = A^c_{.j}$, where $A^c_{.j}$ denotes the jth column of A^c. By Theorem 2.1.1,

$$A \otimes A^c_{.j} = A \otimes A^c \otimes \mathbf{e}_j = \lambda \otimes A^c \otimes \mathbf{e}_j = \lambda \otimes A^c_{.j},$$

with λ the unique eigenvector of A, and the columns of A^c are thus eigenvectors of A. Using the fact that eigenvectors of irreducible matrices have only finite entries (see, for example, Lemma 2.8 in [65]), it follows that A^c has only finite elements. On the one hand, by Corollary 2.1.1, the eigenvector of A is unique. On the other hand, by Theorem 2.1.1, $A^c \otimes x$ is an eigenvector of A for any x. Hence, the projective image of A is a single point (in formula: $\exists v \in \mathbb{PR}^J \ \forall x \in \mathbb{R}^J : \overline{A^c \otimes x} = v$). Applying the second part of the lemma then proves the claim. $\square$

We present a version of Theorem 2.2.4 with a condition that can be directly verified.

Lemma 2.2.3 *Let $\{A(k)\}$ be an i.i.d. sequence of a.s. regular integrable matrices in $\mathbb{R}^{J \times J}_{\max}$ with countable state space. If condition (C) holds, then the statement put forward in Theorem 2.2.4 holds.*

Proof: Let C be as given as in (C) and denote the coupling time of C by c. Because $\{A(k)\}$ is i.i.d. with countable state-space,

$$P\Big(A(N-1) = A(N-2) = \cdots = A(0) = C \Big) > 0\,,$$

implies

$$P\Big(A(cN-1) \otimes A(cN-2) \otimes \cdots \otimes A(0) = C^c \Big) > 0.$$

Since C is primitive, Lemma 2.2.2 implies that C^c is of rank 1 and $\{A(k)\}$ has thus MLP. Hence, Theorem 2.2.4 applies. $\square$

Example 2.2.1 *Consider Example 1.5.5. Matrix D_2 is primitive. Hence, applying Lemma 2.2.3 shows that the Lyapunov exponent of the system exists.*

Remark 2.2.8 *In principle, MLP and condition (C) restrict the class of sequences $\{A(k)\}$ that can be analyzed to those with countable state-space. A* *possible generalization is the following. Suppose that the distribution of $A(k)$ is a mixture of a discrete distribution on a countable state-space, say $\mathcal{A}^c$, and a general distribution on an arbitrary state-space, say $\mathcal{A}^g$. If we require $P(A(N-1) \otimes A(N-2) \otimes \cdots \otimes A(0) \in \mathcal{A}^c) > 0$ in Definition 2.2.1 and $C \in \mathcal{A}^c$ in condition (**C**), respectively, then the results in this section hold for $\{A(k)\}$ with state space $\mathcal{A}^c \cup \mathcal{A}^g$ as well.*

We conclude this section by presenting a generalization of Theorem 2.2.4. As Baccelli and Mairesse show in [11], using the arguments put forward in this section, a limit result can be obtained under a slightly weaker condition than MLP.

Theorem 2.2.5 *Let $\{A(k)\}$ be a stationary and ergodic sequence of integrable and a.s. regular square matrices in $\mathbb{R}^{J \times J}_{\max}$. If there exists $N \in \mathbb{N}$ such that with positive probability $A(N-1) \otimes A(N-2) \otimes \cdots \otimes A(0)$ has a bounded projective image, then the statement put forward in Theorem 2.2.4 holds.*

Proof: By assumption, there exist finite numbers $a, b \in \mathbb{R}$ such that

$$\forall x \in \mathbb{R}^J: \quad ||A(N-1) \otimes A(N-2) \otimes \cdots \otimes A(0) \otimes x||_{\mathbb{P}} \in [a, b].$$

In analogy to (2.25), let T_k denote the time index such that for the kth time a product $A(T_k) \otimes A(T_k - 1) \otimes \cdots \otimes A(T_k - N + 1)$ has been observed whose projective image lies within the interval $[a, b]$; in formula:

$$a \leq ||x(T_k)||_{\max} - ||x(T_k)||_{\min} \leq b$$

for all k. We have assumed that $\{A(k)\}$ is stationary and ergodic, which implies $\lim_{n \to \infty} T_k = \infty$ and $T_k < \infty$ with probability one; for details see Section E.3 in the Appendix. Since $[a, b]$ is compact, the Bolzano-Weierstrass Theorem yields the existence of a subsequence $\{T_{k_n}\}$ of $\{T_k\}$ such that

$$\lim_{n \to \infty} ||x(T_{k_n})||_{\max} - ||x(T_{k_n})||_{\min} = c,$$

for some finite constant c, which implies

$$\lim_{n \to \infty} \frac{1}{k_n} ||x(T_{k_n})||_{\max} = \lim_{n \to \infty} \frac{1}{k_n} ||x(T_{k_n})||_{\min}. \tag{2.29}$$

By Theorem 2.2.1, convergence of the sequences $||x(k)||_{\max}/k$ and $||x(k)||_{\max}/k$ as k tends to infinity is guaranteed. Hence,

$$
\begin{aligned}
\lim_{k \to \infty} \frac{1}{k} ||x(k)||_{\max} &= \lim_{n \to \infty} \frac{1}{k_n} ||x(T_{k_n})||_{\max} \\
&\overset{(2.29)}{=} \lim_{n \to \infty} \frac{1}{k_n} ||x(T_{k_n})||_{\min} \\
&= \lim_{k \to \infty} \frac{1}{k} ||x(k)||_{\min},
\end{aligned}
$$

which proves the claim. $\square$

2.2.3.3 Weak Irreducibility

An approach relaxing the concept of fixed support can be found in [69, 70]. This approach is based on an interpretation of the concept of 'irreducibility' for random matrices which we will explain in the following.

Irreducibility of a matrix A is defined via the communication graph of A, denoted by $\mathcal{G}(A)$. Specifically, A is called irreducible if for any two nodes in $\mathcal{G}(A)$ a path from i to j exists in $\mathcal{G}(A)$. Let $\{A(k)\}$ be a random sequence of $J \times J$ dimensional matrices. The communication graph of a random sequence is itself a random variable and we extend the definition of a path to the sequence $\mathcal{G}(A(k))$ as follows. For any two nodes i, j, a sequence of arcs $\rho = ((i_n, j_n) : 1 \leq n \leq m)$, with $i = i_1$, $j = j_m$ and $j_n = i_{n+1}$ for $1 \leq n < m$, is called a *path of length m* from i to j in $\{A(k)\}$ if (i_n, j_n) is an arc in $\mathcal{G}(A(k + n - 1))$ for $1 \leq n \leq m$, for some $k \in \mathbb{N}$. We say that ρ is a path in $\mathcal{G}(A(k + n - 1) : 1 \leq n \leq m)$.

The *weight of a path* in $\mathcal{G}(A)$ is defined by the sum of the weights of all arcs constituting the path; more formally: let $\rho = ((i_n, j_n) : 1 \leq n \leq m)$ be a path from i to j of length m, then the weight of ρ, denoted by $|\rho|_w$, is given by

$$|\rho|_w = \bigotimes_{n=1}^{m} (A(k + n - 1))_{j_n i_n},$$

with $i = i_1$ and $j = j_m$, for some k.

We now are able to introduce the concept of weak irreducibility: A sequence $\{A(k)\}$ of square matrices is said to be *weakly irreducible* if for any pair of nodes $i, j \in \{1, \ldots, J\}$ a finite number m_{ij} exists such that there is with positive probability a path of length m_{ij} from i to j; in formula: for any i, j, with $1 \leq i, j \leq J$, $m_{ij} \in \mathbb{N}$ exists such that

$$P\left(\left(\bigotimes_{k=0}^{m_{ij}-1} A(k) \right)_{ji} > \varepsilon \right) > 0.$$

Theorem 2.2.6 *Let $\{A(k)\}$ be an i.i.d. sequence of regular, integrable matrices in $\mathbb{R}_{\max}^{J \times J}$ with countable state-space. Assume that $\{A(k)\}$ is weakly irreducible. If there exists at least one node j such that j lies with positive probability on a circuit of length one, then the Lyapunov exponent of $\{A(k)\}$ exists.*

Proof: Consider the collection of numbers m_{ij} for $1 \leq i, j \leq J$. We have assumed that there exits at least one node j^* such that $m_{j^* j^*} = 1$ and the greatest common divisor of the collection of numbers m_{ij}, with $1 \leq i, j \leq J$, is thus equal to one. This implies that a finite number N exists such that each $m \geq N$ can be written as a linear combination of m_{ij}'s, see [26]. Weak analyticity thus implies that for any $m \geq N$ there exists with positive probability a path from any node to any other node; in formula: for any $m \geq N$

$$\forall i, j \in \{1, \ldots, J\} : \quad P\left(\left(\bigotimes_{k=0}^{m-1} A(k) \right)_{ji} > \varepsilon \right) > 0.$$

Let $h \geq N$. Since J is finite, we can choose $j, i \in \{1, \ldots, J\}$ such that a sequence $\{m_n\}$, with $\lim_{n \to \infty} m_n = \infty$, exists for which it holds that

$$\|x(m_n + h)\|_{\min} = x_i(m_n + h) \quad \text{and} \quad \|x(m_n)\|_{\max} = x_j(m_n),$$

for $n \in \mathbb{N}$. By assumption, $\{A(k)\}$ is a weakly irreducible i.i.d. sequence. Hence, we may select a subsequence $\{m_{n_l}\}$ of $\{m_n\}$ such that there is (at least) a fixed path ρ from j to i of length h in $\mathcal{G}(A(m_{n_l+k}) : 0 \leq k < h)$ for any l and

$$w_{ij} \overset{\text{def}}{=} \left(\bigotimes_{m=m_{n_l}}^{m_{n_l}+h-1} A(m) \right)_{ij}$$

is finite. With slight abuse of notation we will identify $\{m_n\}$ and $\{m_{n_l}\}$. This yields

$$\|x(m_n + h)\|_{\min} = x_i(m_n + h)$$
$$= \bigoplus_{k=1}^{J} \left(\bigotimes_{m=m_n}^{m_n+h-1} A(m) \right)_{ik} \otimes x_k(m_n)$$
$$\geq -|w_{ij}| \otimes x_j(m_n)$$
$$= -|w_{ij}| + \|x(m_n)\|_{\max},$$

which establishes the up-crossing property with $M = h$. $\square$

Theorem 2.2.6 provides a sufficient condition for the existence of the Lyapunov exponent completely avoiding the concept of fixed support. The following example illustrates this. Consider $A_1, A_2 \in \mathcal{A}$, with

$$A_1 = \begin{pmatrix} Y_1 & \varepsilon \\ \varepsilon & Y_4 \end{pmatrix} \quad \text{and} \quad A_2 = \begin{pmatrix} \varepsilon & Y_2 \\ Y_3 & \varepsilon \end{pmatrix},$$

for some finite integrable random variables Y_i, $1 \leq i \leq 4$. Let $\{A(k)\}$ be an i.i.d. sequence such that $P(A(k) = A_1) = p > 0$ and $P(A(k) = A_2) = 1 - p > 0$, for $k \geq 0$. Then $\{A(k)\}$ satisfies the condition put forward in Theorem 2.2.6. However, neither does $\{A(k)\}$ have fixed support nor does it satisfy the diagonal condition. Note that the situation in Example 1.5.5 is covered by Theorem 2.2.6, which follows from the fact that D_2 is irreducible and contains one finite element on its diagonal.

As Hong shows in [69, 70], the condition that there is at least one node that lies with positive probability on a circuit of length one is not necessary for Theorem 2.2.6 to hold. Without this simplifying assumption the proof of the theorem becomes however rather technical and the interested reader is referred to [69, 70] for details.

2.3 Stability Analysis of Waiting Times (Type IIa)

A classical result in queuing theory states that if in a G/G/1 queue the expected interarrival time is larger than the expected service time, then the sequence of waiting times converges, independent of the initial condition, to a unique stationary regime. The proof of this result goes back to [81]. In this section, we generalize the classical result on stability of waiting times in the GI/G/1 queue to that of stability of waiting times in open max-plus linear networks. It is worth noting that by virtue of the max-plus formalism we can almost literally copy the proof of the classical result in [81].

We consider the following situation. An open queuing network with J stations is given such that the vector of departure times from the stations, denoted by $x(k)$, follows the recurrence relation

$$x(k+1) = A(k) \otimes x(k) \oplus \tau(k+1) \otimes B(k), \qquad (2.30)$$

with $x(0) = \mathbf{e}$, where $\tau(k)$ denotes the time of the kth arrival to the system. See, equation (1.15) in Section 1.4.2.2 and equation (1.27) in Example 1.5.2, respectively. As usually, we denote by $\sigma_0(k)$ the kth interarrival time, so that the kth arrival of a customer at the network happens at time

$$\tau(k) = \sum_{i=1}^{k} \sigma_0(i), \quad k \geq 1,$$

with $\tau(0) = 0$. Then, $W_j(k) = x_j(k) - \tau(k)$ denotes the time the kth customer arriving to the system spends in the system until completion of service at server j. The vector of kth sojourn times, denoted by $W(k) = (W_1(k), \ldots, W_J(k))$, follows the recurrence relation

$$W(k+1) = A(k) \otimes C(\sigma_0(k+1)) \otimes W(k) \oplus B(k), \quad k \geq 0,$$

with $W(0) = \mathbf{e}$, where $C(h)$ denotes a diagonal matrix with $-h$ on the diagonal and ε elsewhere. See Section 1.4.4 for details. Alternatively, $x_j(k)$ in (2.30) may model the times of the kth beginning of service at station j. With this interpretation of $x(k)$, $W_j(k)$ defined above represents the time spent by the kth customer arriving to the system until beginning of her/his service at j. For example, in the G/G/1 queue $W(k)$ models the waiting time.

In the following we will establish sufficient conditions for $W(k)$ to converge to a unique stationary regime. The main technical assumptions are:

(W1) For $k \in \mathbb{Z}$, let $A(k) \in \mathbb{R}_{\max}^{J \times J}$ be a.s. regular and assume that the maximal Lyapunov exponent of $\{A(k)\}$ exists.

(W2) There exists a fixed number α, with $1 \leq \alpha \leq J$, such that the vector $B^{\alpha}(k) = (B_j(k) : 1 \leq j \leq \alpha)$ has finite elements for any k, and $B_j(k) = \varepsilon$, for $\alpha < j \leq J$ and any k.

(W3) The sequence $\{(A(k), B^{\alpha}(k))\}$ is stationary and ergodic, and independent of $\{\tau(k)\}$, where $\tau(k)$ is given by

$$\tau(k) = \sum_{i=1}^{k} \sigma(i), \quad k \geq 1,$$

with $\tau(0) = 0$ and $\{\sigma(k) : k \in \mathbb{Z}\}$ a stationary and ergodic sequence of positive random variables with mean $\nu \in (0, \infty)$.

In what follows, we establish sufficient conditions for $\{W(k)\}$, with

$$W(k+1) = A(k) \otimes C(\sigma(k+1)) \otimes W(k) \oplus B(k) , \quad k \geq 0 , \qquad (2.31)$$

to have a unique stationary solution.

Provided that $\{A(k)\}$ is a.s. regular and stationary, integrability of $A(k)$ is a sufficient condition for (**W1**), see Theorem 2.2.1. In terms of queuing networks, the main restriction imposed by these conditions stems from the non-negativity of the diagonal of $A(k)$, see Section 2.2 for a detailed discussion and possible relaxations. The part of condition (**W3**) that concerns the arrival stream of the network is, for example, satisfied for Poisson arrival streams.

The proof goes back to [19] and has three main steps. First, we introduce Loynes' scheme for sojourn times. In a second step we show that the Loynes variable converges a.s. to a finite limit. Finally, we show that this limit is the unique stationary solution of equations of type (2.31).

Step 1 (the Loynes's scheme): Let $M(k)$ denote the vector of sojourn times at time zero provided that the sequence of waiting time vectors was started at time $-k$ in $B(-(k+1))$. For $k > 0$, we set

$$\tau(-k) = -\sum_{i=0}^{k-1} \sigma(-i) .$$

By recurrence relation (2.31),

$$M(1) = A(-1) \otimes C(\sigma(0)) \otimes B(-2) \oplus B(-1) .$$

For $M(2)$ we have to replace $B(-2)$ by

$$A(-2) \otimes C(\sigma(-1)) \otimes B(-3) \oplus B(-2) , \qquad (2.32)$$

which yields

$$\begin{aligned} M(2) = {} & A(-1) \otimes C(\sigma(0)) \otimes A(-2) \otimes C(\sigma(-1)) \otimes B(-3) \\ & \oplus A(-1) \otimes C(\sigma(0)) \otimes B(-2) \oplus B(-1) . \end{aligned} \qquad (2.33)$$

By finite induction, we obtain for $M(k)$

$$M(k) = \bigoplus_{j=0}^{k} \bigotimes_{i=1}^{j} A(-i) \otimes C(\sigma(-i+1)) \otimes B(-(j+1)) , \qquad (2.34)$$

where we set the product

$$\bigotimes_{i=1}^{j} A(-i) \otimes C(\sigma(-i+1))$$

to E for $j = 0$.

The sequence $\{M(k)\}$ is called *Loynes sequence*. The above construction implies that $\{M(k)\}$ is monotone increasing in k. To see this, denote for $x, y \in \mathbb{R}_{\max}^J$ the component-wise ordering of x and y by $x \le y$. By calculation,

$$M(k) = \bigoplus_{j=0}^{k} \bigotimes_{i=1}^{j} A(-i) \otimes C(\sigma(-i+1)) \otimes B(-(j+1))$$

$$\le \bigotimes_{i=1}^{k+1} A(-i) \otimes C(\sigma(-i+1)) \otimes B(-(k+1))$$

$$\oplus \bigoplus_{j=0}^{k} \bigotimes_{i=1}^{j} A(-i) \otimes C(\sigma(-i+1)) \otimes B(-(j+1))$$

$$= \bigoplus_{j=0}^{k+1} \bigotimes_{i=1}^{j} A(-i) \otimes C(\sigma(-i+1)) \otimes B(-(j+1))$$

$$= M(k+1),$$

for $k \ge 0$, which proves that $M(k)$ is monotone increasing in k.

The matrix $C(\cdot)$ has the following properties. For any $y \in \mathbb{R}$, $C(y)$ commutes with any matrix $A \in \mathbb{R}_{\max}^{J \times J}$:

$$C(y) \otimes A = A \otimes C(y).$$

Furthermore, for $y, z \in \mathbb{R}$, it holds that

$$C(y) \otimes C(z) = C(z) \otimes C(y) = C(y + z).$$

Specifically,

$$\bigotimes_{i=1}^{j} C(\sigma(-i+1)) = C\left(\bigotimes_{i=1}^{j} \sigma(-i+1) \right) = C(-\tau(-j)).$$

Elaborating on these rules of computation, we obtain

$$\bigotimes_{i=1}^{j} A(-i) \otimes C(\sigma(-i)) \otimes B(-(j+1)) = C(-\tau(-j)) \otimes \bigotimes_{i=1}^{j} A(-i) \otimes B(-(j+1)).$$

Set

$$D(k) = \bigotimes_{i=1}^{k} A(-i) \otimes B(-(k+1)), \quad k \ge 1,$$

and, for $k = 0$, set $D(0) = B(-1)$. Note that $\tau(0) = 0$ implies that $C(-\tau(0)) = E$. Equation (2.34) now reads

$$M(k) = \bigoplus_{j=0}^{k} C(-\tau(-j)) \otimes D(j).$$

Step 2 (pathwise limit): We now show that the limit of $M(k)$ as k tends to ∞ exists and establish a sufficient condition for the limit to be a.s. finite.

Because $M(k)$ is monotone increasing in k, the random variable M, defined by

$$\lim_{k\to\infty} M(k) = \bigoplus_{j\geq 0} C(-\tau(-j)) \otimes D(j)$$
$$\stackrel{\text{def}}{=} M,$$

is either equal to ∞ or finite. The variable M is called *Loynes variable*. In what follows we will derive a sufficient condition for M to be a.s. finite. As a first step towards this result, we study three individual limits.

(i) Under condition (**W1**), a number $\mathbf{a} \in \mathbb{R}$ exists such that, for any $x \in \mathbb{R}^J$,

$$\lim_{k\to\infty} \frac{1}{k} \left\| \bigotimes_{j=1}^{k} A(-j) \otimes x \right\|_{\max} = \mathbf{a} \qquad \text{a.s.}$$

(ii) Under condition (**W3**), the strong law of large numbers (which is a special case of Theorem 2.2.3) implies

$$\lim_{k\to\infty} \frac{1}{k} \|C(-\tau(-k))\|_{\max} = \lim_{k\to\infty} \frac{1}{k}\tau(-k)$$
$$= -\lim_{k\to\infty} \frac{1}{k} \sum_{i=-k+1}^{0} \sigma(i)$$
$$= -\nu \qquad \text{a.s.}$$

(iii) Ergodicity of $\{B^\alpha(k)\}$ (condition (**W3**)) implies that, for $1 \leq j \leq \alpha$, a $b_j \in \mathbb{R}$ exists such that

$$\lim_{k\to\infty} \frac{1}{k} \sum_{i=1}^{k} B_j(-i) = b_j \qquad \text{a.s.},$$

which implies that it holds with probability one that

$$b_j = \lim_{k\to\infty} \frac{1}{k} \sum_{i=1}^{k} B_j(-i)$$
$$= \lim_{k\to\infty} \frac{1}{k} B_j(-k) + \lim_{k\to\infty} \frac{k-1}{k} \frac{1}{k-1} \sum_{i=1}^{k-1} B_j(-i)$$
$$= \lim_{k\to\infty} \frac{1}{k} B_j(-k) + b_j,$$

and thus

$$\lim_{k\to\infty} \frac{1}{k} B_j(-k) = \lim_{k\to\infty} \frac{1}{k} B_j(-(k+1)) = 0 \qquad \text{a.s.},$$

for $j \leq \alpha$. From the above we conclude that

$$\lim_{k \to \infty} \frac{1}{k} \|B(-k)\|_{\max} = 0 \qquad \text{a.s.}$$

From Lemma 1.6.2 it follows that

$$\|C(-\tau(-k)) \otimes D(k)\|_{\max} = \left\| C(-\tau(-k)) \otimes \bigotimes_{i=1}^{k} A(-i) \otimes B(-(k+1)) \right\|_{\max}$$

$$\leq \|C(-\tau(-k))\|_{\max} + \left\| \bigotimes_{i=1}^{k} A(-i) \otimes e \right\|_{\max}$$

$$+ \|B(-(k+1))\|_{\max} .$$

Combining the individual limits (i)-(iii), we obtain

$$\lim_{k \to \infty} \frac{1}{k} \|C(-\tau(-k)) \otimes D(k)\|_{\max} \leq \mathbf{a} - \nu \qquad \text{a.s.}$$

and $\nu > \mathbf{a}$ implies

$$\lim_{k \to \infty} \|C(-\tau(-k)) \otimes D(k)\|_{\max} = -\infty \qquad \text{a.s.} \qquad (2.35)$$

Hence, for k sufficiently large, the vector $C(-\tau(-k)) \otimes D(k)$ has only negative elements and thus doesn't contribute to $M(k)$ (note that $M(k) \geq 0$ by definition). Consequently, $M(k)$ is dominated by the maximum over finitely many vectors $C(-\tau(-k)) \otimes D(k)$ whose elements are all finite. We have thus shown that $\nu > \mathbf{a}$ implies that M is an a.s. finite random variable. In the same vein, one can show that $\nu < \mathbf{a}$ implies $M = \infty$ with probability one.

Step 3 (stationarity and uniqueness): We revisit the construction of $\{M(k)\}$. Under assumption (**W3**), let θ denote an ergodic shift operator such that $A(k) = A \circ \theta^k$, $B(k) = B \circ \theta^k$ and $\sigma(k) = \sigma \circ \theta^k$, for appropriately defined random variables A, B, σ, see Section E.3 in the Appendix. Equation (2.33) thus reads

$$M(2) = A \circ \theta^{-1} \otimes C(\sigma) \otimes M(1) \circ \theta^{-1} \oplus B \circ \theta^{-1}$$

(to see this, note that the expression in (2.32) is equivalent to $M(1) \circ \theta^{-1}$). By finite induction,

$$M(k+1) = A \circ \theta^{-1} \otimes C(\sigma) \otimes M(k) \circ \theta^{-1} \oplus B \circ \theta^{-1}$$

and letting k tend to ∞ in the above equation shows that

$$M = A \circ \theta^{-1} \otimes C(\sigma) \otimes M \oplus B \circ \theta^{-1} .$$

In other words, M is the stationary solution of (2.31).

It remains to be shown that M is the unique limit. Let $M(k, w)$ denote the vector of sojourn times at time 0 provided that the sequence is started at time $-k$ with initial vector $w \in \mathbb{R}^J$, or, more formally, set

$$M(k, w) = \bigotimes_{i=1}^{k} A(-i) \otimes C(\sigma(-i+1)) \otimes w$$
$$\oplus \bigoplus_{j=0}^{k-1} C(-\tau(-j)) \otimes D(j) \, .$$

Because w has only finite elements, we have $\|w\|_{\max} < \infty$. Following the line of argument in step 2 above, it readily follows that

$$\lim_{k \to \infty} \left\| \bigotimes_{i=1}^{k} A(-i) \otimes C(\sigma(-i+1)) \otimes w \right\|_{\max} = -\infty \quad \text{a.s.} \, ,$$

for $\nu > \mathbf{a}$, and

$$\lim_{k \to \infty} \bigotimes_{i=1}^{k} A(-i) \otimes C(\sigma(-i+1)) \otimes w \oplus \bigoplus_{j=0}^{k-1} C(-\tau(-j)) \otimes D(j) = M \quad \text{a.s.}$$

Hence, for any finite initial value w, $M(k, w)$ has the same limit as $M(k)$, which establishes uniqueness. We have thus shown that $M(k, w)$ converges a.s. to a unique stationary limit M, independent of the initial value w.

For $w \in \mathbb{R}^J$, write $W(k, w)$ for the vector of kth system times, initiated at 0 to w. Assumption (**W3**) implies that $M(k, w)$ and $W(k, w)$ are equal in distribution. Hence, M is the unique weak limit of $\{W(k, w)\}$ for arbitrary $w \in \mathbb{R}^J$. We summarize our analysis in the following theorem.

Theorem 2.3.1 *Assume that assumptions* (**W1**), (**W2**) *and* (**W3**) *are satisfied and denote the maximal Lyapunov exponent of* $\{A(k)\}$ *by* $\mathbf{a}$. *If* $\nu > \mathbf{a}$, *then the sequence*

$$W(k+1) = A(k) \otimes C(\sigma(k+1)) \otimes W(k) \oplus B(k)$$

converges with strong coupling to an unique stationary regime W, *with*

$$W = D(0) \oplus \bigoplus_{j \geq 1} C(-\tau(-j)) \otimes D(j) \, ,$$

where $D(0) = B \circ \theta^{-1}$ *and*

$$D(j) = \bigotimes_{i=1}^{j} A(-i) \otimes B(-(j+1)) \, , \quad j \geq 1 \, .$$

Proof: It remains to be shown that the convergence of $\{W(k)\}$ towards W happens with strong coupling. For $w \in \mathbb{R}^J$, let $W(k, w)$ denote the vector of kth

sojourn times, initiated at 0 to w. From the forward construction, see (1.22) on page 20, we obtain

$$W(k+1, w) = \bigotimes_{i=0}^{k} A(i) \otimes C(\sigma(i+1)) \otimes w$$

$$\oplus \bigoplus_{i=0}^{k} \bigotimes_{j=i+1}^{k} A(j) \otimes C(\sigma(j+1)) \otimes B(i) .$$

From the arguments provided in step 2 of the above analysis it follows that

$$\lim_{k \to \infty} \left\| \bigotimes_{i=0}^{k} A(i) \otimes C(\sigma(i+1)) \otimes w \right\|_{max} = -\infty \quad \text{a.s.} ,$$

provided that $\nu > \mathbf{a}$. Hence, there exists an a.s. finite random variable $\beta(w)$, such that

$$\forall k \geq \beta(w) : \quad \left\| \bigotimes_{i=0}^{k} A(i) \otimes C(\sigma(i+1)) \otimes w \right\|_{max} < 0 .$$

In words, after $\beta(w)$ transitions the influence of the initial vector w dies out. We now compare two versions of $\{W(k)\}$. One version is initialized to W, the stationary regime, and the other version is initialized to an arbitrary finite vector w. We obtain that

$$\forall k \geq \max(\beta(w), \beta(W)) : \quad W(k, w) = W(k, W) .$$

Hence, $\{W(k, w)\}$ couples after a.s. finitely many transitions with the stationary version $\{W(k, W)\}$. $\square$

It is worth noting that $\beta(w)$, defined in the proof of Theorem 2.3.1, fails to be a stopping time adapted to the natural filtration of $\{(A(k), B(k)) : k \geq 0\}$. More precisely, $\beta(w)$ is measurable with respect to the σ-field $\sigma((A(k), B(k)) : k \geq 0)$ but, in general, $\{\beta(w) = m\} \notin \sigma((A(k), B(k)) : m \geq k \geq 0)$, for $m \in \mathbb{N}$.

Due to the max-plus formalism, the proof of Theorem 2.3.1 is a rather straightforward extension of the proof of the classical result for the G/G/1 queue. To fully appreciate the conceptual advantage offered by the max-plus formalism, we refer to [6, 13] where the above theorem is shown without using max-plus formalism.

2.4 Harris Recurrent Max-Plus Linear Systems (Type I and Type IIa)

The Markov chain approach to stability analysis of max-plus linear systems presented in this section goes back to [93, 41]. Consider the recurrence relation $x(k+1) = A(k) \otimes x(k)$, $k \geq 0$, and let

$$Z_{j-1}(k) = x_j(k) - x_1(k) , \quad j \geq 2 , \tag{2.36}$$

denote the discrepancy between component $x_1(k)$ and $x_j(k)$ in $x(k)$. The sequence $\{Z(k)\}$ constitutes a Markov chain, as the following theorem shows.

Theorem 2.4.1 *The process $\{Z(k) : k \geq 0\}$ is a Markov chain. Suppose $Z(k) = (z_1, \ldots, z_{J-1})$ for fixed $z_j \in \mathbb{R}$. Then the conditional distribution of $Z_j(k+1)$ given $Z(k) = (z_1, \ldots, z_{J-1})$, is equal to the distribution of the random variable*

$$A_{j+1\,1}(k) \oplus \bigoplus_{i=2}^{J} A_{j+1\,i}(k) \otimes z_{i-1} \; - \; A_{11}(k) \oplus \bigoplus_{i=2}^{J} A_{1i}(k) \otimes z_{i-1} \,,$$

for $1 \leq j \leq J - 1$.

Proof: Note that

$$\begin{aligned}
a \otimes x \oplus b \otimes y - x &= \max(a + x, b + y) - x \\
&= \max(a, b + (y - x)) \\
&= a \oplus b \otimes (y - x) \,.
\end{aligned}$$

Using the above equality, we obtain for $2 \leq j \leq J$:

$$\begin{aligned}
Z_{j-1}(k+1) &= x_j(k+1) - x_1(k+1) \\
&= (A(k) \otimes x(k))_j - (A(k) \otimes x(k))_1 \\
&= A_{j1}(k) \otimes x_1(k) \oplus A_{j2}(k) \otimes x_2(k) \oplus \cdots \oplus A_{jJ}(k) \otimes x_J(k) \; - \\
&\quad A_{11}(k) \otimes x_1(k) \oplus A_{12}(k) \otimes x_2(k) \oplus \cdots \oplus A_{1J}(k) \otimes x_J(k) \\
&= A_{j1}(k) \otimes x_1(k) \oplus A_{j2}(k) \otimes x_2(k) \oplus \cdots \oplus A_{jJ}(k) \otimes x_J(k) - x_1(k) \; - \\
&\quad (A_{11}(k) \otimes x_1(k) \oplus A_{12}(k) \otimes x_2(k) \oplus \cdots \oplus A_{1J}(k) \otimes x_J(k) - x_1(k)) \\
&= A_{j1}(k) \oplus A_{j2}(k) \otimes (x_2(k) - x_1(k)) \oplus \cdots \oplus A_{jJ}(k) \otimes (x_J(k) - x_1(k)) \; \cdot \\
&\quad A_{11}(k) \oplus A_{12}(k) \otimes (x_2(k) - x_1(k)) \oplus \cdots \oplus A_{1J}(k) \otimes (x_J(k) - x_1(k)) \\
&= A_{j1}(k) \oplus A_{j2}(k) \otimes Z_1(k) \oplus \cdots \oplus A_{jJ}(k) \otimes Z_{J-1}(k) \; - \\
&\quad A_{11}(k) \oplus A_{12}(k) \otimes Z_1(k) \oplus \cdots \oplus A_{1J}(k) \otimes Z_{J-1}(k) \,.
\end{aligned}$$

From this expression it follows that the conditional distribution of $Z(k+1)$ given $Z(0), \ldots, Z(k)$ equals the conditional distribution of $Z(k+1)$ given $Z(k)$ and hence the process $\{Z(k) : k \geq 0\}$ is a Markov chain. $\square$

Now define

$$D(k) = x_1(k) - x_1(k-1), \quad k \geq 1 \,.$$

Then, we have

$$x_1(k) = x_1(0) + \sum_{n=1}^{k} D(n), \quad k \geq 1, \tag{2.37}$$

and

$$x_j(k) = x_j(0) + (Z_{j-1}(k) - Z_{j-1}(0)) + \sum_{n=1}^{k} D(n), \quad k \geq 1, j \geq 2. \tag{2.38}$$

Theorem 2.4.2 *For any $k \geq 0$, the distribution of $(D(k + 1), Z(k + 1))$ given $(Z(0), D(1), Z(1), \ldots, D(k), Z(k))$ depends only on $Z(k)$. If $Z(k) = (z_2, \ldots, z_J)$, then the conditional distribution of $D(k + 1)$ given $(Z(0), D(1), Z(1), \ldots, D(k), Z(k))$ is equal to the distribution of the random variable*

$$A_{11}(k) \oplus \bigoplus_{j=2}^{J} A_{1j}(k) \otimes z_{j-1} \, .$$

Proof: We have

$$
\begin{aligned}
D(k+1) &= x_1(k+1) - x_1(k) \\
&= A_{11}(k) \otimes x_1(k) \oplus A_{12}(k) \otimes x_2(k) \oplus \cdots \oplus A_{1J}(k) \otimes x_J(k) \, - x_1(k) \\
&= A_{11}(k) \oplus A_{12}(k) \otimes (x_2(k) - x_1(k)) \oplus \cdots \oplus A_{1J}(k) \otimes (x_J(k) - x_1(k)) \\
&= A_{11}(k) \oplus A_{12}(k) \otimes Z_1(k) \oplus \cdots \oplus A_{1J}(k) \otimes Z_{J-1}(k) \, ,
\end{aligned}
$$

which, together with the previous theorem, yields the desired result. $\square$

If $\{Z(k)\}$ is uniformly ϕ-recurrent and aperiodic (for a definition we refer to the Appendix), then it is ergodic and, as will be shown in the following theorem, a type IIa limit holds. Elaborating on a result from Markov theory for so-called *chain dependent processes*, ergodicity of $\{Z(k)\}$ yields existence of the type I limit and thus of the Lyapunov exponent.

Theorem 2.4.3 *Suppose that the Markov chain $\{Z(k) : k \geq 1\}$ is aperiodic and uniformly ϕ-recurrent, and denote its unique invariant probability measure by π. Then the following holds:*

(i) For $1 \leq i, j \leq J$, $x_i(k) - x_j(k)$ converges weakly to the unique stationary regime π.

(ii) If the elements of $A(k)$ have finite first moments, then a finite number λ exists such that

$$\lim_{k \to \infty} \frac{x_j(k)}{k} = \lambda, \quad j = 1, \ldots, J \, ,$$

almost surely for any finite initial value, and

$$\lambda = \mathbb{E}_\pi[D(1)] \, ,$$

where $\mathbb{E}_\pi$ indicates that the expected value is taken with $Z(0)$ distributed according to π.

Proof: For the proof of the first part of the theorem note that

$$x_i(k) - x_j(k) = Z_{i-1}(k) - Z_{j-1}(k) \, ,$$

for $2 \leq i, j \leq J$, and

$$x_i(k) - x_1(k) = Z_{i-1}(k) \, , \quad x_1(k) - x_i(k) = -Z_{i-1}(k) \, ,$$

for $2 \leq i \leq J$. Hence, weak convergence of $Z(k)$ to a unique stationary regime implies weak convergence of $x_i(k) - x_j(k)$ to a unique stationary regime. Weak convergence of $Z(k)$ to a unique stationary regime follows from uniform ϕ-recurrence and aperiodicity of $Z(k)$, see Section F in the Appendix, and we have thus shown the first part of the theorem.

We now turn to the second part. The process $\{D(k) : k \geq 1\}$ is a so-called chain dependent process and the limit theorem of Griggorescu and Oprisan [55] implies

$$\lim_{k \to \infty} \frac{1}{k} \sum_{n=1}^{k} D(n) = \lambda = \mathbb{E}_\pi[D(1)] \quad \text{a.s.} ,$$

for all initial values x_0. This yields for the limit of $x_1(k)/k$ as k tends to ∞:

$$\lim_{k \to \infty} \frac{1}{k} x_1(k) \overset{(2.37)}{=} \lim_{k \to \infty} \left(\frac{1}{k} x_1(0) + \frac{1}{k} \sum_{n=1}^{k} D(n) \right)$$

$$= \lim_{k \to \infty} \frac{1}{k} \sum_{n=1}^{k} D(n)$$

$$= \lambda \quad \text{a.s.}$$

It remains to be shown that, for $j \geq 2$, the limit of $x_j(k)/k$ as k tends to ∞ equals λ. Suppose that for $j \geq 2$:

$$\lim_{k \to \infty} \frac{1}{k} Z_{j-1}(k) = \lim_{k \to \infty} \frac{1}{k} \left(Z_{j-1}(k) - Z_{j-1}(0) \right) = 0 \quad \text{a.s.} \tag{2.39}$$

With (2.39) it follows from (2.38) that

$$\lim_{k \to \infty} \frac{1}{k} x_j(k) = \lim_{k \to \infty} \frac{1}{k} (Z_{j-1}(k) - Z_{j-1}(0)) + \lambda$$

$$= \lambda \quad \text{a.s.} ,$$

for $j \geq 2$. In what follows we show that (2.39) indeed holds under the conditions of the theorem.

Uniform ϕ-recurrence and aperiodicity of the Markov chain $\{Z(k) : k \geq 1\}$ implies Harris ergodicity. Hence, for $J - 1 \geq j \geq 1$, finite constants c_j exists, such that

$$\lim_{k \to \infty} \frac{1}{k} \sum_{n=1}^{k} Z_j(n) = c_j \quad \text{a.s.}$$

This implies

$$c_j = \lim_{k \to \infty} \frac{1}{k} \sum_{n=1}^{k} Z_j(n)$$

$$= \lim_{k \to \infty} \left(\frac{1}{k} Z_j(k) + \frac{k-1}{k} \frac{1}{k-1} \sum_{n=1}^{k-1} Z_j(n) \right)$$

$$= \lim_{k \to \infty} \frac{1}{k} Z_j(k) + \lim_{k \to \infty} \frac{k-1}{k} \lim_{k \to \infty} \frac{1}{k-1} \sum_{n=1}^{k-1} Z_j(n)$$

$$= \lim_{k \to \infty} \frac{1}{k} Z_j(k) + c_j \,,$$

which yields, for $J - 1 \geq j \geq 1$,

$$\lim_{k \to \infty} \frac{1}{k} Z_j(k) = \lim_{k \to \infty} \frac{1}{k} \Big(Z_j(k) - Z_j(0) \Big) = 0 \quad \text{a.s.}$$

$\square$

Remark 2.4.1 *Let the conditions in Theorem 2.4.3 be satisfied. If, in addition, the elements of $A(k)$ and the initial vector have finite second moments, then*

$$0 \leq \sigma^2 \stackrel{\text{def}}{=} \sum_{n=1}^{\infty} \mathbb{E}_{\pi}[(D(1) - \lambda)(D(n) - \lambda)] < \infty \,,$$

and if $\sigma^2 > 0$, the sequence

$$\frac{(x_1(k), \ldots, x_J(k)) - (k\lambda, \ldots, k\lambda)}{\sigma \sqrt{k}} \,, \quad k \geq 1,$$

converges in distribution to the random vector $(\mathcal{N}, \ldots, \mathcal{N})$, where $\mathcal{N}$ is a standard normal distributed random variable. For details and proof we refer to [93].

Remark 2.4.2 *If the state space of $Z(k)$ is finite, then the convergence in part (i) of Theorem 2.4.3 happens in strong coupling.*

The computational formula for λ put forward in Theorem 2.4.3 is also known as 'Furstenberg's cocycle representation of the Lyapunov exponent;' see [45].

Example 2.4.1 *Consider $x(k)$ as defined in Example 1.5.5, and let $\sigma = 1$ and $\sigma' = 2$. Matrix D_2 is primitive and has (unique) eigenvector $(1, 1, 0, 1)^{\top}$. Let $z(1) = z((1, 1, 0, 1)^{\top}) = (0, -1, 0)^{\top}$. It is easily checked that $\{Z(k)\}$ is a Markov chain on state space $\{z(i) : 1 \leq i \leq 5\}$, with*

$$z(2) = \begin{pmatrix} 0 \\ 0 \\ 0 \end{pmatrix}, \ z(3) = \begin{pmatrix} -1 \\ 0 \\ 0 \end{pmatrix}, \ z(4) = \begin{pmatrix} -1 \\ -2 \\ 0 \end{pmatrix} \text{ and } z(5) = \begin{pmatrix} 0 \\ -1 \\ -1 \end{pmatrix}.$$

Denoting the transition probability of $Z(k)$ from state $z(i)$ to state $z(j)$, for $1 \leq i, j \leq 5$, one obtains the following transition matrix

$$\mathbf{P} = \begin{pmatrix} 1-\theta & \theta & 0 & 0 & 0 \\ 0 & 0 & \theta & 1-\theta & 0 \\ 0 & 0 & \theta & 1-\theta & 0 \\ 0 & 0 & 0 & 0 & 1 \\ 1-\theta & \theta & 0 & 0 & 0 \end{pmatrix}.$$

The chain is aperiodic and since the state space is finite it is uniformly ϕ-recurrent. Moreover, the unique stationary distribution of $Z(k)$ is this given by

$$\pi_{z(1)} = \frac{(1-\theta)^2}{1+\theta(1-\theta)}, \quad \pi_{z(2)} = \frac{\theta(1-\theta)}{1+\theta(1-\theta)}, \quad \pi_{z(3)} = \frac{\theta^2}{1+\theta(1-\theta)},$$

$$\pi_{z(4)} = \frac{\theta(1-\theta)}{1+\theta(1-\theta)} \quad \text{and} \quad \pi_{z(5)} = \frac{\theta(1-\theta)}{1+\theta(1-\theta)}.$$

Applying Theorem 2.4.3, yields $\lambda = \mathbb{E}_\pi[D(1)]$. Evoking Theorem 2.4.2, this expected value can thus be computed as follows:

$$\lambda = \sum_{i=1}^{5} \pi_{z(i)} (1 \oplus 2 \otimes z_2(i))$$

$$= \pi_{z(1)} + 2\pi_{z(2)} + 2\pi_{z(3)} + \pi_{z(4)} + \pi_{z(5)}$$

$$= 1 + \frac{\theta}{1+\theta-\theta^2},$$

for any $\theta \in [0,1]$. For a different example of this kind, see [65].

Example 2.4.2 *Let $\{A(k)\}$, with $A(k) \in \{0,1\}^{2\times 2}$, be an i.i.d. sequence following the distribution $P(A_{ij}(k) = 0) = 1/2 = P(A_{ij}(k) = 1)$ for $1 \leq i, j \leq J$. We turn to the Markov process $\{Z(k)\}$ as defined in (2.36). This process has state space $\{-1, 0, 1\}$. By Theorem 2.4.1, the transition probability of $Z(k)$ is given by*

$$P(Z(k+1) = m \,|\, Z(k) = z)$$
$$= P((A_{21}(k+1) \oplus (A_{22}(k+1) \otimes z)) - A_{11}(k+1) \oplus (A_{12}(k+1) \otimes z) = m),$$

for $m, z \in \{-1, 0, 1\}$, and the transition matrix on $\{Z(k)\}$ can be computed as

$$\mathbf{P} = \begin{pmatrix} \frac{1}{4} & \frac{1}{2} & \frac{1}{4} \\ \frac{3}{16} & \frac{5}{8} & \frac{3}{16} \\ \frac{1}{4} & \frac{1}{2} & \frac{1}{4} \end{pmatrix}.$$

The Markov chain $\{Z(k)\}$ is aperiodic (all elements of $\mathbf{P}$ are positive), uniformly ϕ-recurrent (the state space is finite) and has unique stationary distribution

$$\pi_{-1} = \frac{3}{14}, \quad \pi_0 = \frac{8}{14}, \quad \pi_1 = \frac{3}{14}.$$

From Theorem 2.4.3 together with Theorem 2.4.2 follows that

$$\lambda = \sum_{z \in \{-1,0,1\}} \mathbb{E}[\, A_{11}(1) \oplus (A_{12}(1) \otimes z)\,]\pi_z$$

$$= \frac{3}{14} \times \frac{1}{2} + \frac{8}{14} \times \frac{3}{4} + \frac{3}{14} \times \frac{3}{2}$$

$$= \frac{6}{7}\,.$$

As shown in [93], for this example σ (as defined in Remark 2.4.1) is equal to 33/343.

The above examples are deceitfully simple in the sense that (i) the transition probability (in this case a matrix) of $\{Z(k)\}$ can be calculated easily and (ii) we can deduce that $\{Z(k)\}$ is aperiodic and uniformly ϕ-recurrent from inspecting the transition matrix of $\{Z(k)\}$. In [93], examples with countable state space are discussed. For one example, the elements of $A(k)$ are exponentially distributed with the same parameter; for another example, the elements are assumed to be uniformly distributed over the unit interval. Unfortunately, even when the elements of $A(k)$ are governed by these ostensibly simple distributions, the analysis leads to cumbersome computations. It is mainly for this reason that the Markov chain approach, as presented in this section, will be of avail only in special cases.

2.5 Limits in the Projective Space (Type IIb)

In the previous section, we studied the limit of *differences within $x(k)$*, that is, $x_j(k) - x_1(k)$, for $2 \leq j \leq J$. In what follows, we take a slightly different point of view and consider differences *between $x(k)$ and $x(k-1)$*, that is, $x_j(k) - x_j(k-1)$, for $1 \leq j \leq J$. The basic recurrence relation we study is given by

$$x(k+1) = A(k) \otimes x(k)\,, \quad k \geq 0, \tag{2.40}$$

with $x(0) = x_0 \in \mathbb{R}_{max}^J$ and $A(k) \in \mathbb{R}_{max}^{J \times J}$, for $k \geq 0$.

For the following we use a definition of $Z(k)$ that slightly differs from the definition in Section 2.4. We now let

$$Z(k) = x(k) - x(k-1)\,, \quad k \geq 1, \tag{2.41}$$

denote the component-wise increase of $x(k)$. In particular, the components of $Z(k)$ are given by

$$Z_j(k) = F_j(A(k-1), x(k-1))$$

$$\stackrel{\text{def}}{=} \left(\bigoplus_{i=1}^{J} A_{ji}(k-1) \otimes x_i(k-1) \right) - x_j(k-1)$$

$$= A_{jj}(k-1) \oplus \bigoplus_{\substack{i=1 \\ i \neq j}}^{J} A_{ji}(k-1) \otimes (x_i(k-1) - x_j(k-1))\,, \quad j \geq 1\,.$$

For $\overline{x(k-1)} \in \mathbb{PR}_{\max}^J$, the value of F_j doesn't depend on the representative, that is, for all $X \in \overline{x(k-1)}$ we have $Z_j(k) = F_j(A(k-1), X)$, for $1 \leq j \leq J$, and we write $Z_j(k) = F_j(A(k-1), \overline{x(k-1)})$ to express this fact. For the definition of the modes of convergence used in the following lemma we refer to Section E.4 in the Appendix.

Lemma 2.5.1 *Consider the situation in (2.40) and let $\{A(k) : k \geq 0\}$ be stationary. If $\overline{x(k)} \in \mathbb{PR}^J$ converges weakly to a unique invariant distribution, uniformly over all initial conditions, then $Z(k)$ converges weakly to a unique invariant distribution, uniformly over all initial conditions.*

Proof: Consider the sequence $y(k) = (A(k), \overline{x(k)})$, $k \geq 0$. The sequence $A(k)$ is stationary by assumption with stationary distribution π_A. Let A be distributed according to π_A. If $\overline{x(k)}$ converges weakly to $\overline{x}$, then $y(k)$ converges weakly to $(A, \overline{x})$. Because $F = (F_1, \ldots, F_J)$ defined above is continuous, we obtain from the continuous mapping theorem (see Appendix, Section E.4) the weak convergence of $F(A(k), \overline{x(k)})$. $\square$

In what follows we establish sufficient conditions for weak convergence of $\overline{x(k)}$. By Lemma 2.5.1, this already implies weak convergence of $Z(k)$ which in turn yields type IIb second-order ergodic theorems. As we will show in the following, in many situations, the convergence of $Z(k)$ occurs even in strong coupling. In Section 2.5.1, we will study systems with countable state space and, in Section 2.5.2, we will address the general situation. In Section 2.5.3 we revisit the deterministic setup. Finally, we present a representation of type IIb limits via a renewal type approach in Section 2.5.4.

2.5.1   Countable Models

In this section, we study models with countable state space. Let $\mathcal{A}$ be a finite or countable collection of $J \times J$-dimensional irreducible matrices. We think of $\mathcal{A}$ as the state space of the random sequence $\{A(k)\}$ following a discrete law.

Definition 2.5.1 *Let $\{A(k)\}$, with $A(k) \in \mathcal{A}$, be a random sequence. A matrix $\tilde{A} \in \mathcal{A}$ is called a pattern of $\{A(k)\}$ if a sequence $\tilde{a} = (a_1, \ldots, a_N) \in \mathcal{A}^N$ exists such that*

$$(a) \qquad \tilde{A} = a_N \otimes a_{N-1} \otimes \cdots \otimes a_1$$
$$(b) \qquad P\big(A(N+k) = a_N, \ldots, A(1+k) = a_1 \big) > 0, \quad k \in \mathbb{N}.$$

We call $\tilde{a}$ the sequence constituting $\tilde{A}$.

Note that if $\{A(k)\}$ is i.i.d., then the second condition in the above definition is satisfied if we let $\mathcal{A}$ contain only those possible outcomes of $A(k)$ that have a positive probability. In other words, in the i.i.d. case, the second condition is satisfied if we restrict $\mathcal{A}$ to the support of $A(k)$. Existence of a pattern essentially implies that $\mathcal{A}$ is at most countable, see Remark 2.2.8.

The main technical assumptions we need are the following:

(C1) The sequence $\{A(k)\}$ is i.i.d. with countable state space $\mathcal{A}$.

(C2) Each $A \in \mathcal{A}$ is regular.

(C3) There is a primitive matrix C that is a pattern of $\{A(k)\}$.

Observe that we have already encountered the concept of a pattern - as expressed in condition (C3) - in condition (C) on page 82, although we haven't coined the name 'pattern' for it at that stage.

The following theorem provides a sufficient condition for $\{\overline{x(k)}\}$ to converge in strong coupling.

Theorem 2.5.1 *Let* (C1) - (C3) *be satisfied, then* $\{\overline{x(k)}\}$ *converges with strong coupling to a unique stationary regime for all initial conditions in* $\mathbb{R}^J$. *In particular,* $\overline{x(k)}$ *converges in total variation.*

Proof: Let C be defined as in (C3) and denote the coupling time of C by c. For the sake of simplicity, assume that $C \in \mathcal{A}$, which implies $N = 1$. Set $\tau_0 = 0$ and

$$\tau_{k+1} = \inf\{m \geq \tau_k + c : A(m - i) = C : 0 \leq i \leq c - 1\}, \quad k \geq 0.$$

In words, at time τ_k we have observed for the kth time a sequence of c consecutive occurrences of C. The i.i.d. assumption (C1) implies that $\tau_k < \tau_{k+1} < \infty$ for all k and that $\lim_{k\to\infty} \tau_k = \infty$ with probability one. Let p denote the probability of observing C, then we observe C^c with probability p^c. By construction, the probability of the event $\{\tau_1 = m\}$ is less than or equal to the probability of the event $A(k) \neq C$, $0 \leq k \leq m - c$, and $A(k) = C$, for $k = m - c + 1, \ldots, m$. In other words, for $m \geq c$, it holds that $P(\tau_1 = m) \geq (1 - p)^{m-c}p^c$. Hence,

$$\mathbb{E}[\tau_1] \leq \sum_{m=c}^{\infty} m\,(1 - p)^{m-c}p^c$$
$$= \sum_{m=0}^{\infty} (m + c)\,(1 - p)^m p^c$$
$$= c\,p^c \sum_{m=0}^{\infty} (1 - p)^m + p^c \sum_{m=0}^{\infty} m\,(1 - p)^m$$
$$= \frac{c\,p^c}{p} + \frac{p^c(1 - p)}{p^2}$$
$$< \infty,$$

which implies that $\mathbb{E}[\tau_{k+1} - \tau_k] < \infty$, for $k \in \mathbb{N}$.

At τ_k, $x(\tau_k) \in V(C)$, see Theorem 2.1.1. By condition (C3), C is primitive and, by Corollary 2.1.1, the eigenspace of C is a single point in the projective space (that is, the eigenvector of C is unique). In other words, $\{A(k)\}$ has MLP, see Lemma 2.2.2. By (C2), $\overline{x(k)} \in \mathbb{R}^J$, for any k, and from the above line argument it follows that $\{\overline{x(k)}\}$ is a Harris ergodic Markov chain and regenerates whenever the chain hits the single point $\overline{V(C)}$. This implies that $\{\overline{x(k)}\}$ converges with strong coupling to a unique stationary regime. See Section F in the Appendix. □

What happens if we consider in Theorem 2.5.1 a stationary and ergodic sequence instead of an i.i.d. sequence? The key argument in the proof of Theorem 2.5.1 is that $\{A(k)\}$ has MLP. This is guaranteed by the fact that we observe with positive probability a sequence of occurrences of $A(k)$ such that the partial product over that sequence equals C^c, for some primitive matrix C, where c denotes the coupling time of C, see Lemma 2.2.2. If the coupling time of C is larger than 1, then, under i.i.d. regime, the event that C occurs c times in a row has positive probability. However, this reasoning doesn't apply in the stationary case. To see this, consider the following example. Let $\Omega = \{\omega_1, \omega_2\}$ and $P(\omega_i) = 1/2$, for $i = 1, 2$. Define the shift operator θ by $\theta(\omega_1) = \omega_2$ and $\theta(\omega_2) = \omega_1$. Then θ is stationary and ergodic. Consider the matrices A, B as defined in Example 2.1.1 and let

$$\{A(k, \omega_1)\} = A, B, A, B, \ldots \qquad \{A(k, \omega_2)\} = B, A, B, A, \ldots$$

The sequence $\{A(k)\}$ is thus stationary and ergodic. Furthermore, A, B are primitive matrices whose coupling time is 4 each. But with probability one we never observe a sequence of 4 occurrences in a row of either A or B. Since neither A or B is of rank 1, we cannot conclude that $\{A(k)\}$ has MLP and, consequently, that $\overline{x(k)}$, with $x(k+1) = \bigotimes_{i=0}^{k} A(i) \otimes x_0$, is regenerative. However, if we replace, for example, A by A^m, for $m \geq 4$ (i.e., a matrix of rank 1), then the argument would apply again. For this reason, we require for the stationary and ergodic setup that a matrix of rank 1 exists that is a pattern, so that $\overline{x(k)}$ becomes a regenerative process. Note that the condition 'there exits a pattern of rank 1' is equivalent to the condition '$\{A(k)\}$ has MLP.' The precise statement is given in the following theorem. For a proof we refer to [84].

Theorem 2.5.2 *Let $\{A(k)\}$ be a stationary and ergodic sequence of a.s. regular square matrices. If $\{A(k)\}$ has MLP, then $\{\overline{x(k)}\}$ converges with strong coupling to a unique stationary regime for all initial conditions in $\mathbb{R}^J$. In particular, $\{\overline{x(k)}\}$ converges in total variation.*

2.5.2 General Models

In this section, we consider matrices $A(k)$ the elements of which may follow a distribution that is either discrete or absolutely continuous with respect to the Lebesgue measure, or a mixture of both. For general state-space, the event $\{A(N + k) \otimes \cdots \otimes A(2 + k) \otimes A(1) = \tilde{A}\}$ in Definition 2.5.1 typically has probability zero. For this reason we introduce the following extension of the definition of a pattern. Let $M \in \mathbb{R}_{max}^{J \times J}$ be a deterministic matrix and $\eta > 0$. We denote by $B(M, \eta)$ the open ball with center M and radius η in the supremum norm on $\mathbb{R}^{J \times J}$. More precisely, $A \in B(M, \eta)$ if for all i, j, with $1 \leq i, j \leq J$, it holds that

$$
A_{ij} \begin{cases} \in (M_{ij} - \eta, M_{ij} + \eta) & \text{for } M_{ij} \neq \varepsilon, \\ = \varepsilon & \text{for } M_{ij} = \varepsilon. \end{cases}
$$

With this notation, we can state the fact that a matrix A belongs to the support of a random matrix $\tilde{A}$ by

$$
\forall \eta > 0 \qquad P(\tilde{A} \in B(A, \eta)) > 0.
$$

This includes the case where A is a boundary point of the support. We now state the definition of a pattern for non-countable state-space.

Definition 2.5.2 *Let $\{A(k)\}$ be a random sequence of matrices over $\mathbb{R}_{\max}^{J \times J}$ and let $\tilde{A} \in \mathbb{R}_{\max}^{J \times J}$ be a deterministic matrix. We call $\tilde{A}$ a pattern of $\{A(k)\}$ if a deterministic number N exists such that for any $\eta > 0$ it holds that*

$$
P\left(A(N-1) \otimes A(N-2) \otimes \cdots \otimes A(0) \in B(\tilde{A}, \eta) \right) > 0.
$$

Definition 2.5.2 can be phrased as follows: Matrix $\tilde{A}$ is a pattern of $\{A(k)\}$ if $N \in \mathbb{N}$ exists such that $\tilde{A}$ lies in the support of the random matrix $A(N-1) \otimes A(N-2) \otimes \cdots \otimes A(0)$. The key condition for general state space is the following:

(C4) There exists a (measurable) set of matrices $\mathcal{C}$ such that for any $C \in \mathcal{C}$ it holds that C is a pattern of $\{A(k)\}$ and C is of rank 1. Moreover, a finite number N exists such that

$$
P\left(A(N-1) \otimes A(N-2) \otimes \cdots \otimes A(0) \in \mathcal{C} \right) > 0.
$$

Under condition **(C4)**, the following counterpart of Theorem 2.5.2 for models with general state space can be established; for a proof we refer to [84].

Theorem 2.5.3 *Let $\{A(k)\}$ be a stationary and ergodic sequence of a.s. regular matrices in $\mathbb{R}_{\max}^{J \times J}$. If condition **(C4)** is satisfied, then $\{\overline{x(k)}\}$ converges with strong coupling to a unique stationary regime. In particular, $\{\overline{x(k)}\}$ converges in total variation to a unique stationary regime.*

In Definition 2.5.2, we required that after a fixed number of transitions the pattern lies in the support of the matrix product. The following, somewhat weaker, definition requires that an arbitrarily small η-neighborhood of the pattern can be reached in a finite number of transitions where the number of transitions is deterministic and may depend on η.

Definition 2.5.3 *Let $\{A(k)\}$ be a random sequence of matrices over $\mathbb{R}_{\max}^{J \times J}$ and let $\tilde{A} \in \mathbb{R}_{\max}^{J \times J}$ be a deterministic matrix. We call $\tilde{A}$ an asymptotic pattern of $\{A(k)\}$ if for any $\eta > 0$ a deterministic number N_η exists, such that*

$$
P\left(\overline{A(N_\eta - 1) \otimes A(N_\eta - 2) \otimes \cdots \otimes A(0)} \in \overline{B(\tilde{A}, \eta)} \right) > 0.
$$

Accordingly, we obtain a variant of condition (**C4**).

(**C4**)' There exists a matrix C such that C is an asymptotic pattern of $\{A(k)\}$ and C is of rank 1.

Under condition (**C4**)' only weak convergence of $\{\overline{x(k)}\}$ can be established, whereas (**C4**) even yields total variation convergence. The precise statement is given in the following theorem.

Theorem 2.5.4 *Let $\{A(k)\}$ be a stationary and ergodic sequence of a.s. regular matrices in $\mathbb{R}_{\max}^{J \times J}$. If condition (**C4**)' is satisfied, then $\{\overline{x(k)}\}$ converges with δ-coupling to a unique stationary regime. In particular, $\{x(k)\}$ converges weakly to a unique stationary regime.*

Proof: We only give a sketch of the proof, for a detailed proof see [84]. Suppose that a stationary version $\overline{x} \circ \theta^k$ of $\overline{x(k)}$ exists, where θ denotes a stationary and ergodic shift. We will show that $\overline{x(k)}$ converges with δ-coupling to $\overline{x} \circ \theta^k$. Fix $\eta > 0$ and let τ denote the time of the first occurrence of the pattern. Condition (**C4**)' implies that at time τ the projective distance of the two versions is at most η, in formula:

$$d_{\mathbb{P}}(\overline{x}(\tau), \overline{x} \circ \theta^{\tau}) \leq \eta. \tag{2.42}$$

As Mairesse shows in [84], the projective distance of two sequences driven by the same sequence $\{A(k)\}$ is non-expansive which means that (2.42) already implies

$$\forall k \geq \tau : \quad d_{\mathbb{P}}(\overline{x(k)}, \overline{x} \circ \theta^k) \leq \eta.$$

Hence, for any $\eta > 0$,

$$P\left(d_{\mathbb{P}}(\overline{x(k)}, \overline{x} \circ \theta^k) \leq \eta, \, k \geq \tau\right) = 1.$$

Stationarity of $\{A(k)\}$ implies $\tau < \infty$ a.s. and the above formula can be written

$$\lim_{k \to \infty} P\left(d_{\mathbb{P}}(\overline{x(k)}, \overline{x} \circ \theta^k) \leq \eta\right) = 1.$$

Hence, $\overline{x(k)}$ converges with δ-coupling to a stationary regime. See the Appendix. Uniqueness of the limit follows from the same line of argument. $\square$

We conclude this presentation of convergence results by stating the most general result, namely, that existence of an asymptotic pattern is a necessary and sufficient condition for weak convergence of $\{\overline{x(k)}\}$.

Theorem 2.5.5 *(Theorem 7.4 in [84]) Let $\{A(k)\}$ be a stationary and ergodic sequence on $\mathbb{R}_{\max}^{J \times J}$. A necessary and sufficient condition for $\{x(k)\}$ to converge in δ-coupling (respectively, weakly) to a unique stationary regime is that (**C4**)' is satisfied.*

2.5.3 Periodic Regimes of Deterministic Max-Plus DES

Consider the deterministic max-plus linear system

$$x(k+1) = A \otimes x(k), \quad k \geq 0,$$

with $x(0) = x_0 \in \mathbb{R}^J$ and $A \in \mathbb{R}_{\max}^{J \times J}$ a regular matrix. A periodic regime of period d is a set of vectors $x^1, \ldots, x^d \in \mathbb{R}^J$ such that (i) $\overline{x^i} \neq \overline{x^j}$, for $1 \leq i \neq j \leq d$, and (ii) a finite number μ exists which satisfies

$$x^{i+1} = A \otimes x^i, \quad 1 \leq i \leq d,$$

and $\mu \otimes x^1 = A \otimes x^d$. A consequence of the above definition is that $x^1, \ldots, x^d$ are eigenvectors of A^d and μ is an eigenvalue of A^d. If A is irreducible with cyclicity $\sigma(A)$, then A will possess periodic regimes of period $\sigma(A)$, see Theorem 2.1.1, and $A^{\sigma(A)}$ will have $\sigma(A)$ mutually linear independent eigenvectors.

From a system theoretic point of view, one is interested in the limiting behavior of $x(k)$. More precisely, one is interested in the behaviour of $\overline{x(k)}$ for k large. If A is primitive, $\overline{x(k)}$ converges in a finite number of steps to $\overline{x}$, where x denotes the unique eigenvector of A. In the general situation, however, there are two sources for non-uniqueness of the limiting behavior of $\overline{x(k)}$. First, if A has cyclicity $\sigma(A) > 1$, then $\{\overline{x(k)}\}$ may eventually reach a periodic regime of period $\sigma(A)$. Secondly, even if A has cyclicity one, if the communication graph of A possesses m strongly connected subgraphs, with $m > 1$, then the eigenspace of A is a m-dimensional vector space. See Theorem 2.1.2.

Example 2.5.1 *Consider matrix*

$$A = \begin{pmatrix} 3 & 6 \\ 4 & 4 \end{pmatrix}.$$

A is irreducible with eigenvalue 5 and the critical graph of A consists of the circuit $((1,2),(2,1))$. The critical graph has thus one m.s.c.s. and $\sigma(A) = 2$. It is easily checked that the eigenspace of A is given by

$$V(A) = \left\{ \begin{pmatrix} x_1 \\ x_2 \end{pmatrix} \in \mathbb{R}_{\max}^2 \,\middle|\, \exists a \in \mathbb{R} : \begin{pmatrix} x_1 \\ x_2 \end{pmatrix} = a \otimes \begin{pmatrix} 1 \\ 0 \end{pmatrix} \right\}.$$

Starting in $x(0) \notin V(A)$, will lead to a periodic regime of period 2. For example, taking $x(0) = (0,0)$, yields

$$x(1) = \begin{pmatrix} 6 \\ 4 \end{pmatrix}, \quad x(2) = \begin{pmatrix} 10 \\ 10 \end{pmatrix}, \quad x(3) = \begin{pmatrix} 16 \\ 14 \end{pmatrix}, \quad x(4) = \begin{pmatrix} 20 \\ 20 \end{pmatrix} \quad \cdots$$

In other words, A^2 has eigenvalue 10 and two linear independent eigenvectors, namely

$$\begin{pmatrix} 0 \\ 0 \end{pmatrix}, \quad \begin{pmatrix} 2 \\ 0 \end{pmatrix}.$$

We call the set of all initial conditions x_0 such that $\overline{A^k \otimes x_0}$ eventually reaches $\overline{x}$, for some eigenvector x, (resp. periodic regime $\overline{x^1}, \ldots, \overline{x^d}$) the domain of attraction of x (resp. $x^1, \ldots, x^d$). For example, for the matrix given in Example 2.5.1 above, the vector $x = (0,0)$ lies in the domain of attraction of the periodic regime $(6,4),(10,10)$.

For $J = 3$, Mairesse provides a graphical representation of the domain of attraction in the projective space, see [83] and the extended version [82]. In particular, the eigenvector (resp. periodic regime) in whose domain of attraction an initial value x_0 lies can be deduced from a graphical representation of the eigenspace of A in the projective space.

2.5.4 The Cycle Formula

We revisit the situation in Section 2.2.3.2 and use the notation as introduced therein. Specifically, we assume that $\{A(k)\}$ has MLP. Elaborating on the projective space, (2.26) reads

$$\overline{x(T_k)} = \overline{x}, \quad k \geq 0,$$

for some fixed $x \in \mathbb{R}^J$. This constitutes a regenerative property of $\{\overline{x(k)}\}$. Specifically, the cycles $\{\overline{x(k)} : T_k < n \leq T_{k+1}\}$ constitute an i.i.d. sequence. Moreover, $\{T_k\}$ is a sequence of renewal times for the process $\{x(k) - x(k-1)\}$ as well. Stationarity and ergodicity of $\{A(k)\}$ imply that $\overline{x(k)}$ hits $\overline{x}$ a.s. infinitely often. Hence, $\{x(k) - x(k-1)\}$ is a regenerative process with renewal times $\{T_k\}$, see Section E.9 in the Appendix. Note that

$$\mathbb{E}\left[\sum_{k=T_0+1}^{T_1} \big(x(k) - x(k-1) \big) \right] = \mathbb{E}\left[x(T_1) - x(T_0) \right].$$

Let $\overline{X}$ denote the unique stationary regime of $\{\overline{x(k)}\}$. Provided that $\mathbb{E}[x(T_1) - x(T_0)] < \infty$ and $\mathbb{E}[T_1 - T_0] < \infty$, the limit theorem for regenerative processes yields

$$\lim_{N \to \infty} \frac{1}{N} \sum_{k=1}^{N} \big(x(k) - x(k-1) \big) = \frac{1}{\mathbb{E}[T_1 - T_0]} \mathbb{E}\left[\sum_{k=T_0+1}^{T_1} \big(x(k) - x(k-1) \big) \right] \quad \text{a.s.}$$

Moreover, ergodicity of $\{A(k)\}$ yields

$$\lim_{N \to \infty} \frac{1}{N} \sum_{k=1}^{N} \big(x(k) - x(k-1) \big) = \mathbb{E}[X \circ \theta - X] \quad \text{a.s.},$$

for $X \in \overline{X}$. In particular, for $X \in \overline{X}$, it holds that

$$\mathbb{E}[\overline{X} \circ \theta - \overline{X}] = \mathbb{E}[X \circ \theta - X].$$

We summarize the above analysis in the following lemma.

Lemma 2.5.2 (Cycle Formula) *Let $\{A(k)\}$ be a stationary and ergodic sequence in $\mathbb{R}_{\max}^{J \times J}$ that has MLP. If $\mathbb{E}[x(T_1) - x(T_0)] < \infty$ and $\mathbb{E}[T_1 - T_0] < \infty$, then*

$$\mathbb{E}[\overline{X} \circ \theta - \overline{X}] = \frac{\mathbb{E}[x(T_1) - x(T_0)]}{\mathbb{E}[T_1 - T_0]},$$

where $\overline{X}$ denotes the unique stationary regime of $\{\overline{x(k)}\}$.

Remark 2.5.1 *Note that*

$$\overline{X \circ \theta} = \{Y \mid \exists \alpha : Y = \alpha \otimes (X \circ \theta)\}$$
$$= \{Y \mid \exists \alpha : Y = (\alpha \otimes X) \circ \theta\}$$
$$= \overline{X} \circ \theta$$

and the cycle formula can alternatively be phrased

$$\mathbb{E}[\overline{X} \circ \theta - \overline{X}] = \frac{\mathbb{E}[x(T_1) - x(T_0)]}{\mathbb{E}[T_1 - T_0]}.$$

Remark 2.5.2 *If $\{A(k)\}$ is i.i.d., then in the above theorem the condition that $\{A(k)\}$ has MLP can be replaced by condition (C), see Lemma 2.2.2. Moreover, a simple geometrical trial argument, like the one used in the proof of Theorem 2.5.1, shows that $\mathbb{E}[T_1 - T_0] < \infty$. If, in addition, $A(k)$ is integrable, one can show that $\mathbb{E}[x(T_1) - x(T_0)] < \infty$ holds as well.*

In the following section we will establish sufficient conditions for $\mathbb{E}[\overline{X} \circ \theta - \overline{X}]$ to be equal to the Lyapunov exponent.

2.6   Lyapunov Exponents via Second Order Limits (Type IIb)

The Lyapunov exponent can be defined as a first-order limit, as explained in Section 2.2. However, as we will show in this section, under suitable conditions, the Lyapunov exponent can be obtained by a second-order limit as well. In Section 2.6.1 we establish the general result, whereas in Section 2.6.2 we provide a direct analysis via backward coupling. It is this result that will prove valuable for the analysis provided in Part II. The basic recurrence relation we study is given by

$$x(k+1) = A(k) \otimes x(k), \quad k \geq 0, \tag{2.43}$$

with $x(0) = x_0 \in \mathbb{R}^J$ and $\{A(k)\}$ a stationary sequence of a.s. regular matrices on $\mathbb{R}_{\max}^{J \times J}$.

2.6.1   The Projective Space

Suppose that $\overline{x(k)}$ converges in total variation and let $\overline{X}$ denote the limiting random variable. Goldstein's maximal coupling implies the existence of a random variable N so that for all $k \geq N$

$$\overline{x(k)} = \overline{X \circ \theta^k} \quad \text{a.s.},$$

where, for notational convenience, we have identified the versions of the random variables on the underlying common probability space with the original ones. Let x_0 denote the initial value of the recurrence relation, then we may rephrase the above equation as

$$\overline{A(k) \otimes \cdots \otimes A(0) \otimes x_0} = \overline{X \circ \theta^k}, \quad k \geq N,$$

or, equivalently,

$$\overline{A(0) \otimes A(-1) \otimes \cdots \otimes A(-k) \otimes x_0} = \overline{X}, \quad k \geq N,$$

where $\{A(k) : k = \ldots 1, 0, -1, \ldots\}$ denotes the continuation of the stationary sequence $\{A(k)\}$ to the negative numbers. Hence, for $X \in \overline{X}$ there exists $a \in \mathbb{R}$ so that

$$A(0) \otimes A(-1) \otimes \cdots \otimes A(-k) \otimes x_0 = a \otimes X, \quad k \geq N.$$

This implies, for $k \geq N$,

$$A(1) \otimes A(0) \otimes A(-1) \otimes \cdots \otimes A(-k) \otimes x_0 \;-\; A(0) \otimes \cdots \otimes A(-k) \otimes x_0$$
$$= A(1) \otimes a \otimes X \;-\; a \otimes X$$
$$= A(1) \otimes X \;-\; X,$$

where a.s. regularity of $\{A(k)\}$ and our assumption that $x_0 \in \mathbb{R}^J$ implies that the above differences are well-defined. Taking the limit,

$$\lim_{k \to \infty} A(1) \otimes A(0) \otimes A(-1) \otimes \cdots \otimes A(-k) \otimes x_0 \;-\; A(0) \otimes \cdots \otimes A(-k) \otimes x_0$$
$$= A(1) \otimes A(0) \otimes A(-1) \otimes \cdots \otimes A(-N) \otimes x_0 \;-\; A(0) \otimes \cdots \otimes A(-N) \otimes x_0$$
$$= A(1) \otimes X \;-\; X,$$

for all $X \in \overline{X}$. We introduce the following condition:

(D) *A random variable $Z \in [0, \infty)^J$ exists such that with probability one*

$$\sup_k |A(1) \otimes A(0) \otimes A(-1) \otimes \cdots \otimes A(-k) \otimes x_0 \;-\; A(0) \otimes \cdots \otimes A(-k) \otimes x_0| \leq Z$$

and $\mathbb{E}[Z]$ is finite.

In the next section we will provide sufficient conditions for **(D)**.

Suppose that condition (**D**) is satisfied, applying the dominated convergence theorem then yields

$$\lim_{k\to\infty} \mathbb{E}[x(k+1) - x(k)]$$

$$= \lim_{k\to\infty} \mathbb{E}\Big[A(1) \otimes A(0) \otimes A(-1) \otimes \cdots \otimes A(-k) \otimes x_0 - A(0) \otimes \cdots \otimes A(-k) \otimes x_0 \Big]$$

$$= \mathbb{E}\Big[\lim_{k\to\infty} \Big(A(1) \otimes A(0) \otimes A(-1) \otimes \cdots \otimes A(-k) \otimes x_0 - A(0) \otimes \cdots \otimes A(-k) \otimes x_0 \Big) \Big]$$

$$= \mathbb{E}[A(1) \otimes X - X] < \infty \,.$$

Convergence of $\mathbb{E}[x(k+1) - x(k)]$ implies convergence of the Cesàro-sums (see Section G.1 in the Appendix) and we obtain

$$\lim_{k\to\infty} \mathbb{E}[x(k+1) - x(k)] = \lim_{k\to\infty} \frac{1}{k+1} \sum_{i=1}^{k+1} \mathbb{E}[x(i) - x(i-1)]$$

$$= \lim_{k\to\infty} \mathbb{E}\left[\frac{1}{k+1} \sum_{i=1}^{k+1} (x(i) - x(i-1)) \right]$$

$$= \lim_{n\to\infty} \mathbb{E}\left[\frac{1}{k+1} x(k+1) - \frac{1}{k+1} x(0) \right]$$

$$= \lim_{k\to\infty} \mathbb{E}\left[\frac{1}{k} x(k) \right] \,.$$

We summarize our analysis in the following theorem:

Theorem 2.6.1 *Consider the situation in (2.43). If*

- $\{\overline{x(k)} : k \geq 1\}$ *converges in total variation to* $\overline{x}$,

- $\{A(k)\}$ *is a.s. regular and stationary,*

- *condition* (**D**) *is satisfied,*

then there is an a.s. finite random variable N *so that*

$$\lim_{k\to\infty} \mathbb{E}\left[\frac{x(k)}{k} \right] = \mathbb{E}\left[A(1) \otimes \bigotimes_{i=-N}^{0} A(i) \otimes x_0 - \bigotimes_{i=-N}^{0} A(i) \otimes x_0 \right],$$

for any finite initial value $x_0 \in \mathbb{R}^J$.

Under the conditions in Theorem 2.2.3, $\mathbb{E}[x_j(k)]/k$, $1 \leq j \leq J$, tends to the Lyapunov vector of $\{A(k)\}$ as k tends to ∞. This yields the following representation for the Lyapunov vector:

Lemma 2.6.1 *Consider the situation in (2.43). If*

(i) $\{\overline{x}(k) : k \geq 1\}$ *converges in total variation to* $\overline{x}$,

(ii) *condition (D) is satisfied,*

(iii) $\{A(k)\}$ *is an a.s. regular and stationary sequence of integrable matrices such that*

      − $\{A(k)\}$ *has fixed support,*

      − *any finite element is a.s. non-negative, and*

      − *the elements on the diagonal are a.s. different from ε,*

then there is an a.s. finite random variable N such that

$$\mathbb{E}\left[A(1) \otimes \bigotimes_{i=-N}^{0} A(i) \otimes x_0 - \bigotimes_{i=-N}^{0} A(i) \otimes x_0 \right] = \vec{\lambda},$$

for any integrable initial value $x_0 \in \mathbb{R}^J$, where $\vec{\lambda}$ denotes the Lyapunov vector of $\{A(k)\}$.

Lemma 2.6.1 can be stated in various forms. For example, if we replace condition (iii) by the condition that $\{A(k)\}$ has MLP, then we obtain that the components of the Lyapunov vector are equal, see Theorem 2.2.4.

Recall that we have introduced **e** as the vector with all elements equal to e. For $x \in \mathbb{R}$, the vector with all elements equal to x is then given by $x \otimes \mathbf{e}$. For sequences $\{A(k)\}$ with countable state-space, Lemma 2.6.1 can be phrased as follows:

Lemma 2.6.2 *Consider the situation in (2.43). If*

- *(C1) − (C3) are satisfied, and*

- *condition (D) is satisfied,*

then there is an a.s. finite random variable N so that

$$\mathbb{E}\left[A(1) \otimes \bigotimes_{i=-N}^{0} A(i) \otimes x_0 - \bigotimes_{i=-N}^{0} A(i) \otimes x_0 \right] = \lambda \otimes \mathbf{e},$$

for any integrable initial value x_0, where λ denotes the Lyapunov exponent of $\{A(k)\}$.

Proof: Conditions (C1) − (C3) imply convergence of $\{\overline{x(k)} : k \geq 1\}$ in total variation, see Theorem 2.5.1. By condition (C3), a primitive matrix, say, C exists that is a pattern of $\{A(k)\}$, and we assume, for the sake of simplicity, that $C \in \mathcal{A}$, which implies $N = 1$, Let c denote the coupling time of C. From the i.i.d. assumption it follows that the event $\{A(c-1) = A(c-2) = \cdots = A(0) = C\}$ has positive probability and matrix C therefore satisfies condition (C). By Theorem 2.2.4 we obtain $\lim_{k \to \infty} \mathbb{E}[x_j(k)]/k = \lambda$, for $1 \leq j \leq J$. Hence, the proof of the lemma follows directly from Theorem 2.6.1. $\square$

We conclude this section with a remark on the cycle formula in Section 2.5.4. Under the conditions put forward in the above Lemma it holds that

$$\mathbb{E}[\overline{A(1) \otimes X} - \overline{X}] = \lambda \otimes \mathbf{e}. \tag{2.44}$$

The cycle formula can therefore be rephrased as follows: let the conditions in Lemma 2.6.2 be satisfied and let $\{T_k\}$ denote the time of the kth occurrence of the c-fold concatenation of C, see Section 2.5.4 for a formal definition. A simple geometrical trial argument, like the one used in the proof of Theorem 2.5.1, shows that

$$\mathbb{E}[T_1 - T_0] < \infty. \tag{2.45}$$

Elaborating on the limit theorem for regenerative processes (see Section 2.5.4 for details), (2.45) together with (2.44) implies $\mathbb{E}[x(T_1) - x(T_0)] < \infty$, and the cycle formula reads

$$\lambda \otimes \mathbf{e} = \frac{\mathbb{E}[x(T_1) - x(T_0)]}{\mathbb{E}[T_1 - T_0]}.$$

2.6.2   Backward Coupling

In the previous section, the existence of a coupling time N was shown. In this section, we will provide an explicit construction of N via *backward coupling*. In Markov chain theory, backward coupling, or, coupling from the past, is an approach that allows sampling from the stationary distribution of a finite-state Markov chain. Suppose that we consider a family of Markov chains X^s on a finite state space S, each with the same transition probabilities and with common unique stationary distribution π, but with version X^s starting in state $s \in S$. If we can find a time T in the past such that *all* versions X^s starting, not at time 0, but at time $-T$, have the *same value* at time 0, then this common value is a sample from π, see Theorem 1 in [92]. Intuitively, it is clear why this result holds with such a random time T. Consider a chain starting at $-\infty$ with π. This chain must at time $-T$ pick some value s, and from then on it follows the trajectory from that value. By definition of T, this trajectory reaches at time 0 the same state s' that is reached by X^s no matter what choice of s. Therefore, s' is a sample from π. Propp and Wilson coin the name 'coupling-from-the-past' for this algorithm since in essence $-T$ is a coupling time with the stationary version started at $-\infty$. Based on the same principles, Borovkov and Foss developed in [23, 22] the so-called 'renovating events' approach to stability analysis of stochastically recursive sequences. In particular, the approach to stability analysis via patterns (see Section 2.5) was originally inspired by backward coupling via 'renovating events.'

Elaborating on backward coupling, we combine our results for second-order limits with results for first-order limits in order to represent the Lyapunov exponent (a first-order limit) by the difference of two finite horizon experiments. We follow the line of argument in [7]. The key assumption for our analysis is that $\{A(k)\}$ possesses a pattern $\tilde{A}$ such that $\tilde{A}$ is primitive. The fact that $\{A(k)\}$ admits a pattern resembles a sort of memory loss property of max-plus linear systems. To see this, let $x(k + 1) = A(k) \otimes x(k)$ be a stochastic sequence defined via $\{A(k)\}$ and assume that $\{A(k)\}$ has a pattern with associated matrix $\tilde{A}$ and that $\{A(k)\}$ is a.s. regular. For vectors $x, y \in \mathbb{R}^J$, let $x - y$ denote the component-wise difference, that is, $(x - y)_j = x_j - y_j$. In what follows we consider the limit of $x(k + 1) - x(k)$ as k tends to ∞, where the limit has to be understood component-wise. In order to prove the existence of this limit we will work with a backward coupling argument. For this reason it is more convenient to let the index k run backwards. More precisely, we set

$$A^0_{-m} \stackrel{\text{def}}{=} A(0) \otimes A(-1) \otimes \cdots \otimes A(-m) \stackrel{\text{def}}{=} \bigotimes_{k=-m}^{0} A(k)$$

and

$$x^0_{-m} \stackrel{\text{def}}{=} A^0_{-m} \otimes x_0 = \bigotimes_{k=-m}^{0} A(k) \otimes x_0 \,,$$

with $x^0_0 = x_0 \in \mathbb{R}^J$, that is, x^0_{-m} is the state of the sequence $\{x(k)\}$, started at time $-m$ in x_0, at time 0. The sequence $\{x^0_{-m} : m \geq 0\}$ evolves backwards in time according to

$$x^0_{-(m+1)} = A^0_{-m} \otimes A(-(m + 1)) \otimes x_0 \,.$$

Note that $x(k + 1)$ and x^0_{-k} are equal in distribution. With this notation the second-order limit reads

$$\lim_{k\to\infty} A(1) \otimes x^0_{-k} - x^0_{-k} = \lim_{k\to\infty} \left(A(1) \otimes \bigotimes_{m=-k}^{0} A(m) \otimes x_0 - \bigotimes_{m=-k}^{0} A(m) \otimes x_0 \right).$$

Note that the above differences are well-defined due to the a.s. regularity of $\{A(k)\}$ and our assumption that $x_0 \in \mathbb{R}^J$.

Let condition (**C3**) be satisfied. Suppose that, after going η steps backwards in time, we observe for the first time the $c(\tilde{A})$-fold concatenation of the sequence constituting $\tilde{A}$, the pattern of $\{A(k)\}$. More precisely, let $(a_N, a_{N-1}, \ldots, a_1)$ denote the sequence constituting $\tilde{A}$, that is, $\tilde{A} = a_N \otimes \cdots \otimes a_1$, and let $\tilde{a}$ denote the $c(\tilde{A})$-fold concatenation of the string $(a_N, a_{N-1}, \ldots, a_1)$, which implies that $\tilde{a}$ has $M = c(\tilde{A}) \cdot N$ components. Then,

$$\tilde{A}^{c(\tilde{A})} = \bigotimes_{k=1}^{M} a_k$$

and η is defined by

$$\eta = \inf\{k \geq 0 \,|\, A(-k) = a_1, A(-k + 1) = a_2, \ldots, A(-k + (M - 1)) = a_M\} \,.$$
$$(2.46)$$

In accordance with Theorem 2.1.1, we obtain that the random variable

$$\bigotimes_{k=-(\eta+n)}^{0} A(k) \otimes x_0 , \quad n \geq 0 ,$$

is an eigenvector of $\tilde{A}$, in formula:

$$\bigotimes_{k=-(\eta+n)}^{0} A(k) \otimes x_0 \in V(\tilde{A}) , \quad n \geq 0 .$$

Remark 2.6.1 *The random variable η denotes the index of the matrix that completes the first occurrence of $\tilde{a}$. Since we start counting the elements of the series of matrices from zero, the total number of transitions until this happens is $\eta + 1$.*

Recall that multiplication of a vector $v \in \mathbb{R}_{\max}^J$ with a scalar $\gamma \in \mathbb{R}_{\max}$ is defined by component-wise multiplication: $(\gamma \otimes u)_j = \gamma \otimes u_j$. It can be easily checked that

$$\forall \gamma \in \mathbb{R}_{\max}, v \in \mathbb{R}_{\max}^J : \quad B \otimes v - C \otimes v = B \otimes (\gamma \otimes v) - C \otimes (\gamma \otimes v) , \quad (2.47)$$

for all $B, C \in \mathbb{R}_{\max}^{I \times J}$. We now use the fact that the eigenvector of a primitive matrix is unique (up to scalar multiplication): if $u, v \in V(A)$, then a $\gamma \in \mathbb{R}_{\max}$ exists such that $v = \gamma \otimes u$, see Corollary 2.1.1. Hence, (2.47) implies

$$\forall v, u \in V(A) : \quad B \otimes v - C \otimes v = B \otimes u - C \otimes u , \quad (2.48)$$

for matrices $A, B, C \in \mathbb{R}_{\max}^{J \times J}$. Combining the above arguments, we obtain

$$\lim_{k \to \infty} A(1) \otimes x_{-k}^0 - x_{-k}^0$$

$$= \lim_{k \to \infty} \left(A(1) \otimes \bigotimes_{m=-k}^{0} A(m) \otimes x_0 - \bigotimes_{m=-k}^{0} A(m) \otimes x_0 \right)$$

$$= A(1) \otimes \bigotimes_{m=-\eta+M}^{0} A(m) \otimes \underbrace{\tilde{A}^{c(\tilde{A})} \otimes \bigotimes_{m=-\infty}^{-\eta-1} A(m) \otimes x_0}_{\in V(\tilde{A})}$$

$$- \bigotimes_{m=-\eta+M}^{0} A(m) \otimes \underbrace{\tilde{A}^{c(\tilde{A})} \otimes \bigotimes_{m=-\infty}^{-\eta-1} A(m) \otimes x_0}_{\in V(\tilde{A})}$$

$$\overset{(2.48)}{=} A(1) \otimes \bigotimes_{m=-\eta+M}^{0} A(m) \otimes \underbrace{\tilde{A}^{c(\tilde{A})}}_{\in V(\tilde{A})} \otimes x_0$$

$$- \bigotimes_{m=-\eta+M}^{0} A(m) \otimes \underbrace{\tilde{A}^{c(\tilde{A})}}_{\in V(\tilde{A})} \otimes x_0$$

$$= A(1) \otimes \bigotimes_{m=-\eta}^{0} A(m) \otimes x_0 - \bigotimes_{m=-\eta}^{0} A(m) \otimes x_0$$

$$= A(1) \otimes A_{-\eta}^{0} \otimes x_0 - A_{-\eta}^{0} \otimes x_0 < \infty.$$

Hence, the second-order limit can be represented by a random horizon experiment.

Next, we will show that the above limit representation also holds if we consider expected values. We have assumed that $x_0 \in \mathbb{R}^J$. This together with a.s. regularity of $A(k)$ yields that $x(k) \in \mathbb{R}^J$ a.s. for all k. Let $(\cdot)_j$ denote the projection on the jth component. Applying Lemma 1.6.1 yields

$$\left| \left(A(1) \otimes \bigotimes_{k=-m}^{0} A(k) \otimes x_0 \right)_j - \left(\bigotimes_{k=-m}^{0} A(k) \otimes x_0 \right)_j \right|$$

$$\leq \left\| A(1) \otimes \bigotimes_{k=-m}^{0} A(k) \otimes x_0 \right\|_{\oplus} + \left\| \bigotimes_{k=-m}^{0} A(k) \otimes x_0 \right\|_{\oplus}$$

$$\leq 2 \sum_{k=-m}^{1} \|A(k)\|_{\oplus} + 2\|x_0\|_{\oplus}.$$

From the preceding analysis follows that, for any m,

$$\left| \left(A(1) \otimes \bigotimes_{k=-m}^{0} A(k) \otimes x_0 \right)_j - \left(\bigotimes_{k=-m}^{0} A(k) \otimes x_0 \right)_j \right|$$

$$\leq 2 \sum_{k=-\eta}^{1} \|A(k)\|_{\oplus} + 2\|x_0\|_{\oplus}.$$

Let $A(1)$ be integrable, then $\mathbb{E}[\|A(1)\|_{\oplus}] < \infty$, and assume that $\mathbb{E}[\eta] < \infty$. By construction, for $m \geq 0$, the event $\{\eta = m\}$ is independent of $\{A(-k) : k > m\}$. Provided that $\{A(k)\}$ is i.i.d., Wald's equality (see Section E.8 in the Appendix) yields

$$\mathbb{E}\left[\sum_{k=-\eta}^{1} \|A(n)\|_{\oplus} \right] = \mathbb{E}[\eta + 1]\,\mathbb{E}[\|A(1)\|_{\oplus}] < \infty.$$

Hence, provided that $\mathbb{E}[\eta] < \infty$, we may apply the dominated convergence theorem to the second-order limit and obtain

$$
\lim_{k \to \infty} \mathbb{E}[x(k+1) - x(k)]
$$

$$
= \lim_{k \to \infty} \mathbb{E}\left[A(1) \otimes x^0_{-k} - x^0_{-k}\right]
$$

$$
= \mathbb{E}\left[\lim_{k \to \infty} \left(A(1) \otimes x^0_{-k} - x^0_{-k}\right)\right]
$$

$$
= \mathbb{E}\left[\bigotimes_{k=-\eta}^{1} A(k) \otimes x_0 - \bigotimes_{k=-\eta}^{0} A(k) \otimes x_0\right] < \infty. \tag{2.49}
$$

In particular, the above analysis shows that if $\mathbb{E}[\eta] < \infty$, then $(\mathbf{C1}) - (\mathbf{C3})$ already imply $(\mathbf{D})$, and Lemma 2.6.2 can be phrased as follows:

Theorem 2.6.2 *Let $\{A(k)\}$ be a sequence of integrable matrices. If $(\mathbf{C1})-(\mathbf{C3})$ are satisfied, then the Lyapunov exponent of $\{A(k)\}$, denoted by λ, exists and it holds for any initial vector $x_0 \in \mathbb{R}^J$:*

$$
\lambda \otimes \mathbf{e} = \mathbb{E}\left[\bigotimes_{k=-\eta}^{1} A(k) \otimes x_0 - \bigotimes_{k=-\eta}^{0} A(k) \otimes x_0\right]
$$

$$
= \lim_{k \to \infty} \frac{1}{k} \mathbb{E}\left[\bigotimes_{i=0}^{k-1} A(i) \otimes x_0\right]
$$

$$
= \lim_{k \to \infty} \frac{1}{k} \bigotimes_{i=0}^{k-1} A(i) \otimes x_0 \quad a.s.
$$

Proof: We show that $\mathbb{E}[\eta]$ is finite. By assumption $(\mathbf{C3})$, a primitive matrix, say, C exists that is a pattern, and we assume, for the sake of simplicity, that $C \in \mathcal{A}$, which implies $N = 1$. Let c denote the coupling time of C. Because the state space is discrete and the sequence is i.i.d., the probability of observing C, denoted by p, is larger than 0. If $p = 1$, then $\mathbb{E}[\eta] = c$. In case $0 < p < 1$, we argue as follows. By construction, the probability of the event $\{\eta = m\}$ is less than or equal to the probability of the event that $A(k) \neq C$, $0 \geq k \geq -m + c$, and $A(k) = C$, for $k = -m + c - 1, \ldots, -m$. In other words, for $m \geq c$, it holds that $P(\eta = m) \geq (1 - p)^{m-c} p^c$. This implies

$$
\mathbb{E}[\eta] \leq \sum_{m=c}^{\infty} m \, (1 - p)^{m-c} p^c
$$

$$
= \sum_{m=0}^{\infty} (m + c) \, (1 - p)^m p^c
$$

$$
= c \, p^c \sum_{m=0}^{\infty} (1 - p)^m + p^c \sum_{m=0}^{\infty} m \, (1 - p)^m
$$

$$= \frac{c\,p^c}{p} + \frac{p^c(1-p)}{p^2}$$
$$< \infty\,,$$

which concludes the proof. $\square$

We conclude this section with revisiting the cycle formula in Lemma 2.5.2.

Corollary 2.6.1 *Let* $(\mathbf{C1})-(\mathbf{C3})$ *be satisfied. If* C *is a pattern of* $\{A(k)\}$, *then, for any finite initial vector* $x_0 \in V(C)$,

$$\lambda \otimes \mathbf{e} = \frac{\mathbb{E}[x(\eta) - x_0]}{\mathbb{E}[\eta]}\,,$$

where λ *denotes the Lyapunov exponent of* $\{A(k)\}$.

Part II

Perturbation Analysis

Chapter 3

A Max-Plus Differential Calculus

In this chapter we consider parameter-dependent max-plus linear systems where the parameter, denoted by θ, is a parameter of one of the firing time distributions of the event graph. For example, in a queuing application, θ may be the mean service time at one of the queues. We are interested in sensitivity analysis and optimization of performance measures of max-plus linear systems and we therefore want to find algebraic expressions for gradients of max-plus linear systems. Only in special cases the gradients can be calculated explicitly and in the general situation unbiased gradient estimators are obtained.

Perturbation analysis is an approach to gradient estimation that dates back to the pioneering paper by Ho et al. [67]. Since then there has been great interest in gradient estimation and various approaches have been developed. The following monographs [94, 52, 68, 95, 90, 44] may serve as main references.

We work within the framework of *measure-valued differentiation*. One example of such measure-valued derivatives are *weak derivatives* as introduced by Pflug, see [90] and for an early reference we refer to [89]. Specifically, we introduce $\mathcal{D}$-*derivatives of random matrices* (and vectors), where $\mathcal{D}$-differentiability refers to a concept of differentiability that is defined via a class $\mathcal{D}$ of performance functions. In order to develop our calculus of differentiation, we embed the random matrices into a richer set of objects. This enlarged object space allows us to define *sample-path* $\mathcal{D}$-*derivatives* of random matrices. For these sample-path $\mathcal{D}$-derivatives we provide a calculus of $\mathcal{D}$-differentiation that allows us to calculate derivatives of sums and products (or expressions containing mixtures of sums and products) of random matrices. The calculus resembles the standard calculus of differentiation. For various types of max-plus linear systems, we explicitly calculate the sample-path derivatives. It is worth noting that the obtained sample-path derivatives are unbiased gradient estimators by construction.

This chapter is organized as follows. Section 3.1 gives a short introduction to the theory of measure-valued differentiation. In Section 3.2, we introduce a standard example of the function space $\mathcal{D}$. Section 3.3 introduces $\mathcal{D}$-derivatives of random matrices and develops our calculus of $\mathcal{D}$-differentiation. An algebraic framework for calculating $\mathcal{D}$-derivatives of max-plus matrices is derived in Section 3.4, and our differential calculus is established in Section 3.5. Finally, we present unbiased gradient estimators for various types of max-plus linear systems in Section 3.6.

The algebraic framework for $\mathcal{D}$-derivatives of max-plus matrices as established in Section 3.4 is based on [62]. The theory developed in this chapter however extends the results in [62] to unbounded performance measures.

3.1 Measure-Valued Differentiation

This section provides a short introduction to the theory of measure-valued differentiation. Let (S, d_S) be a separable metric space and let $\mathcal{M} = \mathcal{M}(S, \mathcal{S})$ be the set of finite signed measures on the measurable space $(S, \mathcal{S})$, where $\mathcal{S}$ denotes the Borel field of S. The set of all probability measures on $(S, \mathcal{S})$ is denoted by $\mathcal{M}_1 = \mathcal{M}_1(S, \mathcal{S})$. Let $\mathcal{D}(S)$ be a set of mappings from S to $\mathbb{R}$ and assume that the constant function $g \equiv 1$ is in $\mathcal{D}(S)$.

Consider a family $\{\mu_\theta : \theta \in \Theta\}$ of measures on $(S, \mathcal{S})$, with $\Theta \stackrel{\text{def}}{=} (a, b) \subset \mathbb{R}$. Denote the set of continuous absolutely integrable mappings with respect to μ_θ by $\mathcal{L}^1(\mu_\theta)$ and denote by

$$\mathcal{L}^1(\mu_\theta : \theta \in \Theta) \stackrel{\text{def}}{=} \bigcap_{\theta \in \Theta} \mathcal{L}^1(\mu_\theta)$$

the set of mappings that are absolutely integrable with respect to μ_θ for any $\theta \in \Theta$. Moreover, let $C^b(S)$ denote the set of bounded continuous real-valued functions $g : S \mapsto \mathbb{R}$. To simplify the notation we will write C^b for $C^b(S)$ when it is clear which underlying space is meant. Note that for any μ_θ in $\mathcal{M}$ and for any θ in Θ, we have $C^b \subset \mathcal{L}^1(\mu_\theta)$.

Definition 3.1.1 *We call the mapping* $\mu : \Theta \to \mathcal{M}_1$ *$\mathcal{D}$-differentiable at point θ with $\mathcal{D} \subset \mathcal{L}^1(\mu_\theta : \theta \in \Theta)$ if there exists a finite signed measure $\mu'_\theta \in \mathcal{M}$ such that for any g in $\mathcal{D}$:*

$$\lim_{\Delta \to 0} \frac{1}{\Delta} \left(\int_S g(s) \mu_{\theta + \Delta}(ds) - \int_S g(s) \mu_\theta(ds) \right) = \int_S g(s) \mu'_\theta(ds) \, .$$

If μ_θ is $\mathcal{D}$-differentiable, then μ'_θ is a finite signed measure. Any finite signed measure can be written as difference between two finite positive measures. This representation is called *Hahn-Jordan decomposition*, see Section E.1 in the Appendix. More specifically, a set $S^+ \in \mathcal{S}$ exists such that, for $A \in \mathcal{S}$,

$$[\mu'_\theta]^+(A) \stackrel{\text{def}}{=} \mu'_\theta(A \cap S^+) \geq 0 \, ,$$

$$[\mu'_\theta]^-(A) \stackrel{\text{def}}{=} -\mu'_\theta(A \cap (S \setminus S^+)) \geq 0$$

and

$$\mu'_\theta(A) = [\mu'_\theta]^+(A) - [\mu'_\theta]^-(A).$$

Set

$$c_\theta = [\mu'_\theta]^+(S) \tag{3.1}$$

and

$$\tilde{c}_\theta = [\mu'_\theta]^-(S).$$

Since $[\mu'_\theta]^+$ and $[\mu'_\theta]^-$ are finite measures, c_θ and $\tilde{c}_\theta$ are finite. We may thus introduce probability measures μ^+_θ, μ^-_θ on $(S, \mathcal{F})$ through

$$\mu^+_\theta(A) = \frac{1}{c_\theta}[\mu'_\theta]^+(A) \tag{3.2}$$

and

$$\mu^-_\theta(A) = \frac{1}{\tilde{c}_\theta}[\mu'_\theta]^-(A), \tag{3.3}$$

for $A \in \mathcal{S}$. This yields the following representation of μ'_θ

$$\forall A \in \mathcal{S}: \quad \mu'_\theta(A) = c_\theta\,\mu^+_\theta(A) - \tilde{c}_\theta\,\mu^-_\theta(A).$$

The fact that μ'_θ stems from differentiating a probability measure implies that $c_\theta = \tilde{c}_\theta$. Indeed, for $\mu_\theta \in \mathcal{M}_1$, it holds $\mu_\theta(S) = 1$ for all θ and, since we have assumed that the constant function $g \equiv 1$ is in $\mathcal{D}$, this implies

$$0 = \frac{d}{d\theta}\mu_\theta(S) = \frac{d}{d\theta}\int 1\,\mu_\theta(ds) = \int 1\,\mu'_\theta(ds) = \mu'_\theta(S).$$

Hence, $\mu'_\theta(S) = 0$ for all $\theta \in \Theta$, which yields

$$c_\theta\,\mu^+_\theta(S) = \tilde{c}_\theta\,\mu^-_\theta(S).$$

Since μ^+_θ and μ^-_θ are probability measures, we obtain $c_\theta = \tilde{c}_\theta$. Thus, μ'_θ is completely characterized through the triple $(c_\theta, \mu^+_\theta, \mu^-_\theta)$, with $\mu^+_\theta, \mu^-_\theta \in \mathcal{M}_1$. We call μ^+_θ in (3.2) the *(normalized) positive part* and μ^-_θ in (3.3) the *(normalized) negative part* of μ'_θ, respectively. Note that, by the above construction, a set $A \in \mathcal{S}$ exists such that either $\mu^+_\theta(A) = 0$ or $\mu^-_\theta(S \setminus A) = 0$, in symbols: $\mu^+_\theta \perp \mu^-_\theta$. We now state the formal definition of a $\mathcal{D}$-derivative of a probability measure.

Definition 3.1.2 *Let* $\mathcal{D} \subset \mathcal{L}^1(\mu_\theta : \theta \in \Theta)$. *We call a triple* $(c_{\mu_\theta}, \mu^+_\theta, \mu^-_\theta)$, *with* $\mu^\pm_\theta \in \mathcal{M}_1$ *and* $c_{\mu_\theta} \in \mathbb{R}$, *a* $\mathcal{D}$-derivative *of probability measure* μ_θ *at* θ *if, for all* g *in* $\mathcal{D}$, *it holds true that*

$$\lim_{\Delta \to 0} \frac{1}{\Delta}\left(\int_S g(s)\mu_{\theta+\Delta}(ds) - \int_S g(s)\mu_\theta(ds)\right)$$

$$= c_\theta\left(\int_S g(s)\mu^+_\theta(ds) - \int_S g(s)\mu^-_\theta(ds)\right).$$

Remark 3.1.1 *Definition 3.1.1 is easily extended to finite measures in* $\mathcal{M}$. *However, in order to conclude that* $[\mu'_\theta]^+(S) = [\mu'_\theta]^-(S)$ *we need the additional condition that a number* $d \in \mathbb{R}$ *exists such that*

$$\forall \theta \in \Theta : \quad \mu_\theta(S) = d.$$

If $\mu_\theta \in \mathcal{M}$ *satisfies the above condition, then a* $\mathcal{D}$-*derivative of* μ_θ *can be defined as in Definition 3.1.2.*

An instance of a $\mathcal{D}$-derivative can always be found through the Hahn-Jordan decomposition, see (3.1), (3.2) and (3.3), and this construction is called the *standard construction*. However, this is only one of many representations possible. To see that a $\mathcal{D}$-derivative is not unique, let $(c_\theta, \mu_\theta^+, \mu_\theta^-)$ be a $\mathcal{D}$-derivative of μ_θ, and let γ be a probability measure on $(S, \mathcal{S})$ with $\mathcal{D} \subset \mathcal{L}^1(\gamma)$ and let b be a positive constant. Then

$$\mu'_\theta = (c_\theta + b)\left(\frac{c_\theta}{c_\theta + b}\mu_\theta^+ + \frac{b}{c_\theta + b}\gamma\right) - (c_\theta + b)\left(\frac{c_\theta}{c_\theta + b}\mu_\theta^- + \frac{b}{c_\theta + b}\gamma\right)$$

is also a $\mathcal{D}$-derivative of μ_θ. The $\mathcal{D}$-derivative of a probability measure μ_θ becomes unique if we assume that (a) $\mu_\theta^\pm$ are again probability measures, and (b) $\mu_\theta^+ \perp \mu_\theta^-$. Moreover, c_θ is minimized if $\mu_\theta^+ \perp \mu_\theta^-$.

Suppose that μ_θ is $\mathcal{D}$-differentiable and that μ_θ has ν-density f_θ which is differentiable in θ. If, for any $g \in \mathcal{D}$, interchanging the order of integration and differentiation is justified, we obtain

$$\frac{d}{d\theta}\int_S g(s)f_\theta(s)\,\nu(ds) = \int_S g(s)\frac{d}{d\theta}f_\theta(s)\,\nu(ds).$$

Let

$$c_\theta \stackrel{def}{=} \frac{1}{2}\int_S \left|\frac{d}{d\theta}f_\theta(s)\right|\nu(ds)$$

be finite, then

$$f_\theta^+(s) \stackrel{def}{=} \frac{1}{c_\theta}\max\left(0, \frac{d}{d\theta}f_\theta(s)\right)$$

and

$$f_\theta^-(s) \stackrel{def}{=} \frac{1}{c_\theta}\max\left(0, -\frac{d}{d\theta}f_\theta(s)\right),$$

for $s \in S$, are ν-densities, and for any $g \in \mathcal{D}$ it holds that

$$\frac{d}{d\theta}\int_S g(s)f_\theta(s)\,\nu(ds) = c_\theta\int_S g(s)f_\theta^+(s)\,\nu(ds) - c_\theta\int_S g(s)f_\theta^-(s)\,\nu(ds)$$

(that f_θ^+ and f_θ^- are indeed densities follows from the fact that they are limits of measurable mappings). Consequently, we may obtain μ'_θ as the re-scaled difference between the probability measures

$$\mu_\theta^+(A) = \int_A f_\theta^+(s)\,\nu(ds), \quad \mu_\theta^-(A) = \int_A f_\theta^-(s)\,\nu(ds), \qquad (3.4)$$

for $A \in S$, that is, f_θ^+ is the ν-density of μ_θ^+ and f_θ^- is the ν-density of μ_θ^-. Note that this decomposition is nothing else than the Hahn-Jordan decomposition: differentiability of f_θ implies that

$$S^+ = \left\{ s \in S : \frac{d}{d\theta} f_\theta(s) \geq 0 \right\}$$

is measurable and the measures defined in (3.4) satisfy, for any $A \in S$, $c_\theta \mu_\theta^+(A) = \mu_\theta'(A \cap S^+)$ and $c_\theta \mu_\theta^-(A) = -\mu_\theta'(A \cap (S \setminus S^+))$.

A typical choice for $\mathcal{D}$ is $\mathcal{D} = C^b(S)$, the set of bounded continuous real-valued functions. Indeed, Pflug developed his theory of weak differentiation for this class of performance measures, and C^b-derivatives are called *weak derivatives* in [90].

Next, we give an example of a $\mathcal{D}$-derivative and illustrate the non-uniqueness of the $\mathcal{D}$-derivative.

Example 3.1.1 *Let $S = [0, \infty)$ and let*

$$f_\theta(x) \overset{\text{def}}{=} \theta e^{-\theta x}, \quad x \geq 0$$

be the Lebesgue density of an exponential distribution with rate θ, denoted by μ_θ. Take $\mathcal{D} = C^b([0, \infty))$ and $\Theta = [a, b]$, with $0 < a < b < \infty$, then μ_θ is C^b-differentiable. To see this, note that

$$\sup_{\theta \in [a,b]} \left| \frac{d}{d\theta} f_\theta(x) \right| \leq (1 + bx) e^{-ax},$$

which has finite Lebesgue integral. Applying the dominated convergence theorem, we obtain for any $g \in C^b$

$$\frac{d}{d\theta} \int_0^\infty g(x) f_\theta(x) \, dx = \int_0^\infty g(x) \frac{d}{d\theta} f_\theta(x) \, dx.$$

Note that

$$\frac{1}{2} \int_0^\infty \left| \frac{d}{d\theta} f_\theta(x) \right| \, dx = \frac{1}{\theta e},$$

$$\max \left(\frac{d}{d\theta} f_\theta(x), 0 \right) = 1_{[0,1/\theta]}(x) (1 - \theta x) e^{-\theta x}$$

and

$$\max \left(-\frac{d}{d\theta} f_\theta(x), 0 \right) = 1_{[1/\theta,\infty)}(x) (\theta x - 1) e^{-\theta x}.$$

Introducing densities

$$f_\theta^+(x) \overset{\text{def}}{=} \frac{1}{c_\theta} 1_{[0,1/\theta]}(x) (1 - \theta x) e^{-\theta x} \tag{3.5}$$

and

$$f_\theta^-(x) \overset{\text{def}}{=} \frac{1}{c_\theta} 1_{[1/\theta,\infty)}(x) (\theta x - 1) e^{-\theta x}, \tag{3.6}$$

with

$$c_\theta \stackrel{\text{def}}{=} \frac{1}{\theta\, e}\,,$$

we obtain for any $g \in C^b$

$$\frac{d}{d\theta} \int_{\mathbb{R}_+} g(x) f_\theta(x)\, dx$$

$$= \frac{1}{\theta\, e} \int_0^{1/\theta} g(x)\,(\theta - \theta^2 x)\, e^{1-\theta x}\, dx - \frac{1}{\theta\, e} \int_{1/\theta}^\infty g(x)\,(\theta^2 x - \theta)\, e^{1-\theta x}\, dx$$

and a C^b derivative of μ_θ is given by $(c_\theta,\, f_\theta^+(s)\, ds,\, f_\theta^-(s)\, ds\,)$.
    Sometimes the standard construction is not the most efficient one. Let

$$h_{2,\theta}(x) \stackrel{\text{def}}{=} \theta^2 x e^{-\theta x}\,, \quad x \ge 0,$$

denote the Lebesgue density of the Gamma-$(2, \theta)$-distribution. It is easily checked that

$$\frac{d}{d\theta} f_\theta(x) = \frac{1}{\theta}\Big(f_\theta(x) - h_{2,\theta}(x)\Big),$$

which implies that $(1/\theta,\, f_\theta(s)\, ds,\, h_{2,\theta}(s)\, ds\,)$ is a C^b-derivative of μ_θ, that is, for any $g \in C^b$ is holds that

$$\frac{d}{d\theta} \int_{\mathbb{R}} g(x) f_\theta(x)\, dx = \frac{1}{\theta} \int_{\mathbb{R}} g(x)\, f_\theta(x)\, dx - \frac{1}{\theta} \int_{\mathbb{R}} g(x)\, h_{2,\theta}(x)\, dx\,.$$

Let X_θ and Y_θ be independent samples of the exponential distribution with mean $1/\theta$. Then the above equation can be phrased as follows

$$\frac{d}{d\theta} \mathbb{E}[g(X_\theta)] = \frac{1}{\theta} \mathbb{E}[g(X_\theta)] - \frac{1}{\theta} \mathbb{E}[g(X_\theta + Y_\theta)]\,, \quad g \in C^b\,.$$

In words, the derivative of $\mathbb{E}[g(X_\theta)]$ can be estimated by drawing one extra sample from the exponential distribution.

    In the above example, μ_θ as well as μ_θ^+ and μ_θ^- have Lebesgue densities, that is, the measure as well as its $\mathcal{D}$-derivative are dominated by the same measure. The following example demonstrates that this is not always the case.

Example 3.1.2 *Let $\mathcal{U}_{[0,\theta]}$ be the uniform distribution on the interval $[0, \theta]$ for $0 < \theta \le a$, with $a < \infty$. For any g in C^b it holds that*

$$\frac{d}{d\theta} \int g(x)\, \mathcal{U}_{[0,\theta]}(dx) = \frac{d}{d\theta} \left(\frac{1}{\theta} \int_0^\theta g(x)\, dx \right)$$

$$= \frac{1}{\theta} g(\theta) - \frac{1}{\theta^2} \int_0^\theta g(x)\, dx$$

$$= \frac{1}{\theta} \left(\int g(x)\, \delta_\theta(dx) - \int g(x)\, \mathcal{U}_{[0,\theta]}(dx) \right),$$

where δ_x denotes the Dirac measure in x. Hence, $(1/\theta, \delta_\theta, \mathcal{U}_{[0,\theta]})$ is a C^b-derivative of $\mathcal{U}_{[0,\theta]}$.

We conclude this series of examples with an example of a distribution that fails to be $\mathcal{D}$-differentiable.

Example 3.1.3 *Let δ_θ denote the Dirac measure with mass in $\theta \in \Theta$. For any $g : \Theta \to \mathbb{R}$ that is differentiable with respect to θ at a $\theta_0 \in \Theta$, we obtain*

$$\frac{d}{d\theta}\bigg|_{\theta=\theta_0} \int g(x)\,\delta_\theta(dx) = \frac{d}{d\theta}\bigg|_{\theta=\theta_0} g(\theta)\,.$$

Hence, δ_θ fails to be $\mathcal{D}$-differentiable for any reasonable set $\mathcal{D}$. However, we may construct a set $\hat{\mathcal{D}}$ that artificially generates $\hat{\mathcal{D}}$-differentiability. To see this, fix $c, x, y \in \mathbb{R}$ and let $\hat{\mathcal{D}}$ denote the set of all differentiable mappings $g : \Theta \to \mathbb{R}$, such that

$$\frac{d}{d\theta}\bigg|_{\theta=\theta_0} g(\theta) = c\,,$$

$g(x) = 1$ and $g(y) = 0$. Then δ_θ is $\hat{\mathcal{D}}$-differentiable with $\hat{\mathcal{D}}$-derivative (c, δ_x, δ_y). Indeed, for $g \in \hat{\mathcal{D}}$ it holds that

$$\frac{d}{d\theta}\bigg|_{\theta=\theta_0} \int g(x)\,\delta_\theta(dx) = \frac{d}{d\theta}\bigg|_{\theta=\theta_0} g(\theta)$$

$$= c\left(\int g(u)\,\delta_x(du) - \int g(u)\,\delta_y(du)\right)\,.$$

For $\mathcal{D} = C^b$, Definition 3.1.1 recovers the definition of weak differentiability in [90]. Weak differentiability of a probability measure yields statements about derivatives of performance functions out of the restrictive class of bounded continuous performance functions. The results in [62] elaborate on the theory developed in [90] and thus suffer from the restriction to bounded performance functions too. The theory developed in this chapter extends the results in [90] (and thus in [62]) in such a way that unbounded performance functions can be studied as well. The following theorem is our main tool: it establishes a product rule for $\mathcal{D}$-differentiability. For weak derivatives, such a product rule is claimed in Remark 3.28 of [90] but no proof is provided. For the general setup of $\mathcal{D}$-differentiation, we will provide an explicit proof for the product rule. Before we can state the exact statement, we have to introduce the following definition. A set $\mathcal{D}$ of real-valued mappings defined on a common domain is called *solid* if

(i) for $f, h \in \mathcal{D}$ there exists a mapping $g \in \mathcal{D}$ such that $\max(f, h) \leq g$,

(ii) if $f \in \mathcal{D}$, then for any g with $|g| \leq f$ is holds that $g \in \mathcal{D}$.

The precise statement of the product rule is given in the following theorem.

Theorem 3.1.1 *Let $(S, \mathcal{S})$ and $(Z, \mathcal{Z})$ be measurable spaces. Let $\mu_\theta \in \mathcal{M}_1(S, \mathcal{S})$ be $\mathcal{D}(S)$-differentiable at θ and $\nu_\theta \in \mathcal{M}_1(Z, \mathcal{Z})$ be $\mathcal{D}(Z)$-differentiable at θ with $\mathcal{D}(S)$ and $\mathcal{D}(Z)$ solid.*

Then, the product measure $\mu_\theta \times \nu_\theta \in \mathcal{M}_1(S \times Z)$ is $\mathcal{D}(S, Z)$-differentiable at θ, with

$$\mathcal{D}(S, Z) = \left\{ g \in \mathcal{L}^1(\mu_\theta) \cap \mathcal{L}^1(\nu_\theta) \middle| \exists n : |g(s, z)| \leq \sum_{i=0}^{n} d_i f_i(s) h_i(z) , \right.$$

$$\left. 1 \leq f_i \in \mathcal{D}(S), 1 \leq h_i \in \mathcal{D}(Z), d_i \in \mathbb{R} \right\},$$

and it holds that

$$(\mu_\theta \times \nu_\theta)' = (\mu_\theta' \times \nu_\theta) + (\mu_\theta \times \nu_\theta') .$$

Furthermore, if μ_θ has $\mathcal{D}(S)$-derivative $(c_{\mu_\theta}, \mu_\theta^+, \mu_\theta^-)$ and ν_θ has $\mathcal{D}(Z)$-derivative $(c_{\nu_\theta}, \nu_\theta^+, \nu_\theta^-)$, respectively, then the product measure $\mu_\theta \times \nu_\theta$ is $\mathcal{D}(S, Z)$-differentiable and has $\mathcal{D}(S, Z)$-derivative

$$\left(c_{\mu_\theta} + c_{\nu_\theta} , \right.$$

$$\left. \frac{c_{\mu_\theta}}{c_{\mu_\theta} + c_{\nu_\theta}} \mu_\theta^+ \times \nu_\theta + \frac{c_{\nu_\theta}}{c_{\mu_\theta} + c_{\nu_\theta}} \mu_\theta \times \nu_\theta^+, \frac{c_{\mu_\theta}}{c_{\mu_\theta} + c_{\nu_\theta}} \mu_\theta^- \times \nu_\theta + \frac{c_{\nu_\theta}}{c_{\mu_\theta} + c_{\nu_\theta}} \mu_\theta \times \nu_\theta^- \right) .$$

Proof: We show the first part of the theorem. Let $g \in \mathcal{D}(S, Z)$ be such that

$$|g(s, z)| \leq \sum_{i=0}^{n} d_i f_i(s) h_i(z).$$

We have assumed that $\mathcal{D}(S)$ is solid and condition (i) in the definition of solidness implies the existence of a mapping $f \in \mathcal{D}(S)$ such that $f \geq f_i \geq 1$ for $1 \leq i \leq n$. In the same way, solidness of $\mathcal{D}(Z)$ implies and there exists $h \in \mathcal{D}(Z)$ such that $h \geq h_i \geq 1$ for $1 \leq i \leq n$. This yields

$$|g(s, z)| \leq f(s) h(z) \sum_{i=0}^{n} d_i \|f_i\|_f \|h_i\|_h,$$

where

$$\|f_i\|_f \stackrel{\text{def}}{=} \sup_s \frac{f_i(s)}{f(s)} \quad \text{and} \quad \|h_i\|_h \stackrel{\text{def}}{=} \sup_z \frac{h_i(z)}{h(z)}, \tag{3.7}$$

for $1 \leq i \leq n$ (for a proof note that $|f_i(s)| \leq \|f_i\|_f f(s)$, for any $s \in S$). Hence, it suffices for the proof to consider $g \in \mathcal{D}(S, Z)$ such that $|g(s, z)| \leq f(s) h(z)$ for $f \in \mathcal{D}(S)$ and $h \in \mathcal{D}(Z)$. By calculation,

$$\frac{1}{\Delta} \left(\mu_{\theta+\Delta} \times \nu_{\theta+\Delta} - \mu_\theta \times \nu_\theta \right)$$

$$= \frac{1}{\Delta} \left(\mu_{\theta+\Delta} - \mu_\theta \right) \times \nu_\theta + \mu_\theta \times \frac{1}{\Delta} \left(\nu_{\theta+\Delta} - \nu_\theta \right) \tag{3.8}$$

$$+ \frac{1}{\Delta} \left(\mu_{\theta+\Delta} - \mu_\theta \right) \times \left(\nu_{\theta+\Delta} - \nu_\theta \right) .$$

Let

$$\mathcal{D}_f(S) \stackrel{\text{def}}{=} \{ g \in \mathcal{D}(S) \mid \exists c > 0 : |g(s)| \leq c f(s), \forall s \in S \},$$

or, equivalently,

$$\mathcal{D}_f(S) = \{g \in \mathcal{D}(S) : \|g\|_f < \infty\};$$

and let

$$\mathcal{D}_h(Z) \overset{\text{def}}{=} \{g \in \mathcal{D}(Z) \mid \exists c > 0 : |g(s)| \le ch(s), \forall s \in S\},$$

or, equivalently,

$$\mathcal{D}_h(Z) = \{g \in \mathcal{D}(Z) : \|g\|_h < \infty\}.$$

The remainder of the proof uses properties of the relationship between weak convergence defined by the sets D_f and D_h, respectively, and norm convergence with respect to $\|\cdot\|_f$ and $\|\cdot\|_h$, respectively. These statements are of rather technical nature and proofs can be found in Section E.5 in the Appendix.

By condition (ii) in the definition of solidness, μ_θ is in particular $\mathcal{D}_f$-differentiable, which implies that

$$\lim_{\Delta \to 0} \int_S g(s) \frac{1}{\Delta}(\mu_{\theta+\Delta} - \mu_\theta)(ds) = \int_S g(s)\,\mu_\theta'(ds)$$

for any $g \in \mathcal{D}_f$ and, by Theorem E.5.1 in the Appendix, $\|\mu_{\theta+\Delta} - \mu_\theta\|_f$ tends to zero as Δ tends to zero. For the extension of the definition in (3.7) to signed measures, we refer to Section E.5 in the Appendix. In the same vein, ν_θ is $\mathcal{D}_h$-differentiable which implies that

$$\lim_{\Delta \to 0} \int_Z g(z) \frac{1}{\Delta}(\nu_{\theta+\Delta} - \nu_\theta)(dz) = \int_Z g(z)\,\nu_\theta'(dz)$$

for any $g \in \mathcal{D}_h$ and $\|\nu_{\theta+\Delta} - \nu_\theta\|_h$ tends to zero as Δ tends to zero. Applying Lemma E.5.1 in the Appendix to the individual terms on the right-hand side of (3.8) yields

$$\lim_{\Delta \to 0} \frac{1}{\Delta} \int g(s,z)\big((\mu_{\theta+\Delta} - \mu_\theta) \times \nu_\theta\big)(ds,dz) = \int g(s,z)\big(\mu_\theta' \times \nu_\theta\big)(ds,dz)$$

$$\lim_{\Delta \to 0} \frac{1}{\Delta} \int g(s,z)\big(\mu_\theta \times (\nu_{\theta+\Delta} - \nu_\theta)\big)(ds,dz) = \int g(s,z)\big(\mu_\theta \times \nu_\theta'\big)(ds,dz)$$

and

$$\lim_{\Delta \to 0} \frac{1}{\Delta} \int g(s,z)\big((\mu_{\theta+\Delta} - \mu_\theta) \times (\nu_{\theta+\Delta} - \nu_\theta)\big)(ds,dz) = 0$$

which proves the first part of the theorem.

For the proof of the second part of the theorem, one represents μ_θ' and ν_θ' by the corresponding $\mathcal{D}$–derivatives. Let $(c_{\mu_\theta}, \mu_\theta^+, \mu_\theta^-)$ be a $\mathcal{D}(S)$–derivative of μ_θ and let $(c_{\nu_\theta}, \nu_\theta^+, \nu_\theta^-)$ be a $\mathcal{D}(Z)$–derivative of ν_θ. By the first part of the theorem:

$$(\mu_\theta \times \nu_\theta)' = \mu_\theta' \times \nu_\theta + \mu_\theta \times \nu_\theta'$$
$$= \big(c_{\mu_\theta}\mu_\theta^+ - c_{\mu_\theta}\mu_\theta^-\big) \times \nu_\theta + \mu_\theta \times \big(c_{\nu_\theta}\nu_\theta^+ - c_{\nu_\theta}\nu_\theta^-\big),$$

re–grouping the positive and negative parts yields

$$= c_{\mu_\theta}\mu_\theta^+ \times \nu_\theta + c_{\nu_\theta}\mu_\theta \times \nu_\theta^+ - c_{\mu_\theta}\mu_\theta^- \times \nu_\theta - c_{\nu_\theta}\mu_\theta \times \nu_\theta^-$$

and normalizing the parts in order to obtain probability measures gives

$$= (c_{\mu_\theta} + c_{\nu_\theta})\left(\left(\frac{c_{\mu_\theta}}{c_{\mu_\theta} + c_{\nu_\theta}}\mu_\theta^+ \times \nu_\theta + \frac{c_{\nu_\theta}}{c_{\mu_\theta} + c_{\nu_\theta}}\mu_\theta \times \nu_\theta^+\right)\right.$$
$$\left. - \left(\frac{c_{\mu_\theta}}{c_{\mu_\theta} + c_{\nu_\theta}}\mu_\theta^- \times \nu_\theta + \frac{c_{\nu_\theta}}{c_{\mu_\theta} + c_{\nu_\theta}}\mu_\theta \times \nu_\theta^-\right)\right),$$

which completes the proof.   □

Remark 3.1.2 *By assumption, any $g \in \mathcal{D}$ is continuous. However, it is possible to slightly deviate from the continuity assumption. If g is bounded by some $h \in \mathcal{D}$ and if the set of discontinuities, denoted by D_g, satisfies $\mu_\theta^+(D_g) = 0 = \mu_\theta^-(D_g)$, then the analysis applies to g as well.*

The statement of Theorem 3.1.1 can be rephrased as follows. Let $X_{\mu_\theta} \in S$ have distribution μ_θ and let $X_{\nu_\theta} \in Z$ have distribution ν_θ with X_{μ_θ} independent of X_{ν_θ}. If μ_θ is $\mathcal{D}(S)$-differentiable and ν_θ is $\mathcal{D}(Z)$-differentiable, then random variables $X_{\mu_\theta}^+$, $X_{\mu_\theta}^-$ and $X_{\nu_\theta}^+$, $X_{\nu_\theta}^-$ exist, such that for all g in $\mathcal{D}(S, Z)$:

$$\frac{d}{d\theta}\mathbb{E}\Big[g(X_{\mu_\theta}, X_{\nu_\theta})\Big]$$
$$= \mathbb{E}\Big[c_{\mu_\theta}g(X_{\mu_\theta}^+, X_{\nu_\theta}) + c_{\nu_\theta}g(X_{\mu_\theta}, X_{\nu_\theta}^+) - \Big(c_{\mu_\theta}g(X_{\mu_\theta}^-, X_{\nu_\theta}) + c_{\nu_\theta}g(X_{\mu_\theta}, X_{\nu_\theta}^-)\Big)\Big].$$

In order to make the concept of $\mathcal{D}$-differentiability fruitful for applications, we have to choose $\mathcal{D}$ in such a way that

- it is rich enough to contain interesting performance functions,

- the product of $\mathcal{D}$-differentiable measures is again $\mathcal{D}$-differentiable.

In what follows, we study two examples of $\mathcal{D}$: The space C^b of bounded continuous performance mappings and the space C_p to be introduced presently.

3.2   The Space C_p

Let the measurable space $(S, \mathcal{S})$ be equipped with an upper bound $|| \cdot ||_S$, see Definition 1.6.1. For $p \in \mathbb{N}$, we denote by $C_p(S, || \cdot ||_S)$ the set of all continuous functions $g : S \to \mathbb{R}$ such that

$$|g(x)| \le a_g + b_g||x||^p, \quad x \in S,$$

for finite constants a_g, b_g. Note that $C^b(S, || \cdot ||_S)$ is a solid space and that $C^b(S, || \cdot ||_S) \subset C_p(S, || \cdot ||_S)$ for all $p \ge 0$.

The space $C_p(S, || \cdot ||_S)$ allows us to describe many interesting performance characteristics as the following example illustrates.

Convention: When it is clear what S is, we will simply write C_p instead of $C_p(S, ||\cdot||_S)$.

Example 3.2.1 *For $J \geq 1$, take $S = [0,\infty)^J$, $||\cdot||_S = ||\cdot||_\oplus$ and let $X = (X_1, \ldots, X_J) \in S$ be defined on a probability space $(\Omega, \mathcal{A}, P)$ such that $P^X = \mu$.*

- *Taking $g(x) = \exp(-r\,x_j) \in C_0(S)$, with $r \geq 0$, we obtain the Laplace transform of X through*

$$\mathbb{E}\left[e^{-rX_j}\right] = \int g(x)\,\mu(dx).$$

- *For $g(x) = x_j^p \in C_p$, we obtain the higher-order moments of X through*

$$\mathbb{E}\left[X_j^p\right] = \int g(x)\,\mu(dx), \quad for\ p \geq 1.$$

- *Let $\mathbb{E}[X_i] = a_i$ and $\mathbb{E}[X_j] = a_j$ for specified i and j, with $i \neq j$, and assume that a_i, a_j are finite. Setting*

$$g(x_1, \ldots, x_J) = x_i\,x_j - a_i\,a_j,$$

we obtain from $\mathbb{E}[g(X)]$ the covariance between the ith and j^{th} component of X.

Remark 3.2.1 *In the literature, see for example [15], Taylor series expansions for max-plus linear systems are developed for performance functions $f : [0,\infty) \to [0,\infty)$ such that $f(x) \leq c^f x^\nu$ for all $x \geq 0$, where $\nu \in \mathbb{N}$. This class of performance functions is a true subset of $C_\nu([0,\infty))$. For example, take $f(x) = \sqrt{x}$, then no $c^f \in \mathbb{R}$ and $\nu \in \mathbb{N}$ exist such that $f(x) \leq c^f x^\nu$, whereas $f(x) \leq 1 + x^2$ and thus $f \in C_2([0,\infty))$.*

In what follows we study C_p-differentiability of product measures, that is, we take $\mathcal{D} = C_p$.

Example 3.2.2 *We revisit the situation in Example 3.1.1. Let $f_\theta^\pm(x)$ be given as in (3.5) and (3.6), respectively. Since all higher moments of the exponential distributions are finite, it follows that μ_θ is $C_p([0,\infty), |\cdot|)$-differentiable for any $p \in \mathbb{N}$.*

C_p-spaces have the nice property that, under appropriate conditions, the product of C_p-differentiable measures is again C_p-differentiable and it is this property of C_p-spaces that makes them a first choice for $\mathcal{D}$ when working with $\mathcal{D}$-derivatives. The main technical property needed for such a product rule of C_p-differentiation to hold is established in the following lemma. The statement of the lemma is expressed in terms of the influence of binary mappings on C_p-differentiability.

Lemma 3.2.1 *Let X, Y, S be non-empty sets equipped with upper bounds $||\cdot||_X$, $||\cdot||_Y$ and $||\cdot||_S$, respectively.*

- *Let $h : X \times Y \to S$ denote a binary operation. For $g \in C_p(S)$, let $g_h(x,y) = g(h(x,y))$, for $x \in X$, $y \in Y$. If finite constants c_X, c_Y exist such that for any $x \in X, y \in Y$*

$$||h(x,y)||_S \leq c_X||x||_X + c_Y||y||_Y,$$

then

$$g_h \in \mathcal{C}(X,Y),$$

with

$$\mathcal{C}(X,Y) = \left\{ g : X \times Y \to \mathbb{R} \Big| \exists n : |g(x,y)| \leq \sum_{i=0}^{n} d_i f_i(x) h_i(y), \right.$$
$$\left. f_i \in C_p(X, ||\cdot||_X), h_i \in C_p(Y, ||\cdot||_Y), d_i \in \mathbb{R} \right\}. \quad (3.9)$$

- *If finite constants c_X, c_Y and an upper bound $||\cdot||_{X \times Y}$ on $X \times Y$ exist, such that for any $x \in X, y \in Y$*

$$||(x,y)||_{X \times Y} \leq c_X||x||_X + c_Y||y||_Y,$$

then $C_p(X,Y) \subset \mathcal{C}(X,Y)$, with $\mathcal{C}(X,Y)$ as defined above.

Proof: Let $g \in C_p(S, ||\cdot||_S)$. For $r = h(x,y)$, with $x \in X$ and $y \in Y$, we obtain

$$
\begin{aligned}
|g(r)| &\leq a_g + b_g ||r||_S^p \\
&= a_g + b_g ||h(x,y)||_S^p \\
&\leq a_g + b_g (c_X||x||_X + c_Y||y||_Y)^p \\
&= a_g + b_g \sum_{i=0}^{p} \binom{p}{i} c_X^{p-i} ||x||_X^{p-i} c_Y^i ||y||_Y^i \\
&= d_{p+1} + \sum_{i=0}^{p} d_i ||x||_X^{p-i} ||y||_Y^i,
\end{aligned}
$$

with $d_i \in \mathbb{R}$, for $0 \leq i \leq p+1$. By definition, $||\cdot||_X^{p-i} \in C_p(X)$ and $||\cdot||_Y^i \in C_p(Y)$ for $0 \leq i \leq p$. Hence, $g_h \in \mathcal{C}(X,Y)$ which concludes the proof of the first part of the lemma.

The proof of the second part of the lemma follows from the first part with $S = X \times Y$ and $h(x,y) = (x,y)$. $\square$

An immediate consequence of Lemma 3.2.1 above is a version of Theorem 3.1.1 for C_p-spaces yielding a product rule for C_p-differentiability of measures.

Theorem 3.2.1 *Let $(S, \mathcal{S})$ and $(Z, \mathcal{Z})$ be measurable spaces equipped with upper bounds $||\cdot||_S$ and $||\cdot||_Z$, respectively, and let the product space $S \times Z$ be equipped with upper bound $||\cdot||_{S \times Z}$. If*

- *for any $(s, z) \in S \times Z$ it holds that*

$$\|(s, z)\|_{S \times Z} \le \|s\|_S + \|z\|_Z \,,$$

- *$\mu_\theta \in \mathcal{M}_1(S, \mathcal{S})$ is $C_p(S, \| \cdot \|_S)$-differentiable and $\nu_\theta \in \mathcal{M}_1(Z, \mathcal{Z})$ is $C_p(Z, \| \cdot \|_Z)$-differentiable,*

then $\mu_\theta \times \nu_\theta$ is $C_p(S \times Z, \| \cdot \|_{S \times Z})$-differentiable and it holds

$$(\mu_\theta \times \nu_\theta)' = \mu_\theta' \times \nu_\theta + \mu_\theta \times \nu_\theta' \,.$$

Proof: Let h be the identity on $S \times Z$. Applying Lemma 3.2.1, it follows that $C_p(S \times Z) \subset \mathcal{C}(S, Z)$, with $\mathcal{C}(S, Z)$ as defined in (3.9). We will now apply Theorem 3.1.1. Specifically, we take $\mathcal{D}(S) = C_p(S)$ and $\mathcal{D}(Z) = C_p(Z)$, which yields $\mathcal{D}(S, Z) = \mathcal{C}(S, Z)$, and the proof follows from Theorem 3.1.1 together with the fact that $C_p(S)$ and $C_p(Z)$ are solid spaces and that $C_p(S \times Z) \subset \mathcal{C}(S, Z)$. $\square$

Combining Theorem 3.1.1 with Lemma 3.2.1 yields a powerful result on C_p-differentiability of binary mappings.

Theorem 3.2.2 *Let $(X, \mathcal{X}), (Y, \mathcal{Y}), (S, \mathcal{S})$ be measurable spaces equipped with upper bounds $\| \cdot \|_X$, $\| \cdot \|_Y$ and $\| \cdot \|_S$, respectively, and let $h : X \times Y \to S$ denote a measurable binary operation. If*

- *finite constants c_X, c_Y exist such that for any $x \in X, y \in Y$*

$$\|h(x, y)\|_S \le c_X \|x\|_X + c_Y \|y\|_Y \,,$$

- *$\mu_\theta \in \mathcal{M}_1(X, \mathcal{X})$ is $C_p(X, \| \cdot \|_X)$-differentiable and $\nu_\theta \in \mathcal{M}_1(Y, \mathcal{Y})$ is $C_p(Y, \| \cdot \|_Y)$-differentiable,*

then it holds for any $g \in C_p(S, \| \cdot \|_S)$ that

$$\frac{d}{d\theta} \int_{X \times Y} g(h(x, y)) \, \mu_\theta \times \nu_\theta(dx, dy)$$

$$= \int_{X \times Y} g(h(x, y)) \, (\mu_\theta' \times \nu_\theta)(dx, dy) + \int_{X \times Y} g(h(x, y)) \, (\mu_\theta \times \nu_\theta')(dx, dy) \,,$$

or, more concisely,

$$\left((\mu_\theta \times \mu_\theta)^h \right)' = \left((\mu_\theta \times \mu_\theta)' \right)^h \,.$$

Proof: Let $g \in C_p(S)$. For $x \in X$ and $y \in Y$, set $g_h(x, y) = g(h(x, y))$. By Lemma 3.2.1, $g_h \in \mathcal{C}(X, Y)$, with $\mathcal{C}(X, Y)$ as defined in (3.9). Moreover, from Theorem 3.1.1 applied to the C_p-spaces $C_p(X, \| \cdot \|_X)$ and $C_p(Y, \| \cdot \|_Y)$, respectively, we obtain that the product measure $\mu_\theta \times \nu_\theta$ is $\mathcal{C}(X, Y)$-differentiable. Since $g_h \in \mathcal{C}(X, Y)$, we obtain

$$\frac{d}{d\theta} \int_{X \times Y} g_h(x, y) \, (\mu_\theta \times \nu_\theta)(dx, dy)$$

$$= \int_{X \times Y} g_h(x, y) \, (\mu_\theta' \times \nu_\theta)(dx, dy) + \int_{X \times Y} g_h(x, y) \, (\mu_\theta \times \nu_\theta')(dx, dy) \,.$$

Rewriting $g_h(x, y)$ as $g(h(x, y))$, we obtain

$$\frac{d}{d\theta} \int_{X \times Y} g_h(x, y) \, (\mu_\theta \times \nu_\theta)(dx, dy)$$

$$= \int_{X \times Y} g(h(x, y)) \, (\mu'_\theta \times \nu_\theta)(dx, dy) + \int_{X \times Y} g(h(x, y)) \, (\mu_\theta \times \nu'_\theta)(dx, dy) \, ,$$

which concludes the proof of the theorem. $\square$

When we study max-plus linear systems, we will consider $h = \otimes, \oplus$ and $\|\cdot\|_\oplus$ as upper bound. This choice satisfies the condition in Theorem 3.2.2 and the theorem thus applies to $\mathbb{R}_{\max}^{I \times J}$. In the following, we will use this richness of the max-plus algebra to establish a calculus for C_p-differentiability.

3.3 $\mathcal{D}$-Derivatives of Random Matrices

For $\Theta = (a, b) \subset \mathbb{R}$, let $(A_\theta \in \mathbb{R}_{\max}^{I \times J} : \theta \in \Theta)$ be a family of random matrices defined on a common probability space.

Definition 3.3.1 *We call $A_\theta \in \mathbb{R}_{\max}^{I \times J}$ $\mathcal{D}$-differentiable if the distribution of A_θ is $\mathcal{D}$-differentiable. Moreover, let $(c_{\mu_\theta}, \mu_\theta^+, \mu_\theta^-)$ be a $\mathcal{D}$-derivative of the distribution of A_θ. Then, the triple $(c_{A_\theta}, A_\theta^+, A_\theta^-)$, with $c_{A_\theta} = c_{\mu_\theta}$, A_θ^+ distributed according to μ_θ^+ and A_θ^- distributed according to μ_θ^-, is called a $\mathcal{D}$-derivative of the random matrix A_θ, and it holds for any $g \in \mathcal{D}$*

$$\frac{d}{d\theta} \mathbb{E}[\, g(A_\theta)\,] = \mathbb{E}\left[\, c_{A_\theta} \Big(g(A_\theta^+) - g(A_\theta^-) \Big) \right] \, .$$

The goal of our analysis is to establish a Leibnitz rule for $\mathcal{D}$-differentiation of the type: if A and B are $\mathcal{D}$-differentiable, then $A \oplus B$ and $A \otimes B$ are $\mathcal{D}$-differentiable, for random matrices A, B of appropriate size. Working with a general set $\mathcal{D}$ has the drawback that the set of performance functions with respect to which the $\oplus$-sum of two random matrices is differentiable is only implicitly given, cf. Theorem 3.1.1 where a precise statement in terms of measures in given. Fortunately, it will turn out that this problem does not arise when we work with C_p-spaces defined via the upper bound $\|\cdot\|_\oplus$. Specifically, we will be able to show that it holds that if $A, B \in \mathbb{R}_{\max}^{J \times I}$ are $C_p(\mathbb{R}_{\max}^{J \times I}, \|\cdot\|_\oplus)$-differentiable, then $A \oplus B$ is $C_p(\mathbb{R}_{\max}^{J \times I}, \|\cdot\|_\oplus)$-differentiable and a similar result will hold for $\otimes$-product of matrices of appropriate size. For this reason, we will present our results for C_p-spaces rather than in the most general setting possible.

Let matrix $A_\theta \in \mathbb{R}_{\max}^{J \times I}$ be a measurable mapping of random variables $X_{\theta,1}, \ldots, X_{\theta,m}$, with $X_{\theta,i} \in \mathbb{R}_{\max}$ for $1 \leq i \leq m$, that is, assume that

$$A_\theta = A(X_{\theta,1}, \ldots, X_{\theta,m}) \, .$$

We call $X_{\theta,1}, \ldots, X_{\theta,m}$ the *input* of A_θ. The following theorem establishes sufficient conditions for the existence of a C_p-derivative of a matrix with input $(X_{\theta,1}, \ldots, X_{\theta,m})$.

Theorem 3.3.1 *Let A_θ have input $X_{\theta,1}, X_2, \ldots, X_m$, with $(X_2, \ldots, X_m)$ independent of θ, and let $X_{\theta,1}$ have cumulative distribution function F_θ such that F_θ has $C_p(\mathbb{R}_{\max}, \|\cdot\|_\oplus)$-derivative*

$$\left(c_{F_\theta}, F_\theta^+, F_\theta^- \right) .$$

If $X_{\theta,1}$ is stochastically independent of $(X_2, \ldots, X_m)$ and if a constant $c \in (0, \infty)$ exists such that

$$\|A(X_{\theta,1}, X_2, \ldots, X_m)\|_\oplus \le c \,\|(X_{\theta,1}, X_2, \ldots, X_m)\|_\oplus ,$$

then A_θ has $C_p(\mathbb{R}_{\max}^{I \times J}, \|\cdot\|_\oplus)$-derivative $(c_{A_\theta}, A_\theta^+, A_\theta^-)$ with $c_{F_\theta} = c_{A_\theta}$ and

- *$A_\theta^+ = A(X_{\theta,1}^+, X_2, \ldots, X_m)$, where $X_{\theta,1}^+$ is distributed according to F_θ^+;*

- *$A_\theta^- = A(X_{\theta,1}^-, X_2, \ldots, X_m)$, where $X_{\theta,1}^-$ is distributed according to F_θ^-.*

    Proof: The mapping A maps $\mathbb{R}_{\max} \times (\mathbb{R}_{\max})^{m-1}$ onto $\mathbb{R}_{\max}^{I \times J}$. Writing $A(X_{\theta,1}, X_2 \ldots, X_m)$ as $h(X_{\theta,1}, (X_2, \ldots, X_m))$, it holds by assumption that

$$\|h(X_{\theta,1}, (X_2, \ldots, X_m))\|_\oplus \le c \,\|(X_{\theta,1}, X_2 \ldots, X_m)\|_\oplus$$

and Corollary 1.6.1 yields

$$\|h(X_{\theta,1}, (X_2, \ldots, X_m))\|_\oplus \le c \,\|X_{\theta,1}\|_\oplus + c \|(X_2, \ldots, X_m)\|_\oplus .$$

Hence, Theorem 3.2.2 applies. Using the fact that the distribution of $(X_2, \ldots, X_m)$ is independent of θ completes the proof. $\square$

    In a queuing application, the entries of matrix $A_\theta(k)$ are typically sums of service times and the condition in the above theorem is satisfied. The following example illustrates this for a specific situation.

Example 3.3.1 *Consider the homogeneous model of the queuing network in Example 1.5.2. Let the interarrival times $\sigma_0(\theta, k)$ be exponentially distributed with mean $1/\theta$, that is, $P(\sigma_0(\theta, k) \le x) = F_\theta(x) = 1 - e^{-\theta x}$. For this model is holds that*

$$A_\theta(k) = A(\sigma_0(\theta, k+1), \sigma_1(k+1), \ldots, \sigma_J(k+1)) ,$$

see Equation (1.26). Assume that the interarrival time $\sigma_0(\theta, k)$ is stochastically independent of the service times $(\sigma_1(k), \ldots, \sigma_J(k))$ and that the service times are independent of θ. In accordance with Example 3.1.1, we see that F_θ is $C_p([0, \infty), \|\cdot\|_\oplus)$-differentiable with $C_p([0, \infty), \|\cdot\|_\oplus)$-derivative $(\theta^{-1}, F_\theta, \Gamma(2, \theta))$, where $\Gamma(2, \theta)$ denotes the Gamma-$(2, \theta)$-distribution. Observe that

$$\|A(\sigma_0(\theta, k), \sigma_1(k), \ldots, \sigma_J(k))\|_\oplus \le (J+1) \,\|(\sigma_0(\theta, k), \sigma_1(k), \ldots, \sigma_J(k))\|_\oplus .$$

Hence, Theorem 3.3.1 applies and we obtain the following $C_p(\mathbb{R}_{\max}^{I \times J}, || \cdot ||_\oplus)$-*derivative of* $A_\theta(k)$

$$A_\theta^+(k) = A_\theta(k),$$

$c_{A_\theta} = 1/\theta$ *and*

$$A_\theta^-(k) = A(\sigma_0^-(\theta, k+1), \sigma_1(\theta, k+1), \ldots, \sigma_J(k+1)),$$

where $\sigma_0^-(k+1)$ *has distribution* $\Gamma(2, \theta)$.

If it causes no confusion, we will suppress θ in order to simplify the notation and write A *in lieu of* A_θ.

The following lemma states a first result on the $\mathcal{D}$-differentiability of products and sums, respectively, of random matrices.

Lemma 3.3.1 *If* A, $B \in \mathbb{R}_{\max}^{I \times J}$ *are stochastically independent and* $C_p(\mathbb{R}_{\max}^{I \times J}, || \cdot ||_\oplus)$-*differentiable, then for all* $g \in C_p(\mathbb{R}_{\max}^{I \times J}, || \cdot ||_\oplus)$

$$\frac{d}{d\theta} \mathbb{E}_\theta[g(A \oplus B)]$$
$$= \mathbb{E}_\theta \left[c_A \, g(A^+ \oplus B) + c_B \, g(A \oplus B^+) - \left(c_A \, g(A^- \oplus B) + c_B \, g(A \oplus B^-) \right) \right].$$

Furthermore, if $A \in \mathbb{R}_{\max}^{I \times J}$ *is* $C_p(\mathbb{R}_{\max}^{I \times J}, || \cdot ||_\oplus)$-*differentiable and* $B \in \mathbb{R}_{\max}^{J \times K}$ *is* $C_p(\mathbb{R}_{\max}^{J \times K}, || \cdot ||_\oplus)$-*differentiable and stochastically independent of* A, *then for all* $g \in C_p(\mathbb{R}_{\max}^{I \times K}, || \cdot ||_\oplus)$

$$\frac{d}{d\theta} \mathbb{E}_\theta[g(A \otimes B)]$$
$$= \mathbb{E}_\theta \left[c_A g(A^+ \otimes B) + c_B \, g(A \otimes B^+) - \left(c_A \, g(A^- \otimes B) + c_B \, g(A \otimes B^-) \right) \right].$$

Proof: By Lemma 1.6.1, the upper bound $|| \cdot ||_\oplus$ satisfies the condition in Theorem 3.2.2 for the operations $\oplus$ and $\otimes$ as well. Switching from random matrices to their distributions, applying Theorem 3.2.2 and switching back to C_p-derivatives proves the lemma. $\square$

Lemma 3.3.1 provides the means of calculating the derivative of $\mathbb{E}[g(A \oplus B)]$. Unfortunately, it does not answer the question regarding what the $\mathcal{D}$-derivative of $A \oplus B$ looks like nor if it exists at all. This is due to the fact that there exists no (c, C^+, C^-), such that

$$\mathbb{E}_\theta \left[c \left(g(C^+) - g(C^-) \right) \right]$$
$$= \mathbb{E}_\theta \left[c_A \, g(A^+ \oplus B) + c_B \, g(A \oplus B^+) - \left(c_A \, g(A^- \oplus B) + c_B \, g(A \oplus B^-) \right) \right].$$

But to establish the $\mathcal{D}$-differentiability of $A \oplus B$ we require such an object (c, C^+, C^-). Suppose that we could give meaning to the equations

$$c_A \, A^+ \oplus B + c_B \, A \oplus B^+ = C^+ \tag{3.10}$$

and

$$c_A \, A^- \oplus B + c_B \, A \oplus B^- = C^- \qquad (3.11)$$

and suppose further that g is linear; then we would obtain from Lemma 3.3.1

$$\frac{d}{d\theta} \mathbb{E}_\theta[g(A \oplus B)] = \mathbb{E}_\theta[g(C^+) - g(C^-)].$$

Hence, the $\mathcal{D}$-derivative of $A \oplus B$ would be $c_{A \oplus B} = 1$, $(A \oplus B)^+ = C^+$ and $(A \oplus B)^- = C^-$. As already said, Equations (3.10) and (3.11) have no meaning in $\mathbb{R}_{\max}{}^{I \times J}$. Furthermore, g is by no means linear.

In the following section we will embed $\mathbb{R}_{\max}{}^{I \times J}$ into a richer object space, called $M^{I \times J}$, where $M^{I \times J}$ will be the set of all finite sequences of triples (c, A, B), such that $c \in \mathbb{R}$ and $A, B \in \mathbb{R}_{\max}^{I \times J}$. Thus, the $\mathcal{D}$-derivative (c, A^+, A^-) of a matrix $A \in \mathbb{R}_{\max}^{I \times J}$ will be an element of $M^{I \times J}$. In particular,

- we define $\oplus$-sums and $\otimes$-products on $M^{J \times J}$ in such a way that the semi-ring $\mathbb{R}_{\max}^{J \times J}$ is a proper sub-structure of the (later defined) structure $\mathcal{M}^{J \times J}$ over $M^{J \times J}$;

- all real-valued mappings g on $\mathbb{R}_{\max}^{I \times J}$ can be extended to $M^{I \times J}$;

- on $M^{I \times J}$ we can define a binary operation '+' and scalar multiplication by real numbers in such a way that all real-valued mappings g on $\mathbb{R}_{\max}^{I \times J}$ are linear on $M^{I \times J}$.

Hence, Equations (3.10) and (3.11) have solutions in $M^{I \times J}$. Since the extension of $g \in \mathcal{D}$ to $M^{I \times J}$ is linear in $M^{I \times J}$, we can then calculate

$$\frac{d}{d\theta} \mathbb{E}_\theta[g(A \oplus B)] = \mathbb{E}_\theta\Big[c_A \, g(A^+ \oplus B) + c_B \, g(A \oplus B^+)$$
$$- \Big(c_A \, g(A^- \oplus B) + c_B \, g(A \oplus B^-)\Big)\Big]$$
$$= \mathbb{E}_\theta\Big[g(c_A \, A^+ \oplus B + c_B \, A \oplus B^+)$$
$$- g(c_A \, A^- \oplus B + c_B \, A \oplus B^-)\Big)\Big]$$
$$= \mathbb{E}_\theta\Big[g(C^+) - g(C^-)\Big],$$

i.e., $(1, C^+, C^-)$ is the $\mathcal{D}$-derivative of $A \oplus B$ in $M^{I \times J}$. It will turn out that simple rules of $\mathcal{D}$-differentiation exist in $M^{I \times J}$. In other words, $M^{I \times J}$ is a suitable space for calculating $\mathcal{D}$-derivatives of complex functions of random matrices. With a simple trick, called *randomization*, we are even able to project objects in $M^{I \times J}$ on random elements in $\mathbb{R}_{\max}^{I \times J}$. Moreover, this projection leads to an unbiased gradient estimation algorithm for random matrices (which will be discussed in Section 3.5).

3.4   An Algebraic Tool for $\mathcal{D}$-Derivatives of Random Matrices

As was stated in the previous section, we will work in applications with C_p-derivatives rather than with general $\mathcal{D}$-derivatives. However, the construction of the algebraic extension of $\mathbb{R}_{\max}^{I \times J}$ is independent of the set $\mathcal{D}$ with respect to which we define the derivatives and we will use the term $\mathcal{D}$-derivative in this section (since this generality comes at no costs).

In the following we construct $M^{I \times J}$ and develop a calculus which enables us to calculate $\mathcal{D}$-derivatives of functions of random matrices. We take as $M^{I \times J}$ the set of all finite sequences of triples (c, A, B), with $c \in \mathbb{R}$ and $A, B \in \mathbb{R}_{\max}^{I \times J}$. A generic element $\alpha \in M^{I \times J}$ is then given by

$$\alpha = ((c_1, A_1, B_1), (c_2, A_2, B_2) \ldots, (c_{n_\alpha}, A_{n_\alpha}, B_{n_\alpha})),$$

where $n_\alpha < \infty$ is called the *length* of α. If α is of length one, that is, $n_\alpha = 1$, we call it *elementary*. Observe that the $\mathcal{D}$-derivative (c_A, A^+, A^-) of a matrix A is an elementary element of $M^{I \times J}$.

On $M^{I \times J}$ we introduce the binary operation '+' as concatenation of strings. For example, let $\alpha \in M^{I \times J}$ be given by

$$\alpha = (\alpha_i : 1 \le i \le n_\alpha),$$

with α_i elementary, then

$$\alpha = \sum_{i=1}^{n_\alpha} \alpha_i = \alpha_1 + \alpha_2 + \cdots + \alpha_{n_\alpha}.$$

More generally, for $\alpha, \beta \in M^{I \times J}$ application of the '+' operator yields

$$\alpha + \beta = (\alpha_1, \ldots, \alpha_{n_\alpha}, \beta_1, \ldots, \beta_{n_\beta})$$

$$= \sum_{i=1}^{n_\alpha} \alpha_i + \sum_{j=1}^{n_\beta} \beta_j. \tag{3.12}$$

For $\alpha = (c^\alpha, A^\alpha, B^\alpha)$ and $\beta = (c^\beta, A^\beta, B^\beta)$ elementary in $M^{I \times J}$ we set

$$\alpha \oplus \beta = (c^\alpha \cdot c^\beta, A^\alpha \oplus A^\beta, B^\alpha \oplus B^\beta),$$

where $x \cdot y$ denotes conventional multiplication in $\mathbb{R}$, and for $\alpha = (c^\alpha, A^\alpha, B^\alpha) \in M^{I \times J}$, $\beta = (c^\beta, A^\beta, B^\beta) \in M^{J \times K}$ we define

$$\alpha \otimes \beta = (c^\alpha \cdot c^\beta, A^\alpha \otimes A^\beta, B^\alpha \otimes B^\beta).$$

These definitions are extended to general α, β as follows. The $\oplus$-sum is given by

$$\alpha \oplus \beta = \sum_{j=1}^{n_\beta} \alpha_1 \oplus \beta_j + \sum_{j=1}^{n_\beta} \alpha_2 \oplus \beta_j + \cdots + \sum_{j=1}^{n_\beta} \alpha_{n_\alpha} \oplus \beta_j$$

$$= \sum_{i=1}^{n_\alpha} \sum_{j=1}^{n_\beta} \alpha_i \oplus \beta_j,$$

for $\alpha, \beta \in M^{I \times J}$, that is, $\alpha \oplus \beta$ is the concatenation of all elementary $\oplus$-sums, which implies $n_{\alpha \oplus \beta} = n_\alpha \cdot n_\beta$. For the $\otimes$-product we set

$$
\alpha \otimes \beta = \sum_{j=1}^{n_\beta} \alpha_1 \otimes \beta_j + \sum_{j=1}^{n_\beta} \alpha_2 \otimes \beta_j + \cdots + \sum_{j=1}^{n_\beta} \alpha_{n_\alpha} \otimes \beta_j
$$
$$
= \sum_{i=1}^{n_\alpha} \sum_{j=1}^{n_\beta} \alpha_i \otimes \beta_j ,
$$

for $\alpha \in M^{I \times J}$ and $\beta \in M^{J \times K}$, that is, $\alpha \otimes \beta$ is the concatenation of all elementary $\otimes$-products, which implies $n_{\alpha \otimes \beta} = n_\alpha \cdot n_\beta$. In particular, for $\alpha \in M^{I \times J}$ and $x \in M^J \overset{\text{def}}{=} M^{J \times 1}$ the matrix-vector product $\alpha \otimes x$ is defined.

Set $\mathcal{E}^{I \times J} = (1, \mathcal{E}(I, J), \mathcal{E}(I, J))$, then $\mathcal{E}^{I \times J}$ is the neutral element of $\oplus$ in $M^{I \times J}$. The element $\mathcal{E}^{I \times J}$ is unique in the sense that for all $\alpha \in M^{I \times J}$: $n_{\alpha \oplus \mathcal{E}^{I \times J}} = n_\alpha$. Furthermore, set $E^{J \times J} = (1, E(J, J), E(J, J))$, then $E^{J \times J}$ is the neutral element of $\otimes$ in $M^{J \times J}$ and it is unique in the sense that for all $\alpha \in M^{J \times J}$: $n_{\alpha \otimes E^{J \times J}} = n_\alpha$.

We define scalar multiplication as follows. For elementary $\alpha = (c, A, B) \in M^{I \times J}$ we set $r \cdot \alpha = (r \cdot c, A, B)$ and for $\alpha = (\alpha_1, \ldots, \alpha_{n_\alpha}) \in M^{I \times J}$ we set

$$
r \cdot \alpha = \sum_{i=1}^{n_\alpha} r \cdot \alpha_i . \tag{3.13}
$$

We embed $\mathbb{R}_{\max}^{I \times J}$ into $M^{I \times J}$ via a homomorphism τ given by

$$
A^\tau \overset{\text{def}}{=} \tau(A) = (1, A, A) ,
$$

for $A \in \mathbb{R}_{\max}^{I \times J}$. It is easily checked that $(A \oplus B)^\tau = A^\tau \oplus B^\tau$ and $(A \otimes B)^\tau = A^\tau \otimes B^\tau$.

We now define the τ-image of a function $g : \mathbb{R}_{\max}^{I \times J} \to \mathbb{R}$. For $\alpha = ((c_1, A_1, B_1), \ldots, (c_{n_\alpha}, A_{n_\alpha}, B_{n_\alpha})) \in M^{I \times J}$ we set

$$
g^\tau(\alpha) = \sum_{i=1}^{n_\alpha} 1_{c_i \neq \varepsilon} c_i \Big(g(A_i) - g(B_i) \Big) . \tag{3.14}
$$

The mapping $g^\tau(\cdot)$ is called τ-*projection w.r.t.* g *onto* $\mathbb{R}$, or (τ, g)-*projection* for short. For ease of notation, we suppress the superscript τ when this causes no confusion and write $g(\cdot)$ instead of $g^\tau(\cdot)$.

Remark 3.4.1 *For* $A \in \mathbb{R}_{\max}^{J \times J}$, *the τ-projection with respect to any* $g : \mathbb{R}_{\max}^{J \times J} \to \mathbb{R}$ *yields* $g^\tau(\tau(A)) = 0$. *However, we can recover* g *via the τ-projection with respect to* g *through a linear transformation. More precisely, take* $\pi^{J \times J} = (1, E(J, J), \mathcal{E}(J, J)) \in M^{J \times J}$, *then* $\pi^{J \times J} \otimes \tau(A) = (1, A, \mathcal{E}(J, J))$ *and we obtain*

$$
\forall A \in \mathbb{R}_{\max}^{J \times J} : \quad g^\tau \left(\pi^{J \times J} \otimes \tau(A) \right) = g(A) ,
$$

where we assume without loss of generality that $g(\mathcal{E}(J, J)) = 0$.

The definition of addition and scalar multiplication are tailored to make the extension of any real-valued function on $\mathbb{R}_{\max}^{I \times J}$ to $M^{I \times J}$ linear. This is shown in the following lemma.

Lemma 3.4.1 *For $\alpha, \beta \in M^{I \times J}$ and $c_\alpha, c_\beta \in \mathbb{R}$ it holds true that*

$$\forall g \in \mathbb{R}^{\mathbb{R}_{\max}^{I \times J}} : \quad g^\tau (c_\alpha \cdot \alpha + c_\beta \cdot \beta) = c_\alpha \, g^\tau(\alpha) + c_\beta \, g^\tau(\beta) \,.$$

Proof: For $\alpha = ((c_i^\alpha, A_i^\alpha, B_i^\alpha) : 1 \leq i \leq n_\alpha)$, $\beta = ((c_i^\beta, A_i^\beta, B_i^\beta) : 1 \leq i \leq n_\beta)$ and $c_\alpha, c_\beta \in \mathbb{R}$ we obtain

$$
\begin{aligned}
& g^\tau (c_\alpha\, \alpha + c_\beta\, \beta) \\
&= g^\tau \big(c_\alpha((c_1^\alpha, A_1^\alpha, B_1^\alpha), \ldots, (c_{n_\alpha}^\alpha, A_{n_\alpha}^\alpha, B_{n_\alpha}^\alpha)) \\
& \qquad\qquad + c_\beta((c_1^\beta, A_1^\beta, B_1^\beta), \ldots, (c_{n_\beta}^\beta, A_{n_\beta}^\beta, B_{n_\beta}^\beta)) \big) \\
&\overset{(3.13)}{=} g^\tau \big((c_\alpha(c_1^\alpha, A_1^\alpha, B_1^\alpha), \ldots, c_\alpha(c_{n_\alpha}^\alpha, A_{n_\alpha}^\alpha, B_{n_\alpha}^\alpha)) \\
& \qquad\qquad + (c_\beta(c_1^\beta, A_1^\beta, B_1^\beta), \ldots, c_\beta(c_{n_\beta}^\beta, A_{n_\beta}^\beta, B_{n_\beta}^\beta)) \big) \\
&\overset{(3.12)}{=} g^\tau \big(((c_\alpha c_1^\alpha, A_1^\alpha, B_1^\alpha), \ldots, (c_\alpha c_{n_\alpha}^\alpha, A_{n_\alpha}^\alpha, B_{n_\alpha}^\alpha), \\
& \qquad\qquad (c_\beta c_1^\beta, A_1^\beta, B_1^\beta), \ldots, (c_\beta c_{n_\beta}^\beta, A_{n_\beta}^\beta, B_{n_\beta}^\beta)) \big) \\
&\overset{(3.14)}{=} \sum_{i=1}^{n_\alpha} c_\alpha\, c_i^\alpha \left(g(A_i^\alpha) - g(B_i^\alpha) \right) + \sum_{i=1}^{n_\beta} c_\beta\, c_i^\beta \left(g(A_i^\beta) - g(B_i^\beta) \right) \\
&= c_\alpha \sum_{i=1}^{n_\alpha} c_i^\alpha \left(g(A_i^\alpha) - g(B_i^\alpha) \right) + c_\beta \sum_{i=1}^{n_\beta} c_i^\beta \left(g(A_i^\beta) - g(B_i^\beta) \right) \\
&\overset{(3.14)}{=} c_\alpha\, g^\tau(\alpha) + c_\beta\, g^\tau(\beta) \,.
\end{aligned}
$$

□

The operator '+' does the trick to make any $g : \mathbb{R}_{\max}^{I \times J} \to \mathbb{R}$ linear on $M^{I \times J}$. Unfortunately, the structure $\mathcal{M}^{J \times J} = (M^{J \times J}, \oplus, \otimes, +, E^{J \times J}, \mathcal{E}^{J \times J})$ has very poor algebraic properties. For example, the operation $\oplus$ fails to be commutative in $M^{I \times J}$. However, in what follows we will show that most of these properties can be recovered in a 'weak' sense.

On $M^{I \times J}$, the equation $\alpha = \beta$ means that α is element-wise equal to β. We call this the *strong equality* on $M^{I \times J}$. We say that $\alpha, \beta \in M^{I \times J}$ are *equal in the weak $\mathcal{D}$-sense* if and only if

$$\forall g \in \mathcal{D} : \quad \mathbb{E}[g^\tau(\alpha)] = \mathbb{E}[g^\tau(\beta)] \,,$$

in symbols: $\alpha \equiv_\mathcal{D} \beta$, where $\mathcal{D}$ is a non-empty set of mappings from $M^{I \times J}$ onto $\mathbb{R}$ (to simplify the notation, we will write $\alpha \equiv \beta$ when it is clear which set $\mathcal{D}$ is meant). Obviously, strong equality implies weak $\mathcal{D}$-equality. On the other hand we are only interested in results of the type $\forall g \in \mathcal{D} : \mathbb{E}[g^\tau(\ldots)] = \mathbb{E}[g^\tau(\ldots)]$, that is, in all proofs that will follow it is sufficient to work with $\mathcal{D}$-equality on

$M^{I \times J}$. When there can be no confusion about the space $\mathcal{D}$, we use the term 'weak equality' rather than '$\mathcal{D}$-equality.'

We now say that the binary operator f is *weakly commutative* on $M^{I \times J}$ if $\alpha f \beta \equiv \beta f \alpha$ for all $\alpha, \beta \in M^{I \times J}$, and define *weak distributivity, weak associativity* a.s.f. in the same way.

Theorem 3.4.1 (Rules of Weak Computation) *On $M^{I \times J}$, the binary operations $\oplus$ and '+' are weakly associative, weakly commutative and $\oplus$ is weakly distributive with respect to '+'. Furthermore, on $M^{J \times J}$, $\otimes$ is weakly distributive with respect to '+'.*

Proof: Observe that, for $\gamma = (c_1, \ldots, c_{n_\gamma}) \in M^{I \times J}$, $g(\gamma)$ is insensitive with respect to the order of the entries in γ, i.e., for any permutation π on $\{1, \ldots, n_\gamma\}$ it holds true

$$g^\tau((c_1, \ldots c_{n_\gamma})) = g^\tau((c_{\pi(1)}, \ldots c_{\pi(n_\gamma)})) . \tag{3.15}$$

We show weak commutativity of $\oplus$: for $\alpha = (a_1, \ldots, a_{n_\alpha}), \beta = (b_1, \ldots, b_{n_\beta}) \in M^{I \times J}$, $\alpha \oplus \beta$ contains all elementary $\oplus$-sums $a_i \oplus b_j$ for $1 \le i \le n_\alpha$ and $1 \le j \le n_\beta$. Hence, $\alpha \oplus \beta$ and $\beta \oplus \alpha$ only differ in the order of their entries. In accordance with equation (3.15), $g(\cdot)$ is insensitive with respect to the order of the entries which implies $g(\alpha \oplus \beta) = g(\beta \oplus \alpha)$. Weak commutativity of '+' follows the same line of argument as well as the proof of weak associativity of $\oplus$, $\otimes$, '+' and we therefore omit the proofs.

Next we show left-distributivity of $\oplus$ with respect to '+': for $\alpha = (a_1, \ldots, a_{n_\alpha}), \beta = (b_1, \ldots, b_{n_\beta})$ and $\gamma = (c_1, \ldots, c_{n_\gamma}) \in M^{I \times J}$, $\alpha \oplus (\beta + \gamma)$ contains all elementary $\oplus$-sums $a_i \oplus b_j$, for $1 \le i \le n_\alpha$, $1 \le j \le n_\beta$, and $a_i \oplus c_k$, for $1 \le i \le n_\alpha$, and $1 \le k \le n_\gamma$. These are exactly the entries of $(\alpha \oplus \beta) + (\alpha \oplus \gamma)$ and weak left-distributivity follows from (3.15). Weak right-distributivity as well as weak left-, respectively right-, distributivity of $\otimes$ with respect to '+' follow the same line of argument. $\square$

Remark 3.4.2 *On $M^{J \times J}$, $\otimes$ fails to be weakly left or right distributive with respect to $\oplus$. To see this, consider $\alpha, \beta, \gamma \in M^{J \times J}$ with $n_\alpha > 1$ which gives*

$$n_{\alpha \otimes (\beta \oplus \gamma)} = n_\alpha n_\beta n_\gamma < n_\alpha n_\beta n_\alpha n_\gamma = n_{(\alpha \otimes \beta) \oplus (\alpha \otimes \gamma)} .$$

Hence, in general, $\alpha \otimes (\beta \oplus \gamma) \neq (\alpha \otimes \beta) \oplus (\alpha \otimes \gamma)$ and weak left-distributivity fails. For weak right-distributivity we argue in the same way. Consequently, $\mathcal{M}^{J \times J}$ is not a semiring in the weak sense.

So far we have introduced a new structure $\mathcal{M}^{J \times J} = (M^{J \times J}, \oplus, \otimes, +, \mathcal{E}^{J \times J}, E^{J \times J})$ and established its (weak) algebraic properties. We now ask: what is the relationship between the structures $\mathcal{R}_{\max}^{J \times J}$ and $\mathcal{M}^{J \times J}$?

Recall that we embedded $\mathbb{R}_{\max}^{I \times J}$ into $M^{I \times J}$ via the mapping τ. We now call

$$\left(\mathbb{R}_{\max}^{I \times J} \right)^\tau = \{ \tau(A) : A \in \mathbb{R}_{\max}^{I \times J} \}$$

the set of *standard elements* of $M^{I \times J}$ and the elements of $M^{I \times J} \setminus \left(\mathbb{R}_{\max}^{I \times J} \right)^\tau$ the *nonstandard elements*.

Theorem 3.4.2 *The structure*

$$\left(\mathcal{R}_{\max}^{J \times J} \right)^\tau = \left(\left(\mathbb{R}_{\max}^{J \times J} \right)^\tau, \oplus, \otimes, E^\tau, \mathcal{E}^\tau \right)$$

constitutes an idempotent semiring in the weak sense.

Proof: We prove commutativity of $\oplus$ on $\left(\mathbb{R}_{\max}^{J \times J} \right)^\tau$. For $A, B \in \mathbb{R}_{\max}^{J \times J}$ we have

$$\begin{aligned}
A^\tau \oplus B^\tau &= (1, A, A) \oplus (1, B, B) \\
&= (1, A \oplus B, A \oplus B) \\
&= (1, B \oplus A, B \oplus A) \\
&= B^\tau \oplus A^\tau,
\end{aligned}$$

where the last but one equality follows from commutativity of $\oplus$ on $\mathbb{R}_{\max}^{J \times J}$. All other properties are checked in the same manner and the proofs are therefore omitted. $\square$

In accordance with Theorem 3.4.2, any formula over $\mathbb{R}_{\max}^{J \times J}$ is valid over $M^{J \times J}$ if we add 'τ' to all constants. For example, from

$$\forall A, B, C \in \mathbb{R}_{\max}^{J \times J} : \quad (A \oplus B) \otimes C = (A \otimes C) \oplus (B \otimes C)$$

we can conclude

$$\forall A, B, C \in \left(\mathbb{R}_{\max}^{J \times J} \right)^\tau : \quad (A \oplus B) \otimes C \equiv (A \otimes C) \oplus (B \otimes C),$$

that is, all formulae valid over $\mathbb{R}_{\max}^{J \times J}$ are also valid if interpreted over $M^{J \times J}$ (even if they contain variables g out of the specified set $\mathcal{D}$). This is known in algebraic model theory as *Leibnitz principle*.

Nonstandard elements, such as $A^\tau + B^\tau$ for $A, B \in \mathbb{R}_{\max}{}^{J \times J}$, cannot be directly interpreted as random matrices in $\mathbb{R}_{\max}^{J \times J}$. However, we can project them with the help of the (τ, g)-projection onto $\mathbb{R}$. We conclude our study of $\mathcal{M}^{I \times J}$ by giving a purely stochastic way of interpreting the '+'-operator in $M^{I \times J}$.

Lemma 3.4.2 *Let σ be uniformly distributed over $\{0, \dots, k\}$ and independent of everything else. If $A(i) \in M^{I \times J}$ $(0 \le i \le k)$, then*

$$\sum_{i=0}^k A(i) \equiv (k+1) A(\sigma).$$

Proof: Let g be a measurable mapping g from $\mathbb{R}_{\max}^{I \times J}$ on $\mathbb{R}$. Applying Lemma 3.4.1 yields

$$\mathbb{E}\left[g^\tau\left(\sum_{i=0}^{k} A(i)\right)\right] = \mathbb{E}\left[\sum_{i=0}^{k} g^\tau(A(i))\right]$$

$$= (k+1)\sum_{i=0}^{k} \mathbb{E}\left[g^\tau(A(i))\right] P(\sigma = i)$$

$$= \mathbb{E}\left[(k+1)g^\tau(A(\sigma))\right]$$

$$= \mathbb{E}\left[g^\tau((k+1)A(\sigma))\right].$$

$\square$

To simplify the notation we introduce the following convention.

Convention: *From now on we identify the elements of $\left(\mathbb{R}_{\max}^{I \times J}\right)^\tau$ and $\mathbb{R}_{\max}^{I \times J}$. For example, for $A, B \in \mathbb{R}_{\max}^{I \times J}$, the formula 'A + B' has to be read '$A^\tau + B^\tau$'.*

3.5 Rules for C_p-Differentiation of Random Matrices

This section provides rules for C_p-differentiation of $\oplus$-sums and $\otimes$-products of random matrices. Firstly, we will introduce the general $\mathcal{D}$-derivative and then establish results for the special case $\mathcal{D} = C_p(\mathbb{R}_{\max}^{I \times J}, \|\cdot\|_\oplus)$.

If $A_\theta \in \mathbb{R}_{\max}{}^{I \times J}$ has $\mathcal{D}$-derivative $(c_{A_\theta}, A_\theta^+, A_\theta^-)$ at θ, we set

$$A_\theta' = (c_{A_\theta}, A_\theta^+, A_\theta^-).$$

It is easily checked that this implies

$$\frac{d}{d\theta}\mathbb{E}_\theta[g(A_\theta)] = \mathbb{E}_\theta[g(A_\theta')],$$

for $g \in \mathcal{D}$, which motivates the following definition (again we will suppress for ease of notation the subscript θ whenever this causes no confusion).

Definition 3.5.1 *For $A \in \mathbb{R}_{\max}^{I \times J}$ we call $A' = (c_A, A^+, A^-) \in M^{I \times J}$ a $\mathcal{D}$-derivative of A if for all $g \in \mathcal{D}$*

$$\frac{d}{d\theta}\mathbb{E}_\theta[g(A)] = \mathbb{E}_\theta[g(A')].$$

If the left-hand side equals zero for all g, we set $A' = (0, A, A)$.

$\mathcal{D}$-differentiation maps $A \in \mathbb{R}_{\max}^{I \times J}$ on a (nonstandard) element $A' \in M^{I \times J}$. However, the extension of the $g \in \mathcal{D}$ to $M^{I \times J}$, see (3.14), projects A' on $\mathbb{R}$ in such a way that we recover the original definition of the $\mathcal{D}$-derivative of a random matrix, see Definition 3.3.1. The main benefit of this approach is that we may consider A' and A as objects in $M^{I \times J}$ and elaborate on the 'rules of weak computation in $M^{I \times J}$' put forward in Theorem 3.4.1 to perform computations.

Remark 3.5.1 *The $\mathcal{D}$-derivative A' of a matrix A is by no means the sample path derivative of A. For this reason we carefully avoid writing $dA/d\theta$ for the $\mathcal{D}$-derivative. However, A' is a random variable such that, for all $g \in \mathcal{D}$, $g(A')$ yields an unbiased estimator for $d\mathbb{E}_\theta[g(A)]/d\theta$, i.e., we may think of A' as an ersatz derivative in the sense of Brémaud, see [27].*

Example 3.5.1 *Let $\theta \in [0,1]$ and let $X_\theta \in \{D_1, D_2\}$, with $D_1, D_2 \in \mathbb{R}_{\max}^{I \times J}$, be governed by the Bernoulli-(θ)-distribution such that $P(X_\theta = D_1) = \theta = 1 - P(X_\theta = D_2)$. Calculation yields*

$$\frac{d}{d\theta}\mathbb{E}[g(X_\theta)] = \frac{d}{d\theta}\Big(g(D_1)\theta + g(D_2)(1 - \theta)\Big)$$
$$= g(D_1) - g(D_2). \qquad (3.16)$$

Hence, $(1, D_1, D_2)$ is a $\mathcal{D}$-derivative of X_θ, where $\mathcal{D}$ is any set of mappings from $\{D_1, D_2\}$ onto $\mathbb{R}$. The derivative at the boundary points 0 and 1 is obtained as one-sided limit and $(1, D_1, D_2)$ is thus a $\mathcal{D}$-derivative of X_θ on the entire interval $[0,1]$.

In what follows, we work, as before, with $\mathcal{D} = C_p$. We revisit Lemma 3.3.1. For $A, B \in \mathbb{R}_{\max}^{I \times J}$ with C_p-derivative (c_A, A^+, A^-) and (c_B, B^+, B^-), respectively, we obtain from Definition 3.5.1

$$A' = (c_A, A^+, A^-)$$

and

$$B' = (c_B, B^+, B^-).$$

Direct calculation yields

$$c_A\Big(g(A^+ \oplus B) - g(A^- \oplus B)\Big) = g^\tau(A' \oplus B^\tau)$$

and

$$c_B\Big(g(A \oplus B^+) - g(A \oplus B^-)\Big) = g^\tau(A^\tau \oplus B'),$$

where we place the superscript τ to indicate that the objects on the right-hand side of the above equations live on $M^{I \times J}$, whereas the objects on the left-hand side live on $\mathbb{R}_{\max}^{I \times J}$. Lemma 3.3.1 applies and making use of the linearity of g over $M^{I \times J}$, see Lemma 3.4.1, we obtain

$$\frac{d}{d\theta}\mathbb{E}_\theta[g(A \oplus B)] = \mathbb{E}_\theta[g^\tau(A' \oplus B^\tau) + g^\tau(A^\tau \oplus B')]$$
$$= \mathbb{E}_\theta[g^\tau(A' \oplus B^\tau + A^\tau \oplus B')], \qquad (3.17)$$

or, elaborating on the weak equality on $C_p(\mathbb{R}_{\max}^{I \times J}, ||\cdot||_\oplus)$ and suppressing the superscript τ,

$$(A \oplus B)' \equiv A' \oplus B + A \oplus B'. \qquad (3.18)$$

In the same way we conclude

$$(A \otimes B)' \equiv A' \otimes B + A \otimes B'.$$

We summarize our analysis in

Corollary 3.5.1 *(Lemma 3.3.1 revisited) If $A, B \in \mathbb{R}_{\max}^{I \times J}$ are stochastically independent and C_p-differentiable, then*

$$(A \oplus B)' \equiv A' \oplus B + A \oplus B'.$$

Furthermore, if $A \in \mathbb{R}_{\max}^{I \times J}$ and $B \in \mathbb{R}_{\max}^{J \times K}$ are stochastically independent and C_p-differentiable, then

$$(A \otimes B)' \equiv A' \otimes B + A \otimes B'.$$

Next we state a simple but useful consequence of the results obtained so far which justifies the intuition that the C_p-derivative of a random matrix which does not depend on θ is 'zero.'

Corollary 3.5.2 *Let $A, B \in \mathbb{R}_{\max}^{I \times J}$ be stochastically independent and have C_p-derivatives A' and B', respectively. If B does not depend on θ, then*

$$(A \oplus B)' \equiv A' \oplus B,$$

and if A does not depend on θ, we obtain $(A \oplus B)' \equiv A \oplus B'$.

Furthermore, let $A \in \mathbb{R}_{\max}^{I \times J}$ and $B \in \mathbb{R}_{\max}^{J \times K}$ be stochastically independent and C_p-differentiable. If B does not depend on θ, then

$$(A \otimes B)' \equiv A' \otimes B,$$

and if A does not depend on θ, we obtain $(A \otimes B)' \equiv A \otimes B'$.

Proof: Note that, for B independent of θ, we have

$$B' = (0, B, B)$$

which implies (see the definition of the τ-projection in (3.14))

$$g^\tau(A^\tau \oplus B') = 0 = g^\tau(A^\tau \otimes B')$$

and the proof follows from (3.17). The proof of the second part of the corollary follows the same line of argument and is therefore omitted. $\square$

Another result that will prove helpful is that the C_p-derivative of a sum equals the sum of the C_p-derivatives of its components. The precise statement is given in the next lemma.

Lemma 3.5.1 *If $A, B \in \mathbb{R}_{\max}^{J \times I}$ are stochastically independent and C_p-differentiable, then*

$$(A + B)' \equiv A' + B'.$$

Proof: For the proof we elaborate on the fact that $g \in C_p$ becomes linear over $\mathcal{M}^{J \times I}$, see Lemma 3.4.1. In the following we mark the use of this argument by (a). For any $g \in C_p$ we obtain

$$\frac{d}{d\theta} \mathbb{E}_\theta[g(A + B)] \stackrel{(a)}{=} \frac{d}{d\theta} \mathbb{E}_\theta[g(A) + g(B)]$$

$$
\begin{aligned}
&= \frac{d}{d\theta} \mathbb{E}_\theta[g(A)] + \frac{d}{d\theta} \mathbb{E}_\theta[g(B)] \\
&= \mathbb{E}_\theta[g^\tau(A')] + \mathbb{E}_\theta[g^\tau(B')] \\
&\overset{(a)}{=} \mathbb{E}_\theta[g^\tau(A' + B')] \, .
\end{aligned}
$$

$\square$

The next theorem states the basic rules of C_p-differentiation. Due to our calculus we are able to give a purely algebraic proof. Let $A(i) \in \mathbb{R}_{max}^{I \times J}$ $(0 \le i \le k)$. For technical convenience, we set

$$
\bigoplus_{i=j}^{k} A(i) = \mathcal{E}(I, J)
$$

for $j > k$, and for $I = J$, we set

$$
\bigotimes_{i=j}^{k} A(i) = E(J, J)
$$

for $j > k$.

Theorem 3.5.1 *Let $A(i) \in \mathbb{R}_{max}^{I \times J}$ $(0 \le i \le k)$ be mutually independent and C_p-differentiable, then*

$$
\left(\bigoplus_{i=0}^{k} A(i) \right)' \equiv \sum_{j=0}^{k} \bigoplus_{i=j+1}^{k} A(i) \oplus A(j)' \oplus \bigoplus_{i=0}^{j-1} A(i) \, ,
$$

and, for $J = I$,

$$
\left(\bigotimes_{i=0}^{k} A(i) \right)' \equiv \sum_{j=0}^{k} \bigotimes_{i=j+1}^{k} A(i) \otimes A(j)' \otimes \bigotimes_{i=0}^{j-1} A(i) \, .
$$

Proof: We prove only the first part of the theorem since the proof of the second part follows the same line of argument.

We give a proof by induction. For $k = 2$, the proof follows from Lemma 3.5.1. Suppose that the statement of the theorem holds true for k, then it follows from the rules of weak computation in $M^{I \times J}$, see Theorem 3.4.1, that

$$
\left(\bigoplus_{i=0}^{k+1} A(i) \right)' \equiv \left(A(k+1) \oplus \bigoplus_{i=0}^{k} A(i) \right)'
$$

$$
\equiv A(k+1)' \oplus \bigoplus_{i=0}^{k} A(i) + A(k+1) \oplus \left(\bigoplus_{i=0}^{k} A(i) \right)'
$$

$$
\equiv A(k+1)' \oplus \bigoplus_{i=0}^{k} A(i) + \sum_{j=0}^{k} \bigoplus_{i=j+1}^{k} A(i) \oplus A(j)' \oplus \bigoplus_{i=0}^{j-1} A(i)
$$

$$\equiv \sum_{j=0}^{k+1} \bigoplus_{i=j+1}^{k+1} A(i) \oplus A(j)' \oplus \bigoplus_{i=0}^{j-1} A(i) .$$

□

We illustrate the statement in the above theorem with an example.

Example 3.5.2 *Consider the situation in Example 3.5.1. Let $A(k)$ $(k = 1, 2)$ be stochastically independent and Bernoulli-(θ)-distributed over $\{D_1, D_2\}$, with $P(A(k) = D_1) = \theta = 1 - P(A(k) = D_2)$, for $\theta \in [0, 1]$. For the C_p-derivative of $A(k)$ we obtain*

$$A(k)' = (1, D_1, D_2) .$$

Theorem 3.5.1 now implies

$$(A(1) \otimes A(2))' \equiv A(1)' \otimes A(2) + A(1) \otimes A(2)' ,$$

or, more explicitly,

$$
\begin{aligned}
(A(1) \otimes A(2))' &\equiv (1, D_1, D_2) \otimes (1, A(2), A(2)) + (1, A(1), A(1)) \otimes (1, D_1, D_2) \\
&= (1, D_1 \otimes A(2), D_2 \otimes A(2)) + (1, A(1) \otimes D_1, A(1) \otimes D_2) \\
&= ((1, A(1) \otimes D_1, A(1) \otimes D_2), (1, D_1 \otimes A(2), D_2 \otimes A(2))) .
\end{aligned}
$$

Applying the (τ, g)-projection yields

$$
\begin{aligned}
\frac{d}{d\theta} \mathbb{E}_\theta \left[g\left(A(1) \otimes A(2) \right) \right] &= \mathbb{E}_\theta \left[g^\tau \left((A(1) \otimes A(2))' \right) \right] \\
&= \mathbb{E}_\theta \Big[g(A(1) \otimes D_1) + g(D_1 \otimes A(2)) \\
&\qquad - g(A(1) \otimes D_2) - g(D_2 \otimes A(2)) \Big] .
\end{aligned}
$$

The above formula can be phrased by saying that the derivative of $\mathbb{E}_\theta[g(A(1) \otimes A(2))]$ can be obtained from the difference between two experiments. For the first experiment, we consider all possible combinations of replacing the nominal matrix $A(k)$ by D_1, the positive part of the C_p-derivative of $A(k)$. For the second experiment, we consider all possible combinations of replacing the nominal matrix $A(k)$ by D_2, the negative part of the C_p-derivative of $A(k)$.

Notice that $A(k)$ converges in total variation to D_1 as θ tends to 0. Hence, taking the derivative of $A(1) \otimes A(2)$ at zero yields

$$(A(1) \otimes A(2))' \equiv ((1, D_1 \otimes D_1, D_1 \otimes D_2), (1, D_1 \otimes D_1, D_2 \otimes D_1))$$

and

$$
\begin{aligned}
\lim_{\theta \downarrow 0} \frac{d}{d\theta} \mathbb{E}_\theta \left[g\left(A(1) \otimes A(2) \right) \right] &= g(D_1 \otimes D_1) + g(D_1 \otimes D_2) \\
&\qquad - g(D_1 \otimes D_1) - g(D_2 \otimes D_1) \\
&= g(D_1 \otimes D_2) - g(D_2 \otimes D_1) .
\end{aligned}
$$

Differentiation of $\oplus$-sums or $\otimes$-products increases the complexity of the C_p-derivative. However, we may introduce a stochastic concept called *randomization* that reduces the complexity. The basic idea (as already presented in Lemma 3.4.2) is to replace the summation with an expectation with respect to an independently, uniformly distributed random variable, say σ.

Let $A(k) \in \mathbb{R}_{\max}^{I \times J}$, with $k \in \mathbb{N}$, be a sequence of C_p-differentiable random matrices. To simplify the notation we write for $k \in \mathbb{N}$ and $j \leq k$

$$\left(\bigoplus_{i=0}^{k} A(i)\right)'(j) \overset{\text{def}}{=} \bigoplus_{i=j+1}^{k} A(i) \oplus A(j)' \oplus \bigoplus_{i=0}^{j-1} A(i), \qquad (3.19)$$

and, for $A(k) \in \mathbb{R}_{\max}^{J \times J}$, we define the expression $\left(\bigotimes_{i=0}^{k} A(i)\right)'(j)$ in the same way.

Randomization indeed simplifies the presentation of our results as the statement in Theorem 3.5.1 can be rephrased as follows.

Corollary 3.5.3 *If* $A(i) \in \mathbb{R}_{\max}^{I \times J}$ $(0 \leq i \leq k)$ *are mutually independent and* C_p-*differentiable and if* σ *is uniformly distributed over* $\{0, \ldots, k\}$ *independent of everything else, then*

$$\left(\bigoplus_{i=0}^{k} A(i)\right)' \equiv (k+1) \left(\bigoplus_{i=0}^{k} A(i)\right)'(\sigma)$$

and, for $I = J$,

$$\left(\bigotimes_{i=0}^{k} A(i)\right)' \equiv (k+1) \left(\bigotimes_{i=0}^{k} A(i)\right)'(\sigma).$$

Proof: Apply Lemma 3.4.2 to Theorem 3.5.1.

3.6 Gradient Estimation for Max-Plus Linear Stochastic Systems

We consider the max-plus recurrence relation

$$x(k+1) = A(k) \otimes x(k) \oplus B(k), \qquad \text{for } k \geq 0.$$

Using basic algebraic calculus, the above recurrence relation leads to the following closed-form expression

$$x(k+1) = \bigotimes_{i=0}^{k} A(i) \otimes x_0 \oplus \bigoplus_{i=0}^{k} \bigotimes_{j=i+1}^{k} A(j) \otimes B(i), \quad k \geq 0. \qquad (3.20)$$

This gives

$$\mathbb{E}_\theta\Big[g(x(k+1))\,\Big|\,x(0) = x_0\Big]$$

$$= \mathbb{E}_\theta\left[g\left(\bigotimes_{i=0}^{k} A(i) \otimes x_0 \oplus \bigoplus_{i=0}^{k} \bigotimes_{j=i+1}^{k} A(j) \otimes B(i)\right)\right].$$

In what follows, we calculate the C_p-derivative of the above expression, where we will distinguish between homogeneous and inhomogeneous recurrence relations. Recall that recurrence relation (3.20) is called *homogeneous* if $x(k+1) = A(k) \otimes x(k)$, for $k \geq 0$, i.e., $B(k) = (\varepsilon, \ldots, \varepsilon)$ for all k. For example, the closed tandem network of Example 1.5.1 is modeled by a homogeneous recurrence relation. On the other hand, recurrence relation (3.20) is called inhomogeneous if $B(k) \neq (\varepsilon, \ldots, \varepsilon)$ for some $k \in \mathbb{N}$. For example, the max-plus representation (1.27) on page 26 of the open tandem system in Example 1.5.2 is of inhomogeneous type.

3.6.1 Homogeneous Recurrence Relations

Since ε is absorbing for $\otimes$, (3.20) can be simplified for any homogeneous recurrence relation to

$$x(k+1) = \bigotimes_{i=0}^{k} A(i) \otimes x_0, \quad k \geq 0.$$

Let $A(i)$ $(0 \leq i \leq k)$ be mutually independent and C_p-differentiable with C_p-derivative $(c_{A(i)}, A^+(i), A^-(i))$. Corollary 3.5.2 implies

$$x'(k+1) \equiv \left(\bigotimes_{i=0}^{k} A(i) \otimes x_0\right)' \equiv \left(\bigotimes_{i=0}^{k} A(i)\right)' \otimes x_0.$$

Let σ be uniformly distributed over $\{0, \ldots, k\}$ independent of everything else. In accordance with Corollary 3.5.3 we obtain

$$x'(k+1) \equiv (k+1)\left(\bigotimes_{i=0}^{k} A(i)\right)(\sigma) \otimes x_0. \tag{3.21}$$

By calculation,

$$\frac{d}{d\theta}\mathbb{E}_\theta\Big[g(x(k+1))\Big|x(0) = x_0\Big] = \mathbb{E}_\theta\Big[g^\tau(x'(k+1))\Big|x(0) = x_0\Big]$$

$$\stackrel{(3.21)}{=} \mathbb{E}_\theta\left[g^\tau\left((k+1)\left(\bigotimes_{j=0}^{k} A(j)\right)(\sigma) \otimes x_0\right)\right]$$

$$\stackrel{(3.19)}{=} \mathbb{E}_\theta \left[g^\tau \left((k+1) \bigotimes_{j=\sigma+1}^{k} A(j) \otimes A'(\sigma) \otimes \bigotimes_{j=0}^{\sigma-1} A(j) \otimes x_0 \right) \right]$$

$$\stackrel{(3.14)}{=} \mathbb{E}_\theta \left[(k+1) c_{A(\sigma)} \left\{ g \left(\bigotimes_{j=\sigma+1}^{k} A(j) \otimes A^+(\sigma) \otimes \bigotimes_{j=0}^{\sigma-1} A(j) \otimes x_0 \right) \right. \right.$$

$$\left. \left. - g \left(\bigotimes_{j=\sigma+1}^{k} A(j) \otimes A^-(\sigma) \otimes \bigotimes_{j=0}^{\sigma-1} A(j) \otimes x_0 \right) \right\} \right].$$

The above expression has a surprisingly simple interpretation. To see this, we introduce two processes $x_j^+ = \{x_j^+(i) : 0 \le i \le k+1\}$ and $x_j^- = \{x_j^-(i) : 0 \le i \le k+1\}$, with $0 \le j \le k$, defined as follows.

Algorithm 3.6.1 *Choose σ uniformly distributed over $\{0, \dots k\}$ independently of everything else; and construct x_σ^+ as follows. Initialize $x_\sigma^+(0)$ to x_0. For all $i < \sigma$ set*

$$x_\sigma^+(i+1) = A(i) \otimes x_\sigma^+(i), \tag{3.22}$$

whereas for $i = \sigma$ set

$$x_\sigma^+(\sigma+1) = A^+(\sigma) \otimes x_\sigma^+(\sigma). \tag{3.23}$$

Continue with (3.22) until $i = n$. In words: for all transitions, except the σth transition, the dynamic of x_σ^+ is identical to that of the original sequence $\{x(i) : 0 \le i \le k+1\}$. Construct x_σ^- in exactly the same way except for (3.23) which has to be replaced by

$$x_\sigma^-(\sigma+1) = A^-(\sigma) \otimes x_\sigma^-(\sigma).$$

From the construction follows that

$$\mathbb{E}_\theta[(k+1) c_{A(\sigma)} \, g(x_\sigma^+(k+1)) | x_\sigma^+(0) = x_0]$$

$$= \mathbb{E}_\theta \left[(k+1) c_{A(\sigma)} g \left(\bigotimes_{j=\sigma+1}^{k} A(j) \otimes A^+(\sigma) \otimes \bigotimes_{j=0}^{\sigma-1} A(j) \otimes x_0 \right) \right]$$

and

$$\mathbb{E}_\theta[(k+1) c_{A(\sigma)} \, g(x_\sigma^-(k+1)) | x_\sigma^-(0) = x_0]$$

$$= \mathbb{E}_\theta \left[(k+1) c_{A(\sigma)} g \left(\bigotimes_{j=\sigma+1}^{k} A(j) \otimes A^-(\sigma) \otimes \bigotimes_{j=0}^{\sigma-1} A(j) \otimes x_0 \right) \right].$$

Hence,

$$\frac{d}{d\theta} \mathbb{E}_\theta \left[g(x(k+1)) \Big| x(0) = x_0 \right] \tag{3.24}$$

$$= \mathbb{E}_\theta \left[(k+1) c_{A(\sigma)} \left(g(x_\sigma^+(k+1)) - g(x_\sigma^-(k+1)) \right) \Big| x_\sigma^+(0) = x_\sigma^-(0) = x_0 \right],$$

for all $g \in C_p(\mathbb{R}_{\max}^J, \| \cdot \|_\oplus)$. In other words, Algorithm 3.6.1 together with (3.24) provides an unbiased estimator for derivatives of finite max-plus matrix products.

Expressions containing C_p-derivatives can be easily transformed into the initial max-plus setting. Thus, sensitivity analysis and optimization can easily be added to any max-plus based simulation of the system performance.

3.6.2 Inhomogeneous Recurrence Relations

Consider the max-plus recurrence relation given in (1.27) on page 26, describing the sample dynamic of the queuing system in Example 1.5.2. In order to obtain the C_p-derivative of $\hat{x}(k+1)$ we could either transform (1.27) into a closed-form expression like (3.20) and calculate the derivative directly (which will lead to tiresome calculations) or we could transform (1.27) into a homogeneous equation. In what follows we explain the latter approach. To this end, we define the $(J+1) \times (J+1)$-dimensional matrix

$$\tilde{A}(k) = \begin{pmatrix} \hat{A}(k) & B(k) \otimes \tau(k+1) \\ \varepsilon \ldots \varepsilon & e \end{pmatrix}$$

and set

$$x(k) = \begin{pmatrix} \hat{x}(k) \\ e \end{pmatrix},$$

with

$$x(0) = \begin{pmatrix} x_0 \\ e \end{pmatrix}.$$

With the above definitions, recurrence relation (1.27) reads

$$x(k+1) = \tilde{A}(k) \otimes x(k), \quad k \geq 0. \tag{3.25}$$

C_p-differentiability of $\hat{A}(k)$ and $(B(k) \otimes \tau(k+1))$ implies that of $\tilde{A}(k)$ (for a proof follow the line of argument in the proof of Theorem 3.3.1 and use the fact that $\hat{A}(k)$ and $(B(k) \otimes \tau(k+1))$ have common input $\sigma_0(k+1), \ldots, \sigma_J(k+1)$). In particular,

$$\tilde{A}^+(k) = \begin{pmatrix} \hat{A}^+(k) & (B(k) \otimes \tau(k+1))^+ \\ \varepsilon \ldots \varepsilon & e \end{pmatrix}$$

and

$$\tilde{A}^-(k) = \begin{pmatrix} \hat{A}^-(k) & (B(k) \otimes \tau(k+1))^- \\ \varepsilon \ldots \varepsilon & e \end{pmatrix}.$$

Following the same line of argument as in Section 3.6.1 we obtain the following algorithm.

Algorithm 3.6.2 *Initialize $x_\sigma^+(0) = x_0 = x_\sigma^-(0)$; choose σ uniformly distributed over $\{0, \ldots, k\}$ independent of everything else; and set for $i \neq \sigma$*

$$x_\sigma^{+/-}(i+1) = \hat{A}(i) \otimes x_\sigma^{+/-}(i) \oplus B(i) \otimes \tau(i+1)$$

whereas for $i = \sigma$

$$x_\sigma^+(\sigma + 1) = \hat{A}^+(\sigma) \otimes x_\sigma^+(\sigma) \oplus (B(\sigma) \otimes \tau(\sigma + 1))^+$$

and

$$x_\sigma^-(\sigma + 1) = \hat{A}^-(\sigma) \otimes x_\sigma^-(\sigma) \oplus (B(\sigma) \otimes \tau(\sigma + 1))^- \,.$$

The above algorithm yields an unbiased gradient estimator for inhomogeneous max-plus recurrence relations. More specifically, for $g \in C_p(\mathbb{R}_{max}^J, \|\cdot\|_\oplus)$ set $\tilde{g}(x, e) = g(x)$, then it holds that

$$\frac{d}{d\theta}\mathbb{E}_\theta\Big[g(x(k + 1))\Big|x(0) = x_0\Big]$$
$$= \mathbb{E}_\theta\Big[(k + 1)c_{\hat{A}(\sigma)}\Big(\tilde{g}(x_\sigma^+(k + 1)) - \tilde{g}(x_\sigma^-(k + 1))\Big)\Big|x_\sigma^+(0) = x_\sigma^-(0) = x_0\Big]\,.$$

In Example 1.5.4 we have explained how waiting times can be represented via inhomogeneous max-plus recurrence relations. We conclude our presentation with an application of our results to C_p-differentiation of waiting times.

Example 3.6.1 *Consider the situation in Example 1.5.4 again. For the sake of simplicity, assume that θ is a parameter of the interarrival time distribution so that the interarrival time $\sigma_0(k)$ has a C_p-derivative $\sigma_0'(k) = (c, \sigma_0^+(k), \sigma_0^-(k))$. The matrix $A(k)$ and the vector $B(k)$ are independent of θ. Furthermore, $C(\sigma_0(k))$ is C_p-differentiable with $C'(\sigma_0(k)) = (c, C(\sigma_0^+(k)), C(\sigma_0^-(k)))$. Our calculus of C_p-differentiation then implies*

$$\Big(A(k) \otimes C(\sigma_0(k + 1))\Big)' \equiv A(k) \otimes C'(\sigma_0(k + 1))\,.$$

Let $W_\sigma^+(i)$ and $W_\sigma^-(i)$ $(0 \le i \le k + 1)$ be two sequences defined as follows. Initialize $W_\sigma^+(0) = W(0) = W_\sigma^-(0)$; chose σ uniformly distributed over $\{0, \ldots, k\}$ independent of everything else; and set for $i \ne \sigma$

$$W_\sigma^{+/-}(i + 1) = A(i) \otimes C(\sigma_0(i + 1)) \otimes W_\sigma^{+/-}(i) \oplus B(i)\,,$$

whereas for $i = \sigma$

$$W_\sigma^+(i + 1) = A(i) \otimes C(\sigma_0^+(i + 1)) \otimes W_\sigma^+(i) \oplus B(i)$$

and

$$W_\sigma^-(i + 1) = A(i) \otimes C(\sigma_0^-(i + 1)) \otimes W_\sigma^-(i) \oplus B(i)\,.$$

Then for all $g \in C_p(\mathbb{R}_{max}^J, \|\cdot\|_\oplus)$ it holds true that

$$\frac{d}{d\theta}\mathbb{E}_\theta\Big[g(W(k + 1))\Big] = c(k + 1)\mathbb{E}_\theta\Big[g(W_\sigma^+(k + 1)) - g(W_\sigma^-(k + 1))\Big]\,.$$

Chapter 4

Higher-Order $\mathcal{D}$-Derivatives

In this chapter, we extend the concept of $\mathcal{D}$-differentiability to that of higher-order $\mathcal{D}$-differentiability. The key contribution of this chapter will be that we establish a Leibnitz rule of higher-order $\mathcal{D}$-differentiation and that we give an explicit formula for higher-order $\mathcal{D}$-derivatives. The general setup is as in the previous chapter.

This chapter is organized as follows. In Section 4.1, we introduce the concept of higher-order $\mathcal{D}$-differentiation. The basic result for higher-order differentiation in C_p-spaces is established in Section 4.2. Higher-order $\mathcal{D}$-differentiation on $\mathcal{M}^{I \times J}$ is discussed in Section 4.3. Then, we take $\mathcal{D} = C_p$ and prove in Section 4.4 a Leibnitz rule of higher-order C_p-differentiation. Finally, we introduce in Section 4.5 the concept of $\mathcal{D}$-analyticity and show that the $\oplus$-sum and the $\otimes$-product on $\mathbb{R}_{\max}$ preserve C_p-analyticity.

4.1 Higher-Order $\mathcal{D}$-Derivatives

The definition of higher-order $\mathcal{D}$-derivatives is a straightforward generalization of Definition 3.1.1.

Definition 4.1.1 *Consider the mapping $\mu \; : \; \Theta \; \to \; \mathcal{M}_1(S, \mathcal{S})$. Let $\mathcal{D} \subset \mathcal{L}^1(\mu_\theta : \theta \in \Theta)$ and set $\mu_\theta^{(0)} = \mu_\theta$. We call μ_θ n times $\mathcal{D}$-differentiable at θ if a finite signed measure $\mu_\theta^{(n)}$ exists such that for any $g \in \mathcal{D}$:*

$$\frac{d^n}{d\theta^n} \int_S g(s)\, \mu_\theta(ds) \; = \; \int_S g(s)\, \mu_\theta^{(n)}(ds)\,.$$

The definition of an nth order derivative readily follows from the above definition.

Definition 4.1.2 *We call a triple $(c_\theta^{(n)}, \mu_\theta^{(n,+1)}, \mu_\theta^{(n,-1)})$, with $\mu_\theta^{(n,+1)}, \mu_\theta^{(n,-1)} \in \mathcal{M}_1(S, \mathcal{S})$ and $c_\theta^{(n)} \in \mathbb{R}$, an nth order $\mathcal{D}$-derivative of μ_θ at θ if μ_θ is n times*

$\mathcal{D}$-differentiable and if for any $g \in \mathcal{D}$:

$$\int_S g(s)\,\mu_\theta^{(n)}(ds) = c_\theta^{(n)} \left(\int_S g(s)\,\mu_\theta^{(n,+1)}(ds) - \int_S g(s)\,\mu_\theta^{(n,-1)}(ds) \right). \quad (4.1)$$

If the left-hand side of the above equation equals zero for all $g \in \mathcal{D}$, we say that the nth order $\mathcal{D}$-derivative of μ_θ is not significant, whereas it is called significant otherwise.

We denote by $s(\mu_\theta)$ the order of the highest significant $\mathcal{D}$-derivative. Specifically, we set $s(\mu_\theta) = \infty$ if μ_θ is ∞ times $\mathcal{D}$-differentiable and all higher order $\mathcal{D}$-derivatives are significant. In case μ_θ fails to be $\mathcal{D}$-differentiable, we set $s(\mu_\theta) = -1$.

For $\theta \in \Theta$, let $\mu_\theta \in \mathcal{M}_1(S, \mathcal{S})$ be absolutely continuous with respect to $\mu \in \mathcal{M}_1$ and denote the μ-density of μ_θ by f_θ. Assume that f_θ is n times differentiable as a function in θ and suppose that interchanging the order of n fold differentiation and integration is justified for any $g \in \mathcal{D}$, in formula:

$$\forall g \in \mathcal{D} : \qquad \frac{d^n}{d\theta^n} \int_S g(s)\,\mu_\theta(ds) = \int_S g(s)\,\frac{d^n}{d\theta^n} f_\theta(s)\,\mu(ds). \quad (4.2)$$

Set

$$c_\theta^{(n)} = \frac{1}{2} \int_S \left| \frac{d^n}{d\theta^n} f_\theta(s) \right| \mu(ds)$$

and assume that $c_\theta^{(n)} < \infty$. We may then define μ-densities

$$f_\theta^{(n,+1)} = \frac{1}{c_\theta^{(n)}} \max\left(\frac{d^n}{d\theta^n} f_\theta, 0 \right), \qquad f_\theta^{(n,-1)} = \frac{1}{c_\theta^{(n)}} \max\left(-\frac{d^n}{d\theta^n} f_\theta, 0 \right).$$

Equation (4.2) then reads

$$\frac{d^n}{d\theta^n} \int_S g(s)\,\mu_\theta(ds) = c_\theta^{(n)} \left(\int_S g(s)\,f_\theta^{(n,+1)}(s)\,\mu(ds) - \int_S g(s)\,f_\theta^{(n,-1)}(s)\,\mu(ds) \right).$$
$$(4.3)$$

From the densities $f_\theta^{(n,+1)}$ and $f_\theta^{(n,-1)}$ we obtain measures $\mu_\theta^{(n,+1)}$ and $\mu_\theta^{(n,-1)}$, respectively, on $(S, \mathcal{F})$ through

$$\mu_\theta^{(n,+1)}(A) = \int_A f_\theta^{(n,+1)}(s)\,\mu(ds) \quad \text{and} \quad \mu_\theta^{(n,-1)}(A) = \int_A f_\theta^{(n,-1)}(s)\,\mu(ds),$$
$$(4.4)$$

for $A \in \mathcal{F}$. For $n = 1$, we recover the definition of $\mathcal{D}$-differentiability as stated in (4.1). Like for first-order $\mathcal{D}$-derivatives, the above representation of $\mu_\theta^{(n)}$ is the Hahn-Jordan decomposition, where

$$S_{\mu_\theta^{(n)}}^+ = \left\{ s \in S : \frac{d^n}{d\theta^n} f_\theta(s) \geq 0 \right\}.$$

In the following we provide examples of infinitely $\mathcal{D}$-differentiable distributions.

Example 4.1.1 *Let $\mu_\theta(x) = 1 - e^{-\theta x}$ be the exponential distribution on $S = [0, \infty)$, with $\Theta = [\theta_l, \theta_r]$ for $0 < \theta_l < \theta_r < \infty$, see Example 3.1.1.*

We show that μ_θ is ∞ times C_p-differentiable, for any $p \in \mathbb{N}$. The Lebesgue density of μ_θ, denoted by f_θ, is bounded by

$$\sup_{\theta \in \Theta} f_\theta(x) = \theta_r e^{-\theta_l x} \overset{\text{def}}{=} K_f^0(x), \quad x \in [0, \infty).$$

For $n \geq 1$, the nth derivative of $f_\theta(x)$ is given by

$$\frac{d^n}{d\theta^n} f_\theta(x) = (-1)^n x^{n-1} (\theta x - n) e^{-\theta x},$$

which implies, for any $x \in [0, \infty)$,

$$\sup_{\theta \in \Theta} \left| \frac{d^n}{d\theta^n} f_\theta(x) \right| \leq (\theta_r x + n) x^{n-1} e^{-\theta_l x} \overset{\text{def}}{=} K_f^n(x), \qquad (4.5)$$

for $n \geq 1$. All moments of the exponential distribution exist and we obtain, for all n and all p,

$$\int_S |x|^p K_f^n(x) \, dx < \infty.$$

From the dominated convergence theorem it then follows

$$\frac{d^n}{d\theta^n} \int_S g(s) \, \mu_\theta(ds) = \frac{d^n}{d\theta^n} \int_S g(s) \, f_\theta(s) \, ds$$

$$= \int_S g(s) \, \frac{d^n}{d\theta^n} f_\theta(s) \, ds.$$

Writing $d^n f_\theta / d\theta^n$ as

$$\frac{d^n}{d\theta^n} f_\theta(s) = \max\left(\frac{d^n}{d\theta^n} f_\theta(s), 0 \right) - \max\left(-\frac{d^n}{d\theta^n} f_\theta(s), 0 \right),$$

where

$$\max\left(\frac{d^n}{d\theta^n} f_\theta(s), 0 \right) = \begin{cases} 1_{[n/\theta, \infty)}(x) \, (\theta x - n) \, e^{-\theta x} & n \text{ even}, \\ 1_{[0, n/\theta)}(x) \, (n - \theta x) \, e^{-\theta x} & \text{otherwise}, \end{cases}$$

and

$$\max\left(-\frac{d^n}{d\theta^n} f_\theta(s), 0 \right) = \begin{cases} 1_{[0, n/\theta)}(x) \, (n - \theta x) \, e^{-\theta x} & n \text{ even}, \\ 1_{[n/\theta, \infty)}(x) \, (\theta x - n) \, e^{-\theta x} & \text{otherwise}, \end{cases}$$

we obtain the nth C_p-derivative of μ_θ through

$$\mu_\theta^{(n,+1)}(ds) = \frac{1}{c_\theta^{(n)}} \max\left(\frac{d^n}{d\theta^n} f_\theta(s), 0 \right) ds,$$

$$\mu_\theta^{(n,-1)}(ds) = \frac{1}{c_\theta^{(n)}} \max\left(-\frac{d^n}{d\theta^n} f_\theta(s), 0\right) ds$$

and

$$c_\theta^{(n)} = \left(\frac{n}{\theta\, e}\right)^n.$$

Hence, μ_θ is ∞ times C_p-differentiable, for any $p \in \mathbb{N}$, and all higher-order C_p-derivatives are significant, that is, $s(\mu_\theta) = \infty$. The above C_p-derivative is notably the Hahn-Jordan decomposition of $\mu_\theta^{(n)}$. Later in the text we will provide an alternative representation elaborating on the Gamma-(n,θ)-distribution.

Example 4.1.2 *Consider the Bernoulli-(θ)-distribution μ_θ on $X = \{D_1, D_2\} \subset S$ with $\mu_\theta(D_1) = \theta = 1 - \mu_\theta(D_2)$. Following Example 3.5.1, we obtain*

$$\left(1, \delta_{D_1}(\cdot), \delta_{D_2}(\cdot)\right)$$

as a first-order $\mathcal{D}$-derivative of μ_θ, where δ_x denotes the Dirac measure in x and $\mathcal{D}$ is any set of mappings from X to $\mathbb{R}$. Furthermore, all higher-order $\mathcal{D}$-derivatives of μ_θ are not significant. Hence, μ_θ is ∞ times $\mathcal{D}$-differentiable with $s(\mu_\theta) = 1$.

For the exponential distribution with rate θ all higher-order C_p-derivatives exist and are significant, whereas for the Bernoulli-(θ)-distribution all higher-order C_p-derivatives exist and but only the first C_p-derivative is significant. We conclude this series of examples with the uniform distribution on $[0, \theta]$: here only the first C^b-derivative exists.

Example 4.1.3 *We revisit Example 3.1.2. There exists no (reasonable) set $\mathcal{D}$, such that the Dirac measure in θ is $\mathcal{D}$-differentiable, see Example 3.1.3. In particular, the Dirac measure fails to be C^b-differentiable.*

In Example 3.1.2 we have shown that the uniform distribution on $[0, \theta]$, denoted by $\mathcal{U}_{[0,\theta]}$, is C^b-differentiable and we have calculated $\mathcal{U}'_{[0,\theta]}$. In particular, $\mathcal{U}'_{[0,\theta]}$ is a measure with a discrete and a continuous component, where the discrete component has its mass at θ and therefore any representation of $\mathcal{U}'_{[0,\theta]}$ in terms of a scaled difference between two probability measures does. In other words, any C^b-derivative of μ_θ involves the Dirac measure in θ. Twice C^b-differentiability of $\mathcal{U}_{[0,\theta]}$ is equivalent to C^b-differentiability of $\mathcal{U}'_{[0,\theta]}$ and thus involves C^b-differentiability of the Dirac measure in θ. Since the Dirac measure in θ fails to be C^b-differentiable, we conclude that the second-order C^b-derivative of $\mathcal{U}_{[0,\theta]}$ does not exist and likewise any higher-order C^b-derivative of $\mathcal{U}_{[0,\theta]}$. Hence, $\mathcal{U}_{[0,\theta]}$ is once C_p-differentiable and $s(\mathcal{U}_{[0,\theta]}) = 1$.

In what follows, we will establish a Leibnitz rule for higher order $\mathcal{D}$-differentiability. Before we state our lemma on $\mathcal{D}$-differentiability of the product of two $\mathcal{D}$-differentiable measures, we introduce the following multi indices. For

$n \in \mathbb{N}$, $m_1, m_2 \in \mathbb{Z}$, with $m_1 < m_2$, and $\mu_k \in \mathcal{M}_1(S, \mathcal{S})$, with $m_1 \leq k \leq m_2$, we set

$$\mathcal{L}[m_1, m_2; n] = \left\{ (l_{m_1}, \ldots, l_{m_2}) \,\middle|\, l_k \in \{0, \ldots, n\}, l_k \leq s(\mu_k) \text{ and } \sum_{k=m_1}^{m_2} l_k = n \right\},$$

and, for $l \in \mathcal{L}[m_1, m_2; n]$, we introduce the set

$$\mathcal{I}[l] = \left\{ (i_{m_1}, \ldots, i_{m_2}) \,\middle|\, i_k \in \{0, +1, -1\}, i_k = 0 \text{ iff } l_k = 0 \text{ and } \prod_{\substack{i_{m_1}, \ldots, i_{m_2} \\ i_k \neq 0}} i_k = +1 \right\}.$$

For $i \in \mathcal{I}[l]$ we introduce the auxiliary multi index i^- as follows. Let k^* be the highest position of a non-zero element in i, that is, $i_k = 0$ for all $k > k^*$ and $i_{k^*} \in \{-1, +1\}$. We now set

$$i^- = (i_{m_1}, \ldots, i_{m_2})^- = (i_{m_1}, \ldots i_{k^*-1}, -i_{k^*}, i_{k^*+1}, \ldots, i_{m_2}),$$

that is, the multi index i^- is generated out of i by changing the sign of the highest non-zero element. In the following theorem we denote the cardinality of a given set H by $\mathrm{card}(H)$.

Theorem 4.1.1 *Let $(S, \mathcal{S})$ and $(Z, \mathcal{Z})$ be measurable spaces. If $\mu_\theta \in \mathcal{M}_1(S, \mathcal{S})$ is n times $\mathcal{D}(S)$-differentiable and $\nu_\theta \in \mathcal{M}_1(Z, \mathcal{Z})$ n times $\mathcal{D}(Z)$-differentiable for solid spaces $\mathcal{D}(S)$ and $\mathcal{D}(Z)$. Then $\mu_\theta \times \nu_\theta$ is n times $\mathcal{D}(S, Z)$-differentiable, where*

$$\mathcal{D}(S, Z) = \left\{ g \in \mathcal{L}^1(\mu_\theta \times \nu_\theta : \theta \in \Theta) \,\middle|\, \exists m : |g(s, z)| \leq \sum_{i=0}^{m} d_i f_i(s) h_i(z), \right.$$

$$\left. f_i \in \mathcal{D}(S), h_i \in \mathcal{D}(Z), d_i \in \mathbb{R} \right\},$$

and it holds

$$(\mu_\theta \times \nu_\theta)^{(n)} = \sum_{l = (l_0, l_1) \in \mathcal{L}[0, 1; n]} \frac{n!}{l_0! l_1!} \mu_\theta^{(l_0)} \times \nu_\theta^{(l_1)}.$$

Moreover, let μ_θ have n^{th} order $\mathcal{D}$-derivative $(c_{\mu_\theta}, \mu_\theta^{(n,+1)}, \mu_\theta^{(n,-1)})$ and let ν_θ have n^{th} order $\mathcal{D}$-derivative $(c_{\nu_\theta}, \nu_\theta^{(n,+1)}, \nu_\theta^{(n,-1)})$, then an nth order $\mathcal{D}(S, Z)$-derivative of $\mu_\theta \times \nu_\theta$ is given by

$$\left(c_{\mu_\theta \times \nu_\theta}^{(n)}, (\mu_\theta \times \nu_\theta)^{(n,+1)}, (\mu_\theta \times \nu_\theta)^{(n,-1)} \right),$$

with

$$c_{\mu_\theta \times \nu_\theta}^{(n)} = \sum_{l = (l_0, l_1) \in \mathcal{L}[0, 1; n]} \frac{n!}{l_0! l_1!} c_{\mu_\theta}^{(l_0)} \cdot c_{\nu_\theta}^{(l_1)} \cdot \mathrm{card}(\mathcal{I}[l]),$$

$$(\mu_\theta \times \nu_\theta)^{(n,+1)} = \sum_{l=(l_0,l_1)\in\mathcal{L}[0,1;n]} \frac{n!}{l_0!l_1!} \frac{c_{\mu_\theta}^{(l_0)}\cdot c_{\nu_\theta}^{(l_1)}}{c_{\mu_\theta\times\nu_\theta}^{(n)}} \sum_{(i_0,i_1)\in\mathcal{I}[l]} \mu_\theta^{(l_0,i_0)} \times \nu_\theta^{(l_1,i_1)}$$

and

$$(\mu_\theta \times \nu_\theta)^{(n,-1)} = \sum_{l=(l_0,l_1)\in\mathcal{L}[0,1;n]} \frac{n!}{l_0!l_1!} \frac{c_{\mu_\theta}^{(l_0)}\cdot c_{\nu_\theta}^{(l_1)}}{c_{\mu_\theta\times\nu_\theta}^{(n)}} \sum_{(i_0,i_1)\in\mathcal{I}[l]} \mu_\theta^{(l_0,i_0^-)} \times \nu_\theta^{(l_1,i_1^-)},$$

where $\mu_\theta^{(0,0)} = \mu_\theta$, $\nu_\theta^{(0,0)} = \nu_\theta$ and $c_{\mu_\theta}^{(0)} = 1 = c_{\nu_\theta}^{(0)}$.

Proof: We prove the first part of the theorem by induction with respect to n. Theorem 3.1.1 implies that the induction hypothesis holds for $n = 1$. Suppose now that the statement of the theorem holds for $n > 1$. Direct calculation yields, for any $g \in \mathcal{D}(S, Z)$:

$$\frac{d^n}{d\theta^n} \int g(u)\, \mu_\theta \times \nu_\theta(du)$$

$$= \frac{d}{d\theta} \left(\frac{d^{n-1}}{d\theta^{n-1}} \int g(u)\, (\mu_\theta \times \nu_\theta)(du) \right)$$

$$= \frac{d}{d\theta} \left(\sum_{(l_0,l_1)\in\mathcal{L}[0,1;n-1]} \frac{(n-1)!}{l_0!l_1!} \int g(u) \left(\mu_\theta^{(l_0)} \times \nu_\theta^{(l_1)} \right) (du) \right)$$

$$= \sum_{(l_0,l_1)\in\mathcal{L}[0,1;n-1]} \frac{(n-1)!}{l_0!l_1!} \frac{d}{d\theta} \left(\int g(u) \left(\mu_\theta^{(l_0)} \times \nu_\theta^{(l_1)} \right) (du) \right).$$

We assumed that the nth $\mathcal{D}(S)$-derivative of μ_θ and the $\mathcal{D}(Z)$-derivative of ν_θ, respectively, exist and evoking Theorem 3.1.1 again yields

$$= \sum_{(l_0,l_1)\in\mathcal{L}[0,1;n-1]} \frac{(n-1)!}{l_0!l_1!} \left(\int g(u) \left(\mu_\theta^{(l_0+1)} \times \nu_\theta^{(l_1)} \right) (du) \right.$$

$$\left. + \int g(u) \left(\mu_\theta^{(l_0)} \times \nu_\theta^{(l_1+1)} \right) (du) \right)$$

$$= \sum_{(l_0,l_1)\in\mathcal{L}[0,1;n]} \frac{n!}{l_0!l_1!} \int g(u) \left(\mu_\theta^{(l_0)} \times \nu_\theta^{(l_1)} \right) (du). \tag{4.6}$$

Since only the first $s(\mu_\theta)$ derivatives of μ_θ and the first $s(\nu_\theta)$ derivatives of ν_θ are significant, we only have to take into account indices $l = (l_0, l_1)$ such that $l_0 \le s(\mu_\theta)$ and $l_1 \le s(\nu_\theta)$.

In order to prove the second part of the theorem, we consider the positive and negative parts of the higher-order derivatives of μ_θ and ν_θ separately, see

(4.1). The measure on the right-hand side of (4.6) then reads

$$\sum_{(l_0,l_1)\in\mathcal{L}[0,1;n]} \frac{n!}{l_0!l_1!}\mu_\theta^{(l_0)} \times \nu_\theta^{(l_1)}$$

$$= \sum_{(l_0,l_1)\in\mathcal{L}[0,1;n]} \frac{n!}{l_0!l_1!}c_{\mu_\theta}^{(l_0)} \left(\mu_\theta^{(l_0,+1)} - \mu_\theta^{(l_0,-1)}\right) c_{\nu_\theta}^{(l_1)} \left(\nu_\theta^{(l_1,+1)}(z) - \nu_\theta^{(l_1,-1)}\right)$$

rearranging the positive and negative parts yields,

$$= \sum_{(l_0,l_1)\in\mathcal{L}[0,1;n]} \frac{n!}{l_0!l_1!} \left\{ c_{\mu_\theta}^{(l_0)} c_{\nu_\theta}^{(l_1)} \left(\mu_\theta^{(l_0,+1)} \times \nu_\theta^{(l_1,+1)}\right) + \left(\mu_\theta^{(l_0,-1)} \times \nu_\theta^{(l_1,-1)}\right) \right.$$

$$\left. - c_{\mu_\theta}^{(l_0)} c_{\nu_\theta}^{(l_1)} \left(\left(\mu_\theta^{(l_0,+1)} \times \nu_\theta^{(l_1,-1)}\right) + \left(\mu_\theta^{(l_0,-1)}(z) \times \nu_\theta^{(l_1,+1)}\right) \right) \right\}$$

$$= \sum_{(l_0,l_1)\in\mathcal{L}[0,1;n]} \frac{n!}{l_0!l_1!} c_{\mu_\theta}^{(l_0)} c_{\nu_\theta}^{(l_1)} \sum_{(i_0,i_1)\in\mathcal{I}[l]} \left(\mu_\theta^{(l_0,i_0)} \times \nu_\theta^{(l_1,i_1)}\right)$$

$$- \sum_{(l_0,l_1)\in\mathcal{L}[0,1;n]} \frac{n!}{l_0!l_1!} c_{\mu_\theta}^{(l_0)} c_{\nu_\theta}^{(l_1)} \sum_{(i_0,i_1)\in\mathcal{I}[l]} \left(\mu_\theta^{(l_0,i_0^-)} \times \nu_\theta^{(l_1,i_1^-)}\right)$$

$$= c_{\mu_\theta\times\nu_\theta}^{(n)} \left((\mu_\theta \times \nu_\theta)^{(n,+1)} - (\mu_\theta \times \nu_\theta)^{(n,-1)}\right).$$

Note that, for $l \in \{0,1\}$ and $i \in \{+1,0,-1\}$, $\mu_\theta^{(l,i)}$ and $\nu_\theta^{(l,i)}$ are probability measures. That $(\mu_\theta \times \nu_\theta)^{(n,\pm1)}$ are indeed probability measures can be seen as follows:

$$(\mu_\theta \times \nu_\theta)^{(n,\pm1)}(S \times Z) = \sum_{l=(l_0,l_1)\in\mathcal{L}[0,1;n]} \frac{n!}{l_0!l_1!} \frac{c_{\mu_\theta}^{(l_0)} \cdot c_{\nu_\theta}^{(l_1)}}{c_{\mu_\theta\times\nu_\theta}^{(n)}} \sum_{(i_0,i_1)\in\mathcal{I}[l]} 1$$

$$= \frac{1}{c_{\mu_\theta\times\nu_\theta}^{(n)}} \sum_{l=(l_0,l_1)\in\mathcal{L}[0,1;n]} \frac{n!}{l_0!l_1!} c_{\mu_\theta}^{(l_0)} \cdot c_{\nu_\theta}^{(l_1)} \,\mathrm{card}(\mathcal{I}[l])$$

$$= 1.$$

This concludes the proof of the theorem. $\square$

We now turn to n^{th} order $\mathcal{D}$-derivatives of random variables.

Definition 4.1.3 *Let X_θ have distribution μ_θ. We call a random variable X_θ n times $\mathcal{D}$-differentiable if μ_θ is n times $\mathcal{D}$-differentiable. Let μ_θ have n^{th} order $\mathcal{D}$-derivative $(c_{\mu_\theta}, \mu_\theta^{(n,+1)}, \mu_\theta^{(n,-1)})$. We call the triple $(c_{X_\theta}^{(n)}, X_\theta^{(n,+1)}, X_\theta^{(n,-1)})$ an nth order $\mathcal{D}$-derivative of X_θ if $X_\theta^{(n,+1)}$ is distributed according to $\mu_\theta^{(n,+1)}$ and $X_\theta^{(n,-1)}$ according to $\mu_\theta^{(n,-1)}$, respectively, that is, if for any $g \in \mathcal{D}$:*

$$\frac{d^n}{d\theta^n}\mathbb{E}[g(X_\theta)] = c_{X_\theta}^{(n)} \left(\mathbb{E}\left[g\left(X_\theta^{(n,+1)}\right)\right] - \mathbb{E}\left[g\left(X_\theta^{(n,-1)}\right)\right]\right),$$

where $c_{X_\theta}^{(n)} = c_{\mu_\theta}^{(n)}$. If the left-hand side of the above equation equals zero for all $g \in \mathcal{D}$, we take $(0, X_\theta, X_\theta)$ as an nth order derivative and call the nth order

$\mathcal{D}$-derivative not significant, whereas it is called significant otherwise. We set $s(X_\theta) = s(\mu_\theta)$.

We illustrate the above definition with the following example.

Example 4.1.4 *Let $X_\theta \in \mathbb{R}$ be exponentially distributed with mean $1/\theta$ and denote the Lebesgue density of X_θ by f_θ. In Example 4.1.1, we have already calculated*

$$\frac{d^n}{d\theta^n} f_\theta(x) = (-1)^n x^{n-1} (\theta\, x - n)\, e^{-\theta\, x} \, .$$

A straightforward way for obtaining a C_p-derivative of X_θ is to split $df_\theta/d\theta$ into its positive and negative part. Re-scaling these functions leads to densities of X_θ^+ and X_θ^-, respectively, and the re-scaling factor will be equal to c_{X_θ}.

However, as already explained in Example 3.1.1 for the first-order C^b-derivative, a more convenient representation of the C_p-derivative is obtainable from the Gamma-(n,θ)-distribution. To see this, recall that

$$h_{n,\theta}(x) = \frac{1}{(n-1)!} \theta^n\, x^{n-1}\, e^{-\theta x} \, , \quad n \geq 1 \, ,$$

is the Lebesgue density of the Gamma-(n,θ)-distribution. Direct calculation yields:

$$\begin{aligned}
\frac{d^n}{d\theta^n} f_\theta(x) &= (-1)^n x^{n-1} (\theta\, x - n)\, e^{-\theta\, x} \\
&= (-1)^n \left(x^n \theta - n x^{n-1} \right) e^{-\theta\, x} \\
&= (-1)^n \frac{n!}{\theta^n} \left(\frac{1}{n!} x^n \theta^{n+1} - \frac{1}{(n-1)!} x^{n-1} \theta^n \right) e^{-\theta\, x} \\
&= (-1)^n \frac{n!}{\theta^n} \left(h_{n+1,\theta}(x) - h_{n,\theta}(x) \right) .
\end{aligned}$$

Hence,

$$c_{X_\theta}^{(n)} = \frac{n!}{\theta^n} \, ,$$

for n even

$$f_\theta^{(n,+1)}(x) = h_{n+1,\theta}(x) \, , \quad f_\theta^{(n,-1)}(x) = h_{n,\theta}(x)$$

and for n odd

$$f_\theta^{(n,+1)}(x) = h_{n,\theta}(x) \, , \quad f_\theta^{(n,-1)}(x) = h_{n+1,\theta}(x) \, .$$

Let $X_\theta^{(n,\pm 1)}$ have Lebesgue density $f_\theta^{(n,\pm 1)}$, then an instance of an nth order C_p-derivative of X_θ is given by $(n!/\theta^n, X_\theta^{(n,+1)}, X_\theta^{(n,-1)})$.

Samples from the Gamma-(n,θ)-distribution can be obtained by summing n i.i.d. copies of exponentially distributed random variables with mean $1/\theta$. This leads to the following scheme for sampling an nth order C_p-derivative of X_θ.

Let $\{X_\theta(k)\}$ be an i.i.d. sequence of exponentially distributed random variables with mean $1/\theta$, then, for n even, the nth order C_p-derivative of X_θ is given by

$$\left(\frac{n!}{\theta^n}, \sum_{k=1}^{n+1} X_\theta(k), \sum_{k=1}^{n} X_\theta(k) \right)$$

and, for n odd, the nth order C_p-derivative of X_θ reads

$$\left(\frac{n!}{\theta^n}, \sum_{k=1}^{n} X_\theta(k), \sum_{k=1}^{n+1} X_\theta(k) \right).$$

Put another way, for any $g \in C_p$, it holds that

$$\frac{d^n}{d\theta^n} \mathbb{E}[g(X_\theta(1))]$$

$$= (-1)^n \frac{n!}{\theta^n} \left(\mathbb{E}\left[g\left(\sum_{k=1}^{n+1} X_\theta(k) \right) \right] - \mathbb{E}\left[g\left(\sum_{k=1}^{n} X_\theta(k) \right) \right] \right). \quad (4.7)$$

Note that the above representation allows for a recursive estimation of higher-order derivatives: the $(n+1)^{st}$ derivative of $\mathbb{E}[g(X_\theta)]$ can be estimated from the same data as the nth derivative and the additional drawing of one sample from an exponential distribution. Taking g as the identity, Equation (4.7) yields the following well known result:

$$\frac{d^n}{d\theta^n} \mathbb{E}[X_\theta(1)] = (-1)^n \frac{n!}{\theta^n} \mathbb{E}[X_\theta(n+1)]$$

$$= (-1)^n \frac{n!}{\theta^{n+1}}$$

$$= \frac{d^n}{d\theta^n} \left(\frac{1}{\theta} \right).$$

Example 4.1.5 Let $X_\theta \in \{D_1, D_2\}$ be Bernoulli-(θ)-distributed, so that $P(X_\theta = D_1) = \theta = 1 - P(X_\theta = D_2)$, for $\theta \in [0, 1]$. From Example 4.1.2 it follows that $(1, D_1, D_2)$ is a $\mathbb{R}^{(D_1, D_2)}$-derivative of X_θ. Since all other derivatives are non-significant, we obtain

$$\left(c^{(n)}, X_\theta^{(n,+1)}, X_\theta^{(n,-1)} \right) = \begin{cases} (1, D_1, D_2) & \text{for } n = 1, \\ (0, X_\theta, X_\theta) & \text{for } n > 1, \end{cases}$$

for $\theta \in [0, 1]$, where we take sided derivatives at the boundary points 0 and 1.

4.2 Higher-Order Differentiation in C_p-Spaces

As in Chapter 3, we will confine ourselves to C_p as space of performance functions in order to derive sufficient conditions for a Leibnitz rule to hold. This will be done in the following theorem which establishes our Leibnitz rule for higher order C_p-differentiability measures, namely that the product of two n times C_p-differentiable measures is again n times C_p-differentiable. This result will serve as the main technical tool for developing a calculus of higher order C_p-differentiation for matrices and vectors over the max-plus semiring.

Theorem 4.2.1 *Let $(S, \mathcal{S})$ and $(Z, \mathcal{Z})$ be measurable spaces equipped with upper bounds $|| \cdot ||_S$ and $|| \cdot ||_Z$, respectively, and let the product space $S \times Z$ be equipped with an upper bound $|| \cdot ||_{S \times Z}$. If*

- *for any $s \in S$, $z \in Z$, it holds that*

$$||(s, z)||_{S \times Z} \leq ||s||_S + ||z||_Z,$$

- *$\mu_\theta \in \mathcal{M}_1(S, \mathcal{S})$ is n times $C_p(S, || \cdot ||_S)$-differentiable and $\nu_\theta \in \mathcal{M}_1(Z, \mathcal{Z})$ is n times $C_p(Z, || \cdot ||_Z)$-differentiable,*

then $\mu_\theta \times \nu_\theta$ is n times $C_p(S \times Z, || \cdot ||_{S \times Z})$-differentiable and it holds

$$(\mu_\theta \times \nu_\theta)^{(n)} = \sum_{l=(l_0, l_1) \in \mathcal{L}[0,1;n]} \frac{n!}{l_0! l_1!} \mu_\theta^{(l_0)} \times \nu_\theta^{(l_1)}.$$

Specifically, an nth order $C_p(S \times Z, || \cdot ||_{S \times Z})$-derivative of $\mu_\theta \times \nu_\theta$ is given by

$$\left(c_{\mu_\theta \times \nu_\theta}^{(n)}, (\mu_\theta \times \nu_\theta)^{(n,+1)}, (\mu_\theta \times \nu_\theta)^{(n,-1)} \right),$$

with

$$c_{\mu_\theta \times \nu_\theta}^{(n)} = \sum_{l=(l_0, l_1) \in \mathcal{L}[0,1;n]} \frac{n!}{l_0! l_1!} c_{\mu_\theta}^{(l_0)} \cdot c_{\nu_\theta}^{(l_1)} \cdot \mathrm{card}(\mathcal{I}[l]),$$

$$(\mu_\theta \times \nu_\theta)^{(n,+1)} = \sum_{l=(l_0, l_1) \in \mathcal{L}[0,1;n]} \frac{n!}{l_0! l_1!} \frac{c_{\mu_\theta}^{(l_0)} \cdot c_{\nu_\theta}^{(l_1)}}{c_{\mu_\theta \times \nu_\theta}^{(n)}} \sum_{(i_0, i_1) \in \mathcal{I}[l]} \mu_\theta^{(l_0, i_0)} \times \nu_\theta^{(l_1, i_1)}$$

and

$$(\mu_\theta \times \nu_\theta)^{(n,-1)} = \sum_{l=(l_0, l_1) \in \mathcal{L}[0,1;n]} \frac{n!}{l_0! l_1!} \frac{c_{\mu_\theta}^{(l_0)} \cdot c_{\nu_\theta}^{(l_1)}}{c_{\mu_\theta \times \nu_\theta}^{(n)}} \sum_{(i_0, i_1) \in \mathcal{I}[l]} \mu_\theta^{(l_0, i_0^-)} \times \nu_\theta^{(l_1, i_1^-)},$$

where $\mu_\theta^{(0,0)} = \mu_\theta, \nu_\theta^{(0,0)} = \nu_\theta$ and $c_{\mu_\theta}^{(0)} = c_{\nu_\theta}^{(0)} = 1$.

Furthermore, let $(R, \mathcal{R})$ be a measurable space equipped with upper bound $|| \cdot ||_R$ and let $h : S \times Z \to R$ be a measurable mapping such that finite constants c_S and c_Z exist which satisfy

$$||h(s, z)||_R \leq c_S ||s||_S + c_Z ||z||_Z, \quad s \in S, z \in Z.$$

If $\mu_\theta \in \mathcal{M}_1(S, \mathcal{S})$ is n times $C_p(S, || \cdot ||_S)$-differentiable and $\nu_\theta \in \mathcal{M}_1(Z, \mathcal{Z})$ is n times $C_p(Z, || \cdot ||_Z)$-differentiable, then $(\mu_\theta \times \nu_\theta)^h$ is n times $C_p(R, || \cdot ||_R)$-differentiable and the nth order $C_p(R, || \cdot ||_R)$-derivative of $(\mu_\theta \times \nu_\theta)^h$ is given by

$$\left((\mu_\theta \times \nu_\theta)^h \right)^{(n)} = \left((\mu_\theta \times \nu_\theta)^{(n)} \right)^h.$$

Proof: Take $\mathcal{D}(S) = C_p(S, ||\cdot||_S)$ and $\mathcal{D}(Z) = C_p(Z, ||\cdot||_Z)$ and note that $\mathcal{D}(S)$ and $\mathcal{D}(Z)$ are solid. Then Theorem 4.1.1 implies that $\mu_\theta \times \nu_\theta$ is n times differentiable with respect to the set

$$\left\{ g \in \mathcal{L}^1(\mu_\theta \times \nu_\theta : \theta \in \Theta) \middle| \exists m : |g(s,z)| \leq \sum_{i=0}^m d_i f_i(s) h_i(z) , \right.$$

$$\left. f_i \in C_p(S, ||\cdot||_S), h_i \in C_p(Z, ||\cdot||_Z), d_i \in \mathbb{R} \right\} ,$$

which coincides with the set $\mathcal{C}(S, Z)$ defined in Lemma 3.2.1. From the same lemma it follows that $C_p(S \times Z, ||\cdot||_{S \times Z}) \subset \mathcal{C}(S, Z)$, and we thus proved the first part of the theorem.

We turn to the proof of the second part of the theorem. In the proof of the first part of the theorem we have shown that $\mu_\theta \times \nu_\theta$ is actually n times $\mathcal{C}(S, Z)$-differentiable. By Lemma 3.2.1, $g_h(\cdot, \cdot) = g(h(\cdot, \cdot)) \in \mathcal{C}(S, Z)$, for any $g \in C_p(R, ||\cdot||_R)$, and, using this fact, we calculate

$$\frac{d^n}{d\theta^n} \int_R g(r) \, (\mu_\theta \times \nu_\theta)^h (dr) = \frac{d^n}{d\theta^n} \int_{S \times Z} g(h(s,z)) \, (\mu_\theta \times \nu_\theta)(ds, dz)$$

$$= \frac{d^n}{d\theta^n} \int_{S \times Z} g_h(s,z) \, (\mu_\theta \times \nu_\theta)(ds, dz)$$

$$= \int_{S \times Z} g_h(s,z) \, (\mu_\theta \times \nu_\theta)^{(n)}(ds, dz)$$

$$= \int_{S \times Z} g(h(s,z)) \, (\mu_\theta \times \nu_\theta)^{(n)}(ds, dz)$$

$$= \int_R g(r) \left(\mu_\theta \times \nu_\theta)^{(n)} \right)^h (dr) ,$$

which proves the second part of the theorem. $\square$

The following lemma extends the above theorem to n fold products of probability measures.

Lemma 4.2.1 *Let $(S, \mathcal{S})$ be a measurable space equipped with upper bound $||\cdot||_S$ and let, for some $m \in \mathbb{N}$, the product space S^{m+1} be equipped with upper bound $||\cdot||_{S^{m+1}}$. If*

- *for any $k \leq m$ it holds that*

$$||(s_0, \ldots, s_k)||_{S^{k+1}} \leq ||(s_0, \ldots, s_{k-1})||_{S^k} + ||s_k||_S ,$$

- *$\mu_\theta \in \mathcal{M}_1(S, \mathcal{S})$ is n times $C_p(S, ||\cdot||_S)$-differentiable,*

then $\nu_\theta = \mu_\theta^{m+1}$ (where μ_θ^{m+1} denotes the $(m+1)$ fold independent product of μ_θ) is n times $C_p(S^{m+1}, ||\cdot||_{S^{m+1}})$-differentiable and it holds

$$\nu_\theta^{(n)} = \sum_{l = (l_0, \ldots, l_m) \in \mathcal{L}[0, m; n]} \frac{n!}{l_0! \cdots l_m!} \prod_{k=0}^m \mu_\theta^{(l_k)} .$$

Specifically, an instance of the nth order $C_p(S^{m+1}, ||\cdot||_{S^{m+1}})$-derivative of ν_θ ($= \mu_\theta^{m+1}$) is given by

$$\left(c_{\nu_\theta}^{(n)}, \nu_\theta^{(n,+1)}, \nu_\theta^{(n,-1)} \right),$$

with

$$c_{\nu_\theta}^{(n)} = \sum_{l=(l_0,\cdots,l_m)\in\mathcal{L}[0,m;n]} \frac{n!}{l_0!\cdots l_m!} \prod_{k=0}^{m} c_{\mu_\theta}^{(l_k)} \cdot \mathrm{card}(\mathcal{I}[l]),$$

$$\nu_\theta^{(n,+1)} = \sum_{l=(l_0,\ldots,l_m)\in\mathcal{L}[0,m;n]} \frac{n!}{l_0!\cdots l_m!} \frac{\prod_{k=0}^{m} c_{\mu_\theta}^{(l_k)}}{c_{\nu_\theta}^{(n)}} \sum_{(i_0,\ldots,i_m)\in\mathcal{I}[l]} \prod_{k=0}^{m} \mu_\theta^{(l_k,i_k)}$$

and

$$\nu_\theta^{(n,-1)} = \sum_{l=(l_0,\ldots,l_m)\in\mathcal{L}[0,m;n]} \frac{n!}{l_0!\cdots l_m!} \frac{\prod_{k=0}^{m} c_{\mu_\theta}^{(l_k)}}{c_{\nu_\theta}^{(n)}} \sum_{(i_0,\ldots,i_m)\in\mathcal{I}[l]} \prod_{k=0}^{m} \mu_\theta^{(l_k,i_k^-)},$$

where $\mu_\theta^{(0,0)} = \mu_\theta$ and $c_{\mu_\theta}^{(0)} = 1$.

Furthermore, let $(R, \mathcal{R})$ be a measurable space equipped with an upper bound $||\cdot||_R$ and let $h : S^k \to R$, for $1 < k \leq m+1$, be a measurable mapping such that for each k, with $2 \leq k \leq m+1$, finite constants $c_1(k)$ and $c_2(k)$ exist which satisfy

$$||h(s_0,\ldots,s_k)||_R \leq c_1(k)||h(s_0,\ldots,s_{k-1})||_R + c_2(k)||s_k||_S, \quad s \in S^{k+1},$$

and for $k = 1$ a finite constant c exist which satisfies:

$$||h(s_0, s_1)||_R \leq c(||s_0||_S + ||s_k||_S).$$

If $\mu_\theta \in \mathcal{M}_1(S, \mathcal{S})$ is n times $C_p(S, ||\cdot||_S)$-differentiable, then $(\nu_\theta)^h$ is n times $C_p(R, ||\cdot||_R)$-differentiable and the nth order $C_p(R, ||\cdot||_R)$-derivative of ν_θ^h is given by

$$\left(\nu_\theta^h \right)^{(n)} = \left(\nu_\theta^{(n)} \right)^h.$$

Proof: The proof follows from Theorem 4.2.1 by finite induction. $\square$

4.3   Higher-Order $\mathcal{D}$-Differentiation on $\mathcal{M}^{I\times J}$

In this section we introduce the basic concepts of higher-order $\mathcal{D}$-differentiation in the extended space $M^{I\times J}$, defined in Section 3.4. We begin with the formal definition of the nth order $\mathcal{D}$-derivative of a random matrix in $\mathbb{R}_{\max}^{J\times I}$ as an object in $M^{I\times J}$.

Definition 4.3.1 *We call $A = A_\theta \in \mathbb{R}_{\max}^{I \times J}$ n times $\mathcal{D}$-differentiable if the distribution of A is n times $\mathcal{D}$-differentiable. We call*

$$A^{(n)} = \left(c_A^{(n)}, A^{(n,+1)}, A^{(n,-1)} \right) \in \mathcal{M}^{I \times J}$$

n^{th} order $\mathcal{D}$-derivative of A if for any $g \in \mathcal{D}$ it holds that

$$\frac{d^n}{d\theta^n} \mathbb{E}_\theta[g(A)] = \mathbb{E}_\theta[g^\tau(A^{(n)})].$$

If the left-hand side equals zero for all g, we set $A^{(n)} = (0, A, A)$ and we call the nth $\mathcal{D}$-derivative of A not significant, whereas it is called significant otherwise. For a first order $\mathcal{D}$-derivative of A we write either A' or $A^{(1)}$.

We illustrate the above definition with examples.

Example 4.3.1 *Consider the Bernoulli case in Example 4.1.5. Only the first $\mathbb{R}^{(D_1, D_2)}$-derivative is significant. More precisely, we obtain $A^{(n)} = (1, D_1, D_2)$ for $n = 1$ and $A^{(n)} = (0, A, A)$ for $n > 1$ as $\mathbb{R}^{(D_1, D_2)}$-derivative of A. Let $g \in \mathbb{R}^{(D_1, D_2)}$, taking the (τ, g)-projection of $A^{(n)}$ (see (3.14) for the definition of this projection) yields*

$$\frac{d^n}{d\theta^n} \mathbb{E}_\theta[g(A)] = \mathbb{E}_\theta[g^\tau(A^{(n)})] = \begin{cases} g(D_1) - g(D_2) & \text{for } n = 1, \\ 0 & \text{for } n > 1, \end{cases}$$

for $\theta \in [0, 1]$, where we take sided derivatives at the boundary points 0 and 1.

In applications, a random matrix may depend on θ only through one of the input variables. Recall that $X_1, \ldots, X_m \in \mathbb{R}_{\max}$ is called the input of $A \in \mathbb{R}_{\max}^{J \times I}$ when the elements of A are measurable mappings of $(X_1, \ldots, X_m)$. As for the first-order C_p-derivative we now show that higher order C_p-differentiation of a matrix A is completely determined by the higher order C_p-differentiability of the input of A.

Corollary 4.3.1 *Let $A_\theta \in \mathbb{R}_{\max}^{I \times J}$ have input $X_{1,\theta}, X_2, \ldots, X_m$, with $X_{\theta,1}, X_i \in \mathbb{R}_{\max}$, for $2 \leq i \leq m$, and let $X_{\theta,1}$ have n^{th} order $C_p(\mathbb{R}_{\max}, || \cdot ||_\oplus)$-derivative*

$$X_{\theta,1}^{(n)} = \left(c_{X_{\theta,1}}^{(n)}, X_{\theta,1}^{(n,+1)}, X_{\theta,1}^{(n,-1)} \right).$$

If

- *$X_{\theta,1}$ is stochastically independent of $(X_2, \ldots, X_m)$,*

- *$(X_2, \ldots, X_m)$ does not depend on θ, and*

- *a constant $c \in (0, \infty)$ exists, such that*

$$||A(X_{\theta,1}, X_2, \ldots, X_m)||_\oplus \leq c \, ||(X_{\theta,1}, X_2, \ldots, X_m)||_\oplus,$$

then A_θ has n^{th} order $C_p(\mathbb{R}_{\max}^{I \times J}, ||\cdot||_\oplus)$-derivative

$$A_\theta^{(n)} = (c_{A_\theta}^{(n)}, A_\theta^{(n,+1)}, A_\theta^{(n,-1)})$$

with

$$c_{X_{\theta,1}}^{(n)} = c_{A_\theta}^{(n)}, \quad A_\theta^{(n,+1)} = A(X_{\theta,1}^{(n,+1)}, X_2, \ldots, X_m)$$

and

$$A_\theta^{(n,-1)} = A(X_{\theta,1}^{(n,-1)}, X_2, \ldots, X_m).$$

Proof: Using Theorem 4.2.1, the proof follows from the same line of argument as the proof of Theorem 3.3.1 followed from Theorem 3.2.2 together with Corollary 1.6.1. $\square$

Example 4.3.2 *We revisit the situation in Example 3.3.1 (an open tandem queuing system the interarrival times of which depend on θ). In accordance with Example 4.1.1, $\sigma_0(\theta, k)$ is ∞ times $C_p(\mathbb{R}_{\max}, ||\cdot||_\oplus)$-differentiable with n^{th} C_p-derivative*

$$\sigma_0^{(n)}(\theta, k) = \left(c_{\sigma_0(\theta)}^{(n)}, \sigma_0^{(n,+1)}(\theta, k), \sigma_0^{(n,-1)}(\theta, k)\right).$$

The condition on $||A(\sigma_0(\theta, k), \sigma_1(k), \ldots, \sigma_J(k))||_\oplus$ in Corollary 4.3.1 is satisfied. The positive part of the nth order $C_p(\mathbb{R}_{\max}^{J+1 \times J+1}, ||\cdot||_\oplus)$-derivative of $A(k)$ is obtained from $A(k)$ by replacing all occurrences of $\sigma_0(\theta, k+1)$ by $\sigma_0^{(n,+1)}(\theta, k+1)$; and the negative part is obtained from replacing all occurrences of $\sigma_0(\theta, k+1)$ by $\sigma_0^{(n,-1)}(\theta, k+1)$. More formally, we obtain an nth order C_p-derivative of $A(k)$ through

$$A^{(n)}(k) = \left(c_{\sigma_0(\theta)}^{(n)}, A(\sigma_0^{(n,+1)}(\theta, k+1), \sigma_1(k+1), \ldots, \sigma_J(k+1)),\right.$$
$$\left. A(\sigma_0^{(n,-1)}(\theta, k+1), \sigma_1(k+1), \ldots, \sigma_J(k+1))\right),$$

for $k \geq 0$.

Following the representation of higher-order C_p-derivatives of the exponential distribution in Example 4.1.4, we obtain the higher-order C_p-derivatives as follows. Let $\{X_\theta(i)\}$ be a sequence of i.i.d. exponentially distributed random variables with mean $1/\theta$. Samples of higher-order C_p-derivatives can be obtained through the following scheme

$$A\left(\sum_{i=1}^{2n+1} X_\theta(i), \sigma_1(k+1), \ldots, \sigma_J(k+1)\right) = A^{(2n,+1)}(k) = A^{(2n+1,+1)}(k),$$

for $n \geq 0$, where we elaborate on the convention $A^{(0,+1)} = A$, and, for $n \geq 1$,

$$A\left(\sum_{i=1}^{2n} X_\theta(i), \sigma_1(k+1), \ldots, \sigma_J(k+1)\right) = A^{(2n-1,-1)}(k) = A^{(2n,-1)}(k).$$

4.4 Rules of C_p-Differentiation

In this section we establish the Leibnitz rule for higher-order C_p-differentiation. Specifically, the C_p-derivative of the $(m + 1)$ fold product of n times C_p-differentiable probability measures can be found by Lemma 4.2.1 and the following lemma provides an interpretation of this result in terms of random matrices.

Lemma 4.4.1 (Leibnitz rule) *Let* $\{A(k)\}$ *be an i.i.d. sequence of n times C_p-differentiable matrices over* $\mathbb{R}_{\max}^{J \times J}$, *then*

$$
\left(\bigotimes_{k=0}^{m} A(k) \right)^{(n)} \equiv \sum_{l \in \mathcal{L}[0,m;n]} \frac{n!}{l_0! l_1! \dots l_m!} \sum_{i \in \mathcal{I}[l]}
$$

$$
\left(\prod_{k=0}^{m} c_{A(k)}^{(l_k)}, \bigotimes_{k=0}^{m} A^{(l_k, i_k)}(k), \bigotimes_{k=0}^{m} A^{(l_k, i_k^-)}(k) \right),
$$

where $A^{(0,0)}(k) = A(k)$ *and* $c_{A(k)}^{(0)} = 1$. *A similar formula can be obtained for the* n^{th} C_p-*derivative of* $A \oplus B$.

Proof: Let $S = \mathbb{R}_{\max}^{J \times J}$ and set h

$$
h(A_0, A_1, \dots, A_m) = \bigotimes_{k=0}^{m} A_k ,
$$

for $A_k \in \mathbb{R}_{\max}^{J \times J}$, for $0 \leq k \leq m$. In accordance with Lemma 1.6.1, for any $m \geq 1$

$$
\left\| \bigotimes_{k=0}^{m} A_k \right\|_{\oplus} \leq \left\| \bigotimes_{k=0}^{m-1} A_k \right\|_{\oplus} + \|A_m\|_{\oplus}
$$

and Lemma 4.2.1 applies to h.

Switching from probability measures to the appropriate random matrices yields an interpretation of nth order derivative in terms of random variables. Canceling out the normalizing factor concludes the proof of the lemma. More specifically, let μ_θ denote the distribution of $A(k)$ and let $A^{(l_k, i_k)}(k)$ be distributed according to $\mu_\theta^{(l_k, i_k)}$ and let $A^{(l_k, i_k^-)}(k)$ be distributed according to $\mu_\theta^{(l_k, i_k^-)}$, for $l \in \mathcal{L}[0, m; n]$ and $i \in \mathcal{I}[l]$.

Lemma 4.2.1 applies to the $(m + 1)$ fold independent product of μ_θ, denoted by μ_θ^{m+1}, and we obtain for $g \in C_p(\mathbb{R}_{\max}^{J \times J}, \| \cdot \|_\oplus)$

$$\frac{d^n}{d\theta^n} \mathbb{E}\left[g\left(\bigotimes_{k=0}^{m} A(k) \right) \right]$$

$$= c_{\mu_\theta^{m+1}}^{(n)} \sum_{l \in \mathcal{L}[0,m;n]} \frac{n!}{l_0! l_1! \ldots l_m!} \prod_{k=0}^{m} \frac{c_{\mu_\theta}^{(l_k)}}{c_{\mu_\theta^{m+1}}^{(n)}} \sum_{i \in \mathcal{I}[l]}$$

$$\times \left(\mathbb{E}\left[g\left(\bigotimes_{k=0}^{m} A^{(l_k, i_k)}(k) \right) \right] - \mathbb{E}\left[g\left(\bigotimes_{k=0}^{m} A^{(l_k, i_k^-)}(k) \right) \right] \right).$$

The factor $c_{\mu_\theta^{m+1}}^{(n)}$ cancels out and according to Definition 4.1.3 it holds that $c_{A(k)}^{(l_k)} = c_{\mu_\theta}^{(l_k)}$. This gives

$$\frac{d^n}{d\theta^n} \mathbb{E}\left[g\left(\bigotimes_{k=0}^{m} A(k) \right) \right]$$

$$= \sum_{l \in \mathcal{L}[0,m;n]} \frac{n!}{l_0! l_1! \ldots l_m!} \prod_{k=0}^{m} c_{A(k)}^{(l_k)} \sum_{i \in \mathcal{I}[l]}$$

$$\times \left(\mathbb{E}\left[g\left(\bigotimes_{k=0}^{m} A^{(l_k, i_k)}(k) \right) \right] - \mathbb{E}\left[g\left(\bigotimes_{k=0}^{m} A^{(l_k, i_k^-)}(k) \right) \right] \right)$$

and switching from g to g^τ, the canonical extension of g to $\mathcal{M}^{J \times J}$, yields

$$= \sum_{l \in \mathcal{L}[0,m;n]} \frac{n!}{l_0! l_1! \ldots l_m!} \sum_{i \in \mathcal{I}[l]}$$

$$\times \mathbb{E}\left[g^\tau \left(\prod_{k=0}^{m} c_{A(k)}^{(l_k)} , \bigotimes_{k=0}^{m} A^{(l_k, i_k)}(k), \bigotimes_{k=0}^{m} A^{(l_k, i_k^-)}(k) \right) \right];$$

we now elaborate on the fact that for any mapping $g \in C_p(\mathbb{R}_{\max}^{J \times J}, \| \cdot \|_\oplus)$ the corresponding mapping g^τ becomes linear on $\mathcal{M}^{J \times J}$, and we arrive at

$$= \mathbb{E}\left[g^\tau \left(\sum_{l \in \mathcal{L}[0,m;n]} \frac{n!}{l_0! l_1! \ldots l_m!} \sum_{i \in \mathcal{I}[l]} \right. \right.$$

$$\left. \left. \left(\prod_{k=0}^{m} c_{A(k)}^{(l_k)} , \bigotimes_{k=0}^{m} A^{(l_k, i_k)}(k), \bigotimes_{k=0}^{m} A^{(l_k, i_k^-)}(k) \right) \right) \right],$$

which concludes the proof of the lemma. $\square$

With the help of the Leibnitz rule we can explicitly calculate higher-order C_p-derivatives. In particular, applying the (τ, g)-projection to higher-order C_p-derivatives yields unbiased estimators for higher-order C_p-derivatives, see Section 3.6.

Example 4.4.1 *Let $\{A_\theta(k)\}$ be a sequence of i.i.d. Bernoulli-(θ)-distributed matrices on $\mathbb{R}_{\max}^{J \times J}$. Only the first-order C_p-derivative of $A_\theta(k)$ is significant and an nth order C_p-derivative of the product of $A_\theta(k)$ over $0 \leq k \leq m$ reads*

$$\left(\bigotimes_{k=0}^{m} A_\theta(k) \right)^{(n)}$$

$$= \sum_{\substack{l=(l_0,\ldots,l_m)\in\{0,1\}^{m+1} \\ \sum l_k = n}} n! \sum_{i\in\mathcal{I}[l]} \left(1, \bigotimes_{k=0}^{m} A_\theta^{(l_k,i_k)}(k), \bigotimes_{k=0}^{m} A_\theta^{(l_k,i_k^-)}(k) \right).$$

When we consider the derivatives at zero, see Example 4.1.5, we obtain $A_0(k) = D_2$ and, for example, the first-order derivative of $g(\bigotimes_{k=0}^{m} A_0(k) \otimes x_0)$ is given by

$$\frac{d}{d\theta} g \left(\bigotimes_{k=0}^{m} A_0(k) \otimes x_0 \right)$$

$$= \sum_{j=0}^{m} g \left(D_2^{m-j} \otimes D_1 \otimes D_2^j \otimes x_0 \right) - (m+1) g \left(D_2^{m+1} \otimes x_0 \right),$$

whereas the second-order derivative reads

$$\frac{1}{2} \frac{d^2}{d\theta^2} g \left(\bigotimes_{k=0}^{m} A_0(k) \otimes x_0 \right)$$

$$= \sum_{j=0}^{m-1} \sum_{i=0}^{m-1-j} g \left(D_2^{m-i-j-1} \otimes D_1 \otimes D_2^i \otimes D_1 \otimes D_2^j \otimes x_0 \right)$$

$$+ (m+1)m \, g \left(D_2^{m+1} \otimes x_0 \right) - m \sum_{j=0}^{m-1} g \left(D_2^{m-j} \otimes D_1 \otimes D_2^j \otimes x_0 \right)$$

$$- m \sum_{j=1}^{m} g \left(D_2^{m-j} \otimes D_1 \otimes D_2^j \otimes x_0 \right)$$

$$= \sum_{j=0}^{m-1} \sum_{i=0}^{m-1-j} g \left(D_2^{m-i-j-1} \otimes D_1 \otimes D_2^i \otimes D_1 \otimes D_2^j \otimes x_0 \right)$$

$$- m \frac{d}{d\theta} g \left(\bigotimes_{k=0}^{m} A_0(k) \otimes x_0 \right)$$

$$- m \sum_{j=1}^{m-1} g \left(D_2^{m-j} \otimes D_1 \otimes D_2^j \otimes x_0 \right).$$

We conclude this section by establishing an upper bound for the nth order

C_p-derivative. To simplify the notation, we set

$$g\left(\left(\bigotimes_{k=0}^{m} A(k)\right)^{(0)} \otimes x_0\right) \overset{\text{def}}{=} g\left(\bigotimes_{k=0}^{m} A(k) \otimes x_0\right) . \tag{4.8}$$

Lemma 4.4.2 *Let $\{A(k)\}$ be an i.i.d. sequence of n times C_p-differentiable square matrices in $\mathbb{R}_{\max}^{J \times J}$. For any $g \in C_p$ it holds with probability one that*

$$\left| g^\tau\left(\left(\bigotimes_{k=0}^{m} A(k)\right)^{(n)} \otimes x_0\right)\right| \le B_{g,m,\{A(k)\}}(n,p) ,$$

where

$$B_{g,m,\{A(k)\}}(n,p)$$
$$\overset{\text{def}}{=} \sum_{l \in \mathcal{L}[0,m;n]} \frac{n!}{l_0! l_1! \ldots l_m!} \sum_{i \in \mathcal{I}[l]} \prod_{k=0}^{m} c_{A(0)}^{(l_k)}$$
$$\times \left(2a_g + b_g\left(\sum_{k=0}^{m}\left\|A^{(l_k,i_k)}(k)\right\|_\oplus + \|x_0\|_\oplus\right)^p\right.$$
$$\left. + b_g\left(\sum_{k=0}^{m}\left\|A^{(l_k,i_k^-)}(k)\right\|_\oplus + \|x_0\|_\oplus\right)^p\right) .$$

In particular,

$$B_{g,m,\{A(k)\}}(0,p) \overset{\text{def}}{=} a_g + b_g\left(\sum_{k=0}^{m}\|A(k)\|_\oplus + \|x_0\|_\oplus\right)^p .$$

In addition to that, let $A(0)$ have state space $\mathcal{A}$ and set

$$\|\mathcal{A}\|_\oplus \overset{\text{def}}{=} \sup_{A \in \mathcal{A}}\|A\|_\oplus$$

and

$$\mathbf{c}_{A(0)} \overset{\text{def}}{=} \max_{0 \le m \le n} c_{A(0)}^{(m)} .$$

If $x_0 = \mathbf{e}$ and $\|\mathcal{A}\|_\oplus < \infty$, then

$$B_{m,\{A(k)\},g}(n,p)$$
$$\le \sum_{l \in \mathcal{L}[0,m;n]} \frac{n!}{l_0! l_1! \ldots l_m!}(\mathbf{c}_{A(0)})^n \, 2^n\left(a_g + b_g\,(m+1)^p\,\|\mathcal{A}\|_\oplus^p\right)$$

and

$$B_{g,m,\{A(k)\}}(0,p) \le a_g + b_g\,(m+1)^p\,\|\mathcal{A}\|_\oplus^p .$$

Proof: We only establish the upper bound for the case $n > 0$. The case $n = 0$ follows readily from this line of argument. From the Leibnitz rule of higher-order C_p-differentiation, see Lemma 4.4.1, we obtain an explicit representation of the nth order C_p-derivative of $\bigotimes_{k=0}^{m} A(k)$. Using the fact that g is linear on $\mathcal{M}^{J \times J}$, we obtain

$$
g^\tau \left(\left(\bigotimes_{k=0}^{m} A(k) \right)^{(n)} \otimes x_0 \right)
$$

$$
= g^\tau \left(\sum_{l \in \mathcal{L}[0,m;n]} \frac{n!}{l_0! l_1! \dots l_m!} \right.
$$

$$
\left. \sum_{i \in \mathcal{I}[l]} \left(\prod_{k=0}^{m} c_{A(0)}^{(l_k)}, \bigotimes_{k=0}^{m} A^{(l_k, i_k)}(k) \otimes x_0, \bigotimes_{k=0}^{m} A^{(l_k, i_k^-)}(k) \otimes x_0 \right) \right)
$$

$$
= \sum_{l \in \mathcal{L}[0,m;n]} \frac{n!}{l_0! l_1! \dots l_m!}
$$

$$
\sum_{i \in \mathcal{I}[l]} g^\tau \left(\prod_{k=0}^{m} c_{A(0)}^{(l_k)}, \bigotimes_{k=0}^{m} A^{(l_k, i_k)}(k) \otimes x_0, \bigotimes_{k=0}^{m} A^{(l_k, i_k^-)}(k) \otimes x_0 \right)
$$

$$
= \sum_{l \in \mathcal{L}[0,m;n]} \frac{n!}{l_0! l_1! \dots l_m!} \sum_{i \in \mathcal{I}[l]} \prod_{k=0}^{m} c_{A(0)}^{(l_k)}
$$

$$
\left(g \left(\bigotimes_{k=0}^{m} A^{(l_k, i_k)}(k) \otimes x_0 \right) - g \left(\bigotimes_{k=0}^{m} A^{(l_k, i_k^-)}(k) \otimes x_0 \right) \right),
$$

where, for the last equality, we take the (τ, g)-projection, see (3.14). Taking absolute values and using the fact that $g \in C_p$ yields

$$
\left| g^\tau \left(\left(\bigotimes_{k=0}^{m} A(k) \right)^{(n)} \otimes x_0 \right) \right|
$$

$$
\leq \sum_{l \in \mathcal{L}[0,m;n]} \frac{n!}{l_0! l_1! \dots l_m!} \sum_{i \in \mathcal{I}[l]} \prod_{k=0}^{m} c_{A(0)}^{(l_k)}
$$

$$
\times \left(2a_g + b_g \left(\left\| \bigotimes_{k=0}^{m} A^{(l_k, i_k)}(k) \otimes x_0 \right\|_\oplus \right)^p + b_g \left(\left\| \bigotimes_{k=0}^{m} A^{(l_k, i_k^-)}(k) \otimes x_0 \right\|_\oplus \right)^p \right).
$$

Applying Lemma 1.6.1 yields

$$
\left(\left\| \bigotimes_{k=0}^{m} A^{(l_k, i_k)}(k) \otimes x_0 \right\|_\oplus \right)^p \leq \left(\sum_{k=0}^{m} \left\| A^{(l_k, i_k)}(k) \right\|_\oplus + \|x_0\|_\oplus \right)^p
$$

and

$$\left(\left\|\bigotimes_{k=0}^{m} A^{(l_k, i_k^-)}(k) \otimes x_0\right\|_{\oplus}\right)^p \leq \left(\sum_{k=0}^{m} \left\|A^{(l_k, i_k^-)}(k)\right\|_{\oplus} + \|x_0\|_{\oplus}\right)^p.$$

This yields

$$\left| g^\tau \left(\left(\bigotimes_{k=0}^{m} A(k)\right)^{(n)} \otimes x_0\right)\right|$$

$$\leq \sum_{l \in \mathcal{L}[0,m;n]} \frac{n!}{l_0! l_1! \ldots l_m!} \sum_{i \in \mathcal{I}[l]} \prod_{k=0}^{m} c_{A(0)}^{(l_k)}$$

$$\times \left(2a_g + b_g \left(\sum_{k=0}^{m} \left\|\left(A^{(l_k, i_k)}(k)\right)\right\|_{\oplus} + \|x_0\|_{\oplus}\right)^p\right.$$

$$\left. + b_g \left(\sum_{k=0}^{m} \left\|A^{(l_k, i_k^-)}(k)\right\|_{\oplus} + \|x_0\|_{\oplus}\right)^p\right),$$

which completes the proof of the first part of the lemma.

We now turn to the proof of the second part of the lemma. Note that $x_0 = \mathbf{e}$ implies that $\|x_0\|_{\oplus} = 0$. Without loss of generality, we assume that the state space of C_p-derivatives of $A(0)$ is (a subset) of $\mathcal{A}$ (this can always be guaranteed by representing the C_p-derivative via the Hahn-Jordan decomposition). This implies, for any $i \in \{-1, 0, 1\}$ and $m \in \{0, 1, \ldots, n\}$,

$$\max(\|A^{(m,i)}(k)\|_{\oplus}, \|A^{(m,-i)}(k)\|_{\oplus}) \leq \|\mathcal{A}\|_{\oplus}, \quad k \geq 0,$$

and we obtain

$$\left(\sum_{k=0}^{m} \|A^{(l_k, i_k)}(k)\|_{\oplus}\right)^p + \left(\sum_{k=0}^{m} \|A^{(l_k, i_k^-)}(k)\|_{\oplus}\right)^p \leq 2(m+1)^p \left(\|\mathcal{A}\|_{\oplus}\right)^p.$$

For any $l \in \mathcal{L}[0, m; n]$, at most n elements of l are different from zero. For $l_k > 0$, $c_{A(k)}^{(l_k)} \leq \mathbf{c}_{A(0)}$. Hence, for any $l \in \mathcal{L}[0, m; n]$

$$\prod_{k=0}^{m} c_{A(0)}^{(l_k)} \leq \left(\mathbf{c}_{A(0)}\right)^n.$$

Furthermore, for any $l \in \mathcal{L}[0, m; n]$, $\mathcal{I}[l]$ has at most 2^{n-1} elements, see Section G.5 in the Appendix. This completes the proof of the second part of the lemma. $\square$

4.5 $\mathcal{D}$-Analyticity

We begin this section by formally defining $\mathcal{D}$-*analyticity* of measures.

Definition 4.5.1 *We call $\mu_\theta \in \mathcal{M}_1(S, \mathcal{S})$ $\mathcal{D}$-analytic on Θ, or, $\mathcal{D}$-analytic for short, if*

- *all higher-order $\mathcal{D}$-derivatives of μ_θ exist on Θ, and*

- *for all $\theta_0 \in \Theta$ an open neighborhood $U_{\theta_0} \subset \Theta$ of θ_0 exists such that for any $g \in \mathcal{D}$ and any $\theta \in U_{\theta_0}$:*

$$\sum_{n=0}^{\infty} \frac{1}{n!}(\theta - \theta_0)^n \int_S g(s)\, \mu_{\theta_0}^{(n)}(ds) = \int_S g(s)\, \mu_\theta(ds).$$

In the following we establish sufficient conditions for C_p-analyticity of some interesting distributions.

Example 4.5.1 *Let μ_θ be the exponential distribution and denote the Lebesgue density of μ_θ by $f_\theta(x) = \theta \exp(-\theta x)$, for $\theta \in \Theta = (0, \infty)$. Then $f_\theta(x)$ is analytic on $(0, \infty)$. In particular, the domain of convergence of the Taylor series for $f_\theta(x)$ developed at any $\theta_0 \in \Theta$ is $(0, \infty)$. For $\theta_0 \in (0, \infty)$, set $U_{\theta_0}(\delta) = (\delta, 2\theta_0 - \delta)$ for $\theta_0 > \delta > 0$, then, for all $x \in [0, \infty)$:*

$$\sum_{n=0}^{\infty} \left| \frac{1}{n!} \frac{d^n}{d\theta^n}\bigg|_{\theta=\theta_0} f_\theta(x)(\theta - \theta_0)^n \right|$$

$$\leq \sum_{n=0}^{\infty} (\theta_0 x^n + n x^{n-1})\, e^{-\theta_0 x} \frac{1}{n!}|\theta - \theta_0|^n$$

$$\leq e^{-\theta_0 x}(\theta_0 + (\theta_0 - \delta)) e^{(\theta_0 - \delta) x}$$

$$= (2\theta_0 - \delta) e^{-\delta x},$$

where the second inequality follows from the fact that $|\theta - \theta_0| \leq \theta_0 - \delta$. This implies, for any $g \in C_p$,

$$\sum_{n=0}^{\infty} \frac{1}{n!}(\theta - \theta_0)^n \int g(x)\, \mu_{\theta_0}^{(n)}(dx)$$

$$= \sum_{n=0}^{\infty} \frac{1}{n!}(\theta - \theta_0)^n \frac{d^n}{d\theta^n}\bigg|_{\theta=\theta_0} \int g(x)\, \mu_\theta(dx)$$

$$= \sum_{n=0}^{\infty} \frac{1}{n!}(\theta - \theta_0)^n \frac{d^n}{d\theta^n}\bigg|_{\theta=\theta_0} \int g(x)\, f_\theta(x)\lambda(dx)$$

$$= \int g(x) \sum_{n=0}^{\infty} \frac{1}{n!}(\theta - \theta_0)^n \frac{d^n}{d\theta^n}\bigg|_{\theta=\theta_0} f_\theta(x)\lambda(dx)$$

$$= \int g(x)\, f_\theta(x)\lambda(dx)$$

$$= \int g(x)\, \mu_\theta(dx).$$

Hence, μ_θ is C_p-analytic on $(0, \infty)$, for any $p \in \mathbb{N}$, and the Taylor series for μ_θ at θ_0 has at least domain of convergence $U_{\theta_0}(\delta)$, for $\theta_0 > \delta > 0$.

Example 4.5.2 *Let μ_θ be the Bernoulli-(θ)-distribution on $\{D_1, D_2\}$. Then μ_θ is ∞ times $\mathbb{R}^{(D_1, D_2)}$-differentiable and since only the first order $\mathcal{D}$- derivative is significant, μ_θ is $\mathcal{D}$-analytic on $[0, 1]$. Notice that taking one-sided derivatives at the boundary points, μ_θ can be expanded into a Taylor series at, say, $\theta = 0$.*

The following theorem characterizes the set of performance functions with respect to which the product of two analytic measures is analytic.

Theorem 4.5.1 *Let $(S, \mathcal{S})$ and $(Z, \mathcal{Z})$ be measurable spaces. If μ_θ is $\mathcal{D}(S)$-analytic and ν_θ is $\mathcal{D}(Z)$-analytic for solid spaces $\mathcal{D}(S)$ and $\mathcal{D}(Z)$, then $\mu_\theta \times \nu_\theta$ is $\mathcal{D}(S, Z)$-analytic, with*

$$\mathcal{D}(S, Z) = \left\{ g \in \mathcal{L}^1(\mu_\theta \times \nu_\theta : \theta \in \Theta) \middle| \exists n : |g(s, z)| \leq \sum_{i=0}^{n} d_i f_i(s) h_i(z) , \right.$$
$$\left. f_i \in \mathcal{D}(S), h_i \in \mathcal{D}(Z), d_i \in \mathbb{R} \right\} .$$

In particular, if, for $\theta_0 \in \Theta$, the Taylor series for μ_θ has domain of convergence $U_{\theta_0}^\mu$ and the Taylor series for ν_θ has domain of convergence $U_{\theta_0}^\nu$, then the domain of convergence of the Taylor series for the product measure $\mu_\theta \times \nu_\theta$ is at least $U_{\theta_0}^\mu \cap U_{\theta_0}^\nu$.

Proof: The Leibnitz rule of higher-order $\mathcal{D}$-differentiation (see Theorem 4.1.1) implies that all higher order $\mathcal{D}(S, Z)$-derivatives of $\mu_\theta \times \nu_\theta$ exist. More precisely, let the Taylor series for μ_θ at θ_0 have domain of convergence $U_{\theta_0}^\mu$ and let the Taylor series for ν_θ at θ_0 have domain of convergence $U_{\theta_0}^\nu$. Then all higher order $\mathcal{D}(S, Z)$-derivatives of $\mu_\theta \times \nu_\theta$ exist on $U_{\theta_0} = U_{\theta_0}^\mu \cap U_{\theta_0}^\nu$.

For $\theta \in U_{\theta_0}$, with $\theta_0 \in \Theta$, we calculate

$$\sum_{m=0}^{\infty} \frac{1}{m!} (\theta - \theta_0)^m \frac{d^m}{d\theta^m} \int_{S \times Z} g(u) \, (\mu_{\theta_0} \times \nu_{\theta_0})(du)$$

$$= \sum_{m=0}^{\infty} \frac{1}{m!} (\theta - \theta_0)^m \int_{S \times Z} g(u) \, (\mu_{\theta_0} \times \nu_{\theta_0})^{(m)}(du)$$

$$= \sum_{m=0}^{\infty} \frac{1}{m!} (\theta - \theta_0)^m \sum_{n=0}^{m} \binom{m}{n} \int_{S \times Z} g(u) \left(\mu_{\theta_0}^{(n)} \times \nu_{\theta_0}^{(k)} \right) (du)$$

$$= \sum_{m=0}^{\infty} \sum_{k+n=m} \frac{1}{n!} \frac{1}{k!} (\theta - \theta_0)^{n+k} \int_{S \times Z} g(u) \left(\mu_{\theta_0}^{(n)} \times \nu_{\theta_0}^{(k)} \right) (du)$$

$$= \sum_{n=0}^{\infty} \sum_{k=0}^{\infty} \frac{1}{n!} (\theta - \theta_0)^n \frac{1}{k!} (\theta - \theta_0)^k \int_{S \times Z} g(u) \left(\mu_{\theta_0}^{(n)} \times \nu_{\theta_0}^{(k)} \right) (du) .$$

Set

$$\int_S g(s,z)\,\mu_{\theta_0}^{(n)}(ds) \;=\; h_g(z)$$

for $z \in Z$, and observe that $g \in \mathcal{D}(S,Z)$ implies $h_g(\cdot) \in \mathcal{D}(Z)$. Applying Fubini's theorem (see Section E.1 in the Appendix), yields

$$\sum_{n=0}^{\infty}\sum_{k=0}^{\infty}\frac{1}{n!}(\theta-\theta_0)^n\frac{1}{k!}(\theta-\theta_0)^k\int_{S\times Z}g(u)\left(\mu_{\theta_0}^{(n)}\times\nu_{\theta_0}^{(k)}\right)(du)$$

$$=\sum_{n=0}^{\infty}\frac{1}{n!}(\theta-\theta_0)^n\sum_{k=0}^{\infty}\frac{1}{k!}(\theta-\theta_0)^k\int_Z\underbrace{\left(\int_S g(s,z)\,\mu_{\theta_0}^{(n)}(ds)\right)}_{=h_g(z),\,h_g\in\mathcal{D}(Z)}\nu_{\theta_0}^{(k)}(dz)$$

and, by $\mathcal{D}(Z)$-analyticity of ν_θ,

$$=\sum_{n=0}^{\infty}\frac{1}{n!}(\theta-\theta_0)^n\int_Z\left(\int_S g(s,z)\,\mu_{\theta_0}^{(n)}(ds)\right)\nu_\theta(dz)$$

$$=\sum_{n=0}^{\infty}\frac{1}{n!}(\theta-\theta_0)^n\int_S\left(\int_Z g(s,z)\,\nu_\theta(dz)\right)\mu_{\theta_0}^{(n)}(ds)\,.$$

Using the fact that

$$\int_Z g(\cdot,z)\,\nu_\theta(dz) \in \mathcal{D}(S)\,,$$

the proof follows from the $\mathcal{D}(S)$-analyticity of μ_θ. $\square$

For C_p-spaces, we are able to characterize binary mappings that preserve C_p-analyticity.

Theorem 4.5.2 *Let $(S,\mathcal{S})$ and $(Z,\mathcal{Z})$ be measurable spaces equipped with upper bounds $\|\cdot\|_S$ and $\|\cdot\|_Z$, respectively, and let the product space $S \times Z$ be equipped with an upper bound $\|\cdot\|_{S\times Z}$. If*

- *for any $s \in S$, $z \in Z$, it holds*

$$\|(s,z)\|_{S\times Z} \le \|s\|_S + \|z\|_Z\,,$$

- *$\mu_\theta \in \mathcal{M}_1(S,\mathcal{S})$ is $C_p(S,\|\cdot\|_S)$-analytic and $\nu_\theta \in \mathcal{M}_1(Z,\mathcal{Z})$ is $C_p(Z,\|\cdot\|_Z)$-analytic,*

then $\mu_\theta \times \nu_\theta$ is $C_p(S \times Z,\|\cdot\|_{S\times Z})$-analytic. In particular, if, for $\theta_0 \in \Theta$, the Taylor series for μ_θ has domain of convergence $U_{\theta_0}^\mu$ and the Taylor series for ν_θ has domain of convergence $U_{\theta_0}^\nu$, then the domain of convergence of the Taylor series for the product measure $\mu_\theta \times \nu_\theta$ is $U_{\theta_0}^\mu \cap U_{\theta_0}^\nu$.

Furthermore, let $(R,\mathcal{R})$ be a measurable space equipped with upper bound $\|\cdot\|_R$ and let $h : S \times Z \to R$ be a measurable mapping, such that finite constants c_S and c_Z exist and for any $s \in S, z \in Z$:

$$\|h(s,z)\|_R \le c_S\,\|s\|_S + c_Z\,\|z\|_Z\,.$$

If $\mu_\theta \in \mathcal{M}_1(\mathcal{S}, \mathcal{S})$ is $C_p(S, \|\cdot\|_S)$-analytic and $\nu_\theta \in \mathcal{M}_1(\mathcal{Z}, \mathcal{Z})$ is $C_p(Z, \|\cdot\|_Z)$-analytic, then $(\mu_\theta \times \nu_\theta)^h$ is $C_p(R, \|\cdot\|_R)$-analytic.

Proof: The proof of the theorem follows from the same line of argument as the proof of Theorem 4.2.1 and is therefore omitted. $\square$

4.6 $\mathcal{D}$-Analyticity on $\mathcal{M}^{I \times J}$

$\mathcal{D}$-analyticity of a random matrix over the max-plus semiring is defined as follows.

Definition 4.6.1 We call $A_\theta \in \mathbb{R}_{\max}^{J \times I}$ $\mathcal{D}$-analytic on Θ if the distribution of A_θ is $\mathcal{D}$-analytic on Θ, that is, if

- all higher-order $\mathcal{D}$-derivatives of A_θ exist on Θ, and

- for all $\theta_0 \in \Theta$, an open neighborhood $U_{\theta_0} \subset \Theta$ of θ_0 exists such that for any $g \in \mathcal{D}$ and all $\theta \in U_{\theta_0}$:

$$\sum_{n=0}^{\infty} \frac{1}{n!} (\theta - \theta_0)^n \mathbb{E}_{\theta_0}[g^\tau(A^{(n)})] = \mathbb{E}_\theta[g(A)].$$

$\mathcal{D}$-analyticity of a random matrix A_θ implies analyticity of the expected value of $g(A_\theta)$ as function of θ for any $g \in \mathcal{D}$ as the following lemma shows.

Lemma 4.6.1 If $A_\theta \in \mathbb{R}_{\max}^{J \times I}$ is $\mathcal{D}$-analytic on Θ, $\mathbb{E}_\theta[g(A)]$ is analytic on Θ for all $g \in \mathcal{D}$. Furthermore, if, for $\theta_0 \in \Theta$, the domain of convergence of the Taylor series for A_θ is U_{θ_0}, then the domain of convergence of the Taylor series for $\mathbb{E}_\theta[g(A)]$ is also U_{θ_0}.

Proof: Let A_θ have distribution μ_θ. $\mathcal{D}$-analyticity of A_θ implies $\mathcal{D}$-analyticity of μ_θ. Hence, for any $g \in \mathcal{D}$ it holds

$$\sum_{n=0}^{\infty} \frac{1}{n!} (\theta - \theta_0)^n \mathbb{E}_{\theta_0}[g^\tau(A^{(n)})] = \sum_{n=0}^{\infty} \frac{1}{n!} (\theta - \theta_0)^n \frac{d^n}{d\theta^n} \mathbb{E}_{\theta_0}[g(A)]$$

$$= \sum_{n=0}^{\infty} \frac{1}{n!} (\theta - \theta_0)^n \frac{d^n}{d\theta^n} \int g(s)\, \mu_{\theta_0}(ds)$$

$$= \int g(s)\, \mu_\theta(ds)$$

$$= \mathbb{E}_\theta[g(A)],$$

for any $\theta \in U_{\theta_0}$. $\square$

We now establish sufficient conditions for $\mathcal{D}$-analyticity of some interesting classes of random variables.

Example 4.6.1 Let $A_\theta \in \mathbb{R}^{1 \times 1}$ be exponentially distributed with Lebesgue density $f_\theta(x) = \theta \exp(-\theta x)$, for $x \geq 0$, and let $\Theta = (0, \infty)$. From Example 4.5.1 it follows that A_θ is C_p-analytic on $(0, \infty)$. The Taylor series for $\mathbb{E}_\theta[g(A)]$ at θ_0 has at least domain of convergence $U_{\theta_0}(\delta) = (\delta, 2\theta_0 - \delta)$, for all $\theta_0 \in (0, \infty)$, with $0 < \delta < \theta_0$.

Example 4.6.2 Let A_θ be Bernoulli-(θ)-distributed on $\{D_1, D_2\} \subset \mathbb{R}_{\max}^{J \times I}$. By Example 4.5.2, A_θ is $\mathbb{R}^{(D_1, D_2)}$-analytic on $[0, 1]$ and the Taylor series for $\mathbb{E}_\theta[g(A)]$ at $\theta \in [0, 1]$ has the domain of convergence is $[0, 1]$.

For applications, we work with C_p-analyticity. The following corollary establishes an immediate consequence of the definition of C_p-analyticity that is useful in many practical situations for deciding whether a matrix over the max-plus semiring is analytic. Recall that $X_1, \ldots, X_m \in \mathbb{R}_{\max}$ is called the input of $A \in \mathbb{R}_{\max}^{J \times I}$ if the elements of A are measurable mappings of $(X_1, \ldots, X_m)$.

Corollary 4.6.1 Let $A_\theta \in \mathbb{R}_{\max}^{I \times J}$ have input $X_{\theta,1}, X_2 \ldots, X_m$, with $X_{\theta,1}, X_i \in \mathbb{R}_{\max}$, for $2 \leq i \leq m$, and let $X_{\theta,1}$ be $C_p(\mathbb{R}_{\max}, || \cdot ||_\oplus)$-analytic. If

- $X_{\theta,1}$ is stochastically independent of $(X_2, \ldots, X_m)$,

- $(X_2, \ldots, X_m)$ does not depend on θ and the elements X_i, $2 \leq i \leq m$, have finite p^{th} moment,

- a finite positive constant c exists, such that

$$||A(X_{\theta,1}, X_2, \ldots, X_m)||_\oplus \leq c\, ||(X_{\theta,1}, X_2, \ldots, X_m)||_\oplus\,,$$

then A_θ is $C_p(\mathbb{R}_{\max}^{I \times J}, || \cdot ||_\oplus)$-analytic and the domain of convergence of the Taylor series for A_θ and $X_{\theta,1}$ coincide.

Proof: To abbreviate the notation, set

$$h_g(x) = \mathbb{E}[\, g(A(X_{\theta,1}, X_2, \ldots, X_m)) \,|\, X_{\theta,1} = x\,]\,.$$

The mapping $h(\cdot)$ is independent of θ and lies in $C_p(\mathbb{R}_{\max}, || \cdot ||_\oplus)$. To see this, note that for $g \in C_p(\mathbb{R}_{\max}^{I \times J}, || \cdot ||_\oplus)$ it holds

$$
\begin{aligned}
|\, h_g(x)\,| &\leq \mathbb{E}[\, a_g + b_g\, (||A(X_{\theta,1}, X_2, \ldots, X_m)||_\oplus)^p \,|\, X_{\theta,1} = x\,] \\
&\leq \mathbb{E}[\, a_g + b_g\, c^p (||(x, X_2, \ldots, X_m)||_\oplus)^p \,|\, X_{\theta,1} = x\,] \\
&\leq a_g + b_g\, c^p \mathbb{E}\left[\left.\left(x + \sum_{i=2}^m |X_i|\right)^p \,\right|\, X_{\theta,1} = x\right].
\end{aligned}
$$

We have assumed that $X_2, \ldots, X_m$ have finite p^{th} moment and that they are stochastically independent of $X_{\theta,1}$. Hence, the expression on the right-hand side of the above inequality seen as a function in x lies in $C_p(\mathbb{R}_{\max}, || \cdot ||_\oplus)$.

$C_p(\mathbb{R}_{\max}, ||\cdot||_\oplus)$-analyticity of $X_{\theta,1}$ is equivalent to that of the distribution on $X_{\theta,1}$, denoted by μ_θ. Direct calculation yields:

$$\sum_{n=0}^{\infty} \frac{1}{n!}(\theta - \theta_0)^n \left.\frac{d^n}{d\theta^n}\right|_{\theta=\theta_0} \mathbb{E}_\theta[g(A)]$$

$$= \sum_{n=0}^{\infty} \frac{1}{n!}(\theta - \theta_0)^n \left.\frac{d^n}{d\theta^n}\right|_{\theta=\theta_0} \mathbb{E}[g(A(X_{\theta,1}, X_2, \ldots, X_m))]$$

$$= \sum_{n=0}^{\infty} \frac{1}{n!}(\theta - \theta_0)^n \left.\frac{d^n}{d\theta^n}\right|_{\theta=\theta_0} \int h_g(x)\,\mu_\theta(dx)$$

$$= \int h_g(x)\,\mu_\theta(dx)$$

$$= \mathbb{E}[g(A(X_{\theta,1}, X_2, \ldots, X_m))]$$

$$= \mathbb{E}_\theta[g(A)],$$

which concludes the proof. $\square$

Remark 4.6.1 *Note that Corollary 4.6.1 requires the first p moments of the input variables $X_2, \ldots, X_m$ to be finite. This is in contrast to Theorem 3.3.1 and Corollary 4.3.1, where no condition on the moments of the input variables is imposed. To see why a stronger condition is required in the setup of Corollary 4.6.1, let μ_θ denote the distribution of $X_{\theta,1}$ and let ν denote the distribution of $(X_2, \ldots, X_m)$. Following the line of proof for Theorem 3.3.1 and Corollary 4.3.1, respectively, we would calculate as follows:*

$$\sum_{n=0}^{\infty} \frac{1}{n!}(\theta - \theta_0)^n \left.\frac{d^n}{d\theta^n}\right|_{\theta=\theta_0} \mathbb{E}_\theta[g(A)]$$

$$= \sum_{n=0}^{\infty} \frac{1}{n!}(\theta - \theta_0)^n \left.\frac{d^n}{d\theta^n}\right|_{\theta=\theta_0} \int g(A(x,y))\,\mu_\theta(dx)\,\nu(dy)$$

$$= \sum_{n=0}^{\infty} \frac{1}{n!}(\theta - \theta_0)^n \int g(A(x,y))\,\mu_{\theta_0}^{(n)}(dx)\,\nu(dy).$$

However, for the proof of the corollary we then still have to show that

$$\sum_{n=0}^{\infty} \frac{1}{n!}(\theta - \theta_0)^n \int g(A(x,y))\,\mu_{\theta_0}^{(n)}(dx)\,\nu(dy) = \int g(A(x,y))\,\mu_{\theta_0}(dx)\,\nu(dy),$$

which can be guaranteed if

$$\int g(A(\cdot,y))\,\nu(dy) \in C_p(\mathbb{R}_{\max}, ||\cdot||_\oplus)$$

and if μ_θ is $C_p(\mathbb{R}_{\max}, ||\cdot||_\oplus)$-analytic at θ_0. It is exactly for this purpose that the moment condition on the input is required.

Example 4.6.3 *We revisit the situation in Example 4.3.2 (an open tandem queuing system whose interarrival times depend on θ introduced in Example 1.5.2). In accordance with Example 4.6.1, $\sigma_0(\theta, k)$ is $C_p(\mathbb{R}_{\max}, \| \cdot \|_\oplus)$-analytic. The condition on $\|A(\sigma_0(\theta, k), \sigma_1(k), \ldots, \sigma_J(k))\|_\oplus$ in Corollary 4.6.1 is satisfied, hence, $A(k)$ is $C_p(\mathbb{R}_{\max}^{J+1 \times J+1}, \| \cdot \|_\oplus)$-analytic.*

The following corollary summarizes our analysis by showing that C_p-analyticity is preserved under finite $\otimes$-multiplication or $\oplus$-addition.

Corollary 4.6.2 (Product rule of C_p-analyticity over $\mathbb{R}_{\max}$) *If $A, B \in \mathbb{R}_{\max}^{J \times J}$ are stochastically independent and $C_p(\mathbb{R}_{\max}^{J \times J}, \| \cdot \|_\oplus)$-analytic on Θ, then $A \otimes B$ and $A \oplus B$ are C_p-analytic on Θ.*

In particular, if, for $\theta_0 \in \Theta$, the Taylor series for A at θ_0 has domain of convergence $U_{\theta_0}^A$ and the Taylor series for B at θ_0 has domain of convergence $U_{\theta_0}^B$, then the domain of convergence of the Taylor series for $A \oplus B$ at θ_0, respectively $A \otimes B$, is $U_{\theta_0}^A \cap U_{\theta_0}^B$.

Proof: By Lemma 1.6.1, we may apply Theorem 4.5.2 where we take $\| \cdot \|_\oplus$ as upper bound and the $\otimes$-product and the $\oplus$-sum of matrices, respectively, as mapping h. $\square$

An immediate consequence of Corollary 4.6.2 is that if $A(k) \in \mathbb{R}_{\max}^{J \times J}$ is an i.i.d. sequence of C_p-analytic random matrices on Θ, then

$$x(k+1) = A(k) \otimes x(k), \quad k \geq 0,$$

where $x(0) = x_0$, is C_p-analytic on Θ for any k. Furthermore, Lemma 4.6.1 implies that $\mathbb{E}_\theta[g(x(k+1)]$ is analytic on Θ for any $g \in C_p$ and $k \in \mathbb{N}$. In addition to that if, for $\theta_0 \in \Theta$, $A(0)$ has domain of convergence $U_{\theta_0}^A$, then $x(k+1)$ has domain of convergence $U_{\theta_0}^A$.

Chapter 5

Taylor Series Expansions

This chapter addresses analyticity of performance measures, say $J(\theta)$, such as completion times, waiting times or the throughput (that is, the inverse Lyapunov exponent), of max-plus linear systems. Specifically, this chapter studies Taylor series expansions for $J(\theta)$ with respect to θ. First results on analyticity of stochastic networks were given by Zazanis [104] who studied analyticity of performance measures of stochastic networks fed by a Poisson arrival stream with respect to the intensity of the arrival stream. Baccelli and Schmidt [17] considered the case in which the network is max-plus linear. Their approach was further developed in [15] and [16]. For applications of their results to waiting times, see [71] and [97]. The results mentioned above are restricted to the case of open networks, where θ is the intensity of the arrival stream. Taylor series expansions for closed networks are addressed in [7] and [8]. Strictly speaking, the aforementioned papers study Maclaurin series, that is, they only consider Taylor series at zero.

In this chapter we establish sufficient conditions for analyticity of max-plus linear stochastic systems. In particular,

1. for open systems, we do not require the arrival stream to be of Poisson type;

2. our analysis applies to open and closed systems as well;

3. the parameter with respect to which the Taylor series is developed, may be a parameter of the distribution of any input variable of the max-plus system;

4. at any point of analyticity we establish lower bounds for the domain of convergence of the Taylor series, which is in contrast to the study of Maclaurin series predominant in the literature.

In some special cases the obtained derivatives can be calculated analytically and in the general case the formulae obtained have a simple interpretation as unbiased estimation algorithm.

Our approach is closely related to Markov chain analysis. Cao obtained in [30] a Maclaurin series for steady-state performance functions of finite-state Markov chains. This result has been extend to general state space Markov chains in [64]. Although the types of systems considered in the aforementioned papers are different from the ones treated here, the approach is closely related to ours.

The chapter is organized as follows. Section 5.1 studies Taylor series expansions for finite horizon performance indices. Section 5.2 deals with Taylor series expansions for random horizon performance indices. Section 5.3 is devoted to Taylor series expansions for the Lyapunov exponent. Finally, we address Taylor series expansions for stationary waiting times in Section 5.4. Throughout this chapter we will equip $\mathbb{R}_{max}$ with upper bound $\|\cdot\|_\oplus$ and we simply write C_p instead of $C_p(\mathbb{R}_{max}^{J\times I}, \|\cdot\|_\oplus)$.

5.1   Finite Horizon Experiments

In this section we establish conditions under which deterministic horizon performance indices of max-plus linear systems can be written as Taylor series. A precise description of the problem is the following:

The Deterministic Horizon Problem: We study sequences $x(k) = x_\theta(k)$, $k \in \mathbb{N}$, following

$$x(k+1) = A(k) \otimes x(k) \oplus B(k), \quad k \geq 0.$$

with $x(0) = x_0$, $A(k) = A_\theta(k) \in \mathbb{R}_{max}^{J\times J}$, $B(k) = B_\theta(k) \in \mathbb{R}_{max}^J$ and $\theta \in \Theta \subset \mathbb{R}$. For a given performance function $g : \mathbb{R}_{max}^J \to \mathbb{R}$, we seek sufficient conditions for the analyticity of

$$\mathbb{E}_\theta[g(x(k+1))|x(0) = x_0]. \qquad (5.1)$$

These conditions will depend on the type of performance function and the particular way in which the matrix $A(k)$, respectively the vector $B(k)$, depends on θ.

The section is organized as follows. Section 5.1.1 states the general result on Taylor series expansions for finite max-plus performance indices. In Section 5.1.2, we address analyticity of transient waiting times in non-autonomous systems (that is, open queuing networks). Finally, in Section 5.1.3, we present a scheme for approximating performance characteristics of max-plus linear systems, called *variability expansion*.

5.1.1   The General Result

Corollary 4.6.2 provides the means to solve the deterministic horizon problem. The precise statement reads as follows.

Corollary 5.1.1 *If $A(k) \in \mathbb{R}_{\max}^{J \times J}$ and $B(k) \in \mathbb{R}_{\max}^{J}$ $(0 \leq k)$ are two i.i.d. sequences of random matrices and vectors, respectively, which are C_p-analytic on Θ and mutually independent, then*

$$x(k+1) = A(k) \otimes x(k) \oplus B(k), \quad k \geq 0,$$

with $x(0) = x_0$, is C_p-analytic on Θ for all k. Moreover, $\mathbb{E}_\theta[g(x(k+1))]$ is analytic on Θ for all $g \in C_p$.

If, for $\theta_0 \in \Theta$, the Taylor series for $A(0)$ at θ_0 has domain of convergence $U_{\theta_0}^A$ and the Taylor series for $B(0)$ at θ_0 has domain of convergence $U_{\theta_0}^B$, then the Taylor series for $x(k+1)$ at θ_0 has domain of convergence $U_{\theta_0}^A \cap U_{\theta_0}^B$.

Proof: Analyticity of $x(k+1)$ follows from Corollary 4.6.2 via induction with respect to k; and analyticity of $\mathbb{E}_\theta[g(x(k+1))]$ is an immediate consequence of Lemma 4.6.1.$\square$

Corollary 5.1.1 is illustrated with the following example.

Example 5.1.1 *We revisit the situation in Example 4.6.3. Let $p \in \mathbb{N}$. By Corollary 4.6.1, the transition matrix $A(k)$ is C_p-analytic on $(0, \infty)$. We now make the additional assumption that the service times and the interarrival times are mutually independent and that interarrival times as well as the service times at the servers are identical distributed. The sequence $\{A(k)\}$ is thus i.i.d. Therefore, $x(k+1)$, with $x(k+1) = A(k) \otimes x(k)$, for $k \geq 0$, is C_p-analytic on $(0, \infty)$ and, for $g \in C_p$, the Taylor series for $\mathbb{E}_\theta[g(x(k+1))]$ at $\theta_0 \in (0, \infty)$ has at least domain of convergence $U_{\theta_0}(\delta) = (\delta, 2\theta_0 - \delta)$, for $\theta_0 > \delta > 0$.*

Note that we cannot apply our theory when we consider recurrence relation (1.27) on page 26 in Example 1.5.2, since $B(k) \otimes \tau(k+1)$ and $\hat{A}(k)$ fail to be stochastically independent. However, we can conveniently work with the homogeneous variant of the model and avoid this problem.

Recall that for any C_p-differentiable matrix A the largest order of a significant C_p-derivative of A is denoted by $s(A)$.

Theorem 5.1.1 *Let $A(k)$ be an i.i.d. sequence of matrices in $\mathbb{R}_{\max}^{J \times J}$ that are $(h+1)$ times C_p-differentiable on a neighborhood $U_{\theta_0} \subset \Theta$ of $\theta_0 \in \Theta$. Then, for any $g \in C_p$:*

$$\mathbb{E}_{\theta_0 + \Delta}\left[g\left(\bigotimes_{k=0}^{m} A(k) \otimes x_0\right)\right]$$

$$= \sum_{n=0}^{h} \frac{\Delta^n}{n!} \mathbb{E}_{\theta_0}\left[g^\tau\left(\left(\bigotimes_{k=0}^{m} A(k)\right)^{(n)} \otimes x_0\right)\right] + r_{h+1}(\theta_0, \Delta),$$

for $\theta_0 + \Delta \in U_{\theta_0}$, where $r_{h+1}(\theta_0, \Delta) = 0$ for $h > (m+1)\, s(A(0))$. Provided that $x_0 = \mathbf{e}$, it holds for $h \leq (m+1)\, s(A(0))$ that

$$|r_{h+1}(\theta_0, \Delta)| \leq R_{h+1}(\theta_0, \Delta)$$
$$\stackrel{\text{def}}{=} \frac{1}{h!} \int_{\theta_0}^{\theta_0 + \Delta} (\theta_0 + \Delta - t)^h \, \mathbb{E}_t \left[B_{g,m,\{A(k)\}}(h+1, p) \right] dt \, .$$

Proof: The product $\bigotimes_{k=0}^{m} A(k)$ is $(h+1)$ times C_p-differentiable, see Lemma 4.4.1. Hence,

$$\mathbb{E}_{\theta_0 + \Delta} \left[g \left(\bigotimes_{k=0}^{m} A(k) \otimes x_0 \right) \right]$$

can be written as Taylor polynomial of degree h the remainder term of which is given through

$$\frac{1}{h!} \int_{\theta_0}^{\theta_0 + \Delta} (\theta_0 + \Delta - t)^h \, \frac{d^{h+1}}{d\theta^{h+1}} \bigg|_{\theta = t} \mathbb{E}_\theta \left[g \left(\bigotimes_{k=0}^{m} A(k) \otimes x_0 \right) \right] dt \, ,$$

see equation (G.2) in Section G.4 in the Appendix. In accordance with Lemma 4.4.2, the $(h+1)^{st}$ C_p-derivative of the above product is bounded by $B_{g,m,\{A(k)\}}(h+1, p)$ and inserting this bound into the above expression for the remainder term concludes the proof of the theorem. $\square$

Example 5.1.2 *In the Bernoulli case, $A(k)$ is C_p-analytic on $[0,1]$, for $k \in \mathbb{N}$ and any $p \in \mathbb{N}$. Provided that $\{A(k)\}$ constitutes an i.i.d. sequence,*

$$\mathbb{E}_\theta \left[g \left(\bigotimes_{k=0}^{m} A(k) \otimes x_0 \right) \right]$$

is analytic on $[0,1]$ for all $g \in C_p$ and all $x_0 \in \mathbb{R}^J$. The domain of convergence of the Taylor series is $[0,1]$. When we consider the Taylor series at zero, the first terms of the series are given in Example 4.4.1.

Consider the system in Example 1.5.5. Let $\sigma = 1$, $\sigma' = 2$, then $\|D_1\|_\oplus = \|D_2\|_\oplus = \max(\sigma, \sigma') = 2$ and the second part of Lemma 4.4.2 yields

$$\left| \frac{d^{h+1}}{d\theta^{h+1}} \bigg|_{\theta = t} \mathbb{E}_\theta \left[g \left(\bigotimes_{k=0}^{m} A(k) \otimes x_0 \right) \right] \right| \leq \sum_{l \in \mathcal{L}[0,m;h+1]} (h+1)! \, 2^{h+1} \left(a_g + b_g (m+1)^p \, 2^p \right),$$

where we have used the fact that (i) only the first order C_p-derivative of $A(k)$ is significant, (ii) $c_{A(k)} = 1$, and (iii) $\|A\|_\oplus = 2$. As shown in Section G.5 in the Appendix, it holds that

$$\sum_{l \in \mathcal{L}[0,m;h+1]} (h+1)! = (m+1)^{h+1}$$

and the remainder term of the Taylor polynomial of degree h is for $h \leq m + 1$ given by

$$\frac{1}{h!} \int_{\theta_0}^{\theta_0 + \Delta} (\theta_0 + \Delta - t)^h \, (m+1)^{h+1} \, 2^{h+1} (a_g + b_g (m+1)^p 2^p) \, dt \,,$$

$$= 2^{h+1} \frac{(m+1)^{h+1}}{(h+1)!} \, (a_g + b_g (m+1)^p 2^p) \, \Delta^{h+1}$$

and the remainder term equals zero for $h > m + 1$.

5.1.2 Analyticity of Waiting Times

In this section we consider open max-plus linear systems like the one in Example 1.5.3; we use the notation introduced in Section 1.5.2. More precisely, we consider max-plus linear models for the beginning of service times in queuing networks. These models typically have the form

$$x(k+1) = A(k) \otimes x(k) \oplus B(k) \otimes \tau(k+1) \,, \quad k \geq 0 \,, \tag{5.2}$$

with $x(0) = x_0 \in \mathbb{R}^J$, $\{A(k)\}$ a sequence of i.i.d. matrices in $\mathbb{R}_{\max}^{J \times J}$, $\{B(k)\}$ a sequence of i.i.d. vectors in $\mathbb{R}_{\max}^J$ and

$$\tau(k) = \sum_{i=1}^{k} \sigma_0(i) \,,$$

where $\tau(k)$ denotes the kth arrival epoch (recall that $\sigma_0(k)$ denotes the kth interarrival time). Provided that the system is initially empty, the time the kth customer arriving at the network spends in the system until beginning of her/his service at station j is given by

$$W_j(k) = x_j(k) - \tau(k) \,, \quad k \geq 1 \,.$$

Recall, that if $x(k)$ in (5.2) models the vector of kth departure times at the stations, then $W_j(k)$ defined above represents the time spend by the kth customer arriving at the system until her/his departure from station j.

For our analysis it is more convenient to include the source into the state-vector, that is, we consider

$$x(k+1) = A(k) \otimes x(k) \,, \quad k \geq 0 \,, \tag{5.3}$$

where $\{A(k)\}$ is an appropriately defined sequence of i.i.d. matrices in $\mathbb{R}_{\max}^{J+1 \times J+1}$ and $x(0) = e$, see Section 1.4.3. From this we recover $(W(k))_j$ through

$$W_j(k) = x_j(k) - x_0(k) \,, \quad 1 \leq j \leq J \,. \tag{5.4}$$

Note that, for $x \in \mathbb{R}^J$, $\|x\|_\oplus = \max_j |x_j|$. To unify notation, we write $\|\cdot\|_\oplus$ instead of $|\cdot|$ when $J = 1$. For $g \in C_p([0,\infty)^J, \|\cdot\|_\oplus)$, we define the mapping $g_W : [0,\infty) \to \mathbb{R}$ by

$$g_W(x_0, \ldots, x_J) \stackrel{\text{def}}{=} g(x_1 - x_0, \ldots, x_J - x_0) \,.$$

Note that if $g \in C_p([0,\infty)^J, ||\cdot||_\oplus)$, then $g_W \in C_p([0,\infty)^{J+1}, ||\cdot||_\oplus)$. This stems from the fact that, for $(x_0, \ldots, x_J) \in [0,\infty)^{J+1}$:

$$||g_W(x_0, \ldots, x_J)||_\oplus = ||g(x_1 - x_0, \ldots, x_J - x_0)||_\oplus$$
$$\leq a_g + b_g \left(||(x_1 - x_0, \ldots, x_J - x_0)||_\oplus \right)^p$$
$$\leq a_g + b_g \left(||(x_0, \ldots, x_J)||_\oplus \right)^p .$$

The main assumption we need for the following is:

(W) The matrix $A(k)$ in (5.3) is a.s. regular and any finite element is non-negative. The initial state is $x_0 = \mathbf{e}$.

Condition **(W)** implies $x(k) \in [0,\infty)^{J+1}$, for $k \geq 0$. Hence, for any $g \in C_p([0,\infty)^J, ||\cdot||_\oplus)$ it holds that

$$g(W(k)) = g_W(x(k)) \quad \text{and} \quad g_W \in C_p([0,\infty)^{J+1}, ||\cdot||_\oplus) . \tag{5.5}$$

By Corollary 4.6.1 together with Corollary 5.1.1, we obtain the following result for waiting times.

Lemma 5.1.1 *Let $x(k)$ and $W(k)$ be defined as in (5.3) and (5.4), respectively, and assume that condition* **(W)** *is satisfied. Let $A(k) = A_\theta(k)$ have input $X_{\theta,1}(k), X_2(k) \ldots, X_m(k)$, with $X_{\theta,1}(k), X_i(k) \in \mathbb{R}_{\max}$, for $2 \leq i \leq m$, and let $X_{\theta,1}(k)$ be $C_p(\mathbb{R}_{\max}, ||\cdot||_\oplus)$-analytic. If*

- $\{(X_{\theta,1}(k), X_2(k), \ldots, X_m(k))\}$ *is an i.i.d. sequence,*

- $X_{\theta,1}(k)$ *is stochastically independent of $(X_2(k), \ldots, X_m(k))$,*

- $(X_2(k), \ldots, X_m(k))$ *does not depend on θ and the entries $X_i(k)$, $2 \leq i \leq m$, have finite p^{th} moment,*

- *a finite constant c exists such that for any $x \in \mathbb{R}_{\max}^m$*

$$||A(x_1, \ldots, x_m)||_\oplus \leq c ||(x_1, \ldots, x_m)||_\oplus ,$$

then $W(k)$ is $C_p([0,\infty)^J, ||\cdot||_\oplus)$-analytic. In particular, if, for $\theta_0 \in \Theta$, the Taylor series for $X_{\theta,1}(k)$ has domain of convergence U_{θ_0}, then the domain of convergence of the Taylor series for $W(k)$ is U_{θ_0} as well.

We illustrate Lemma 5.1.1 with the following example.

Example 5.1.3 *Consider the open queuing system in Example 1.5.2 and suppose that for some $j \in \{0, \ldots, J\}$ service time $\sigma_j(\theta, k)$ depends on θ, whereas all other service times and (in case $j = 0$ the interarrival times) are independent of θ and have finite p^{th} moment. Furthermore, assume that $\{\sigma_j(k)\}$ is i.i.d. for any j and that the sequences are mutually independent. We consider the homogeneous model for departure times from the queues as given in (1.25). The* *matrix $A_\theta(k)$ given in (1.24) has input $\sigma_j(\theta, k+1)$ and $\sigma_i(k+1)$, for $0 \leq i \leq J$, with $i \neq j$. Hence,*

$$A_\theta(k) = A(\sigma_0(k+1), \ldots, \sigma_{j-1}(k+1),$$
$$\sigma_j(\theta, k+1), \sigma_{j+1}(k+1), \ldots, \sigma_J(k+1)).$$

Let $\sigma_j(\theta, k)$ be exponentially distributed with mean $1/\theta$. In accordance with Example 4.6.1, $\sigma_j(\theta, k)$ is $C_p(\mathbb{R}_{\max}, \| \cdot \|_\oplus)$-analytic. The condition on $\|A(\sigma_0(k), \ldots, \sigma_{j-1}(k), \sigma_j(\theta, k), \sigma_{j+1}(k), \ldots, \sigma_J(k))\|_\oplus$ in Lemma 5.1.1 is satisfied. $W(k)$ is thus $C_p([0, \infty)^J, \| \cdot \|_\oplus)$-analytic and, for any $g \in C_p([0, \infty)^J, \| \cdot \|_\oplus)$, the Taylor series for $\mathbb{E}_\theta[g(W(k))]$ at θ_0 has at least domain of convergence $(\delta, 2\theta_0 - \delta)$, for any $\theta_0 \in (0, \infty)$ with $0 < \delta < \theta_0$, see Example 4.6.1.

If $j = 0$, then the arrival process is Poisson with rate θ and we obtain a Taylor series expansion with respect to the rate of the Poisson process. Under additional assumptions on the sequences $\{A(k) : k \geq 0\}$ Baccelli et al. show in [15] that an analytic continuation of $\mathbb{E}_\theta[W_j(k)]$ to the complex plane exists which is analytic in zero. Moreover, provided that the service times are deterministic, they explicitly calculate the remainder term of this series expansion.

Remark 5.1.1 *Lemma 5.1.1 applies to functions g that evaluate several waiting times simultaneously. For example, taking $g_W(x_0, \ldots, x_J) = g(x_i - x_0, x_j - x_0)$, for $i \neq j$, leads to the evaluation of the correlation between $W_i(k)$ and $W_j(k)$.*

Lemma 5.1.1 applies to general renewal processes and thereby extends the result in [15], where analyticity of $\mathbb{E}_\theta[W_j(k)]$ is shown under the assumption that the arrival process is a Poisson process with intensity θ.

In the remainder of this section, we give an explicit representation of the Taylor series for $W(m)$, for $m \geq 0$. Let the conditions in Lemma 5.1.1 be in force; in particular, $X_{\theta,1}(k)$ is the only input variable that depends on θ. In order to simplify the notation we write $A(k) = A_\theta(k) = A(k, X_{\theta,1}(k))$. We assume that $\{A(k)\}$ constitutes an i.i.d. sequence. For $l \in \mathcal{L}[0, m-1; n]$ and $i \in \mathcal{I}[l]$ let $x^{(l,i)}(k)$ follow the recurrence relation

$$x^{(l,i)}(k+1) = A(k, X_{\theta,1}^{(l_k, i_k)}(k)) \otimes x^{(l,i)}(k), \quad 0 \leq k < m,$$

with $x^{(l,i)}(0) = e$, and, for $1 \leq j \leq J$, define the waiting times by

$$W_j^{(l,i)}(k) = x_j^{(l,i)}(k) - x_0^{(l,i)}(k), \quad 1 \leq k \leq m,$$

c.f. equation (5.4). In words, for generating $x^{(l,i)}(k+1)$ replace all occurrences of $X_{\theta,1}(k)$ in $A(k)$ by $X_{\theta,1}^{(l_k, i_k)}(k)$.

Let $X_{\theta,1}(k)$ be C_p-analytic and denote the domain of convergence of the Taylor series for $X_{\theta,1}(k)$ at θ_0 by U_{θ_0}. Hence, for any $g \in C_p$ and any $\theta \in U_{\theta_0}$

we obtain:

$$\mathbb{E}_\theta[g(W(m))] = \sum_{n=0}^{\infty} \sum_{l \in \mathcal{L}[0,m-1;n]} \frac{(\theta - \theta_0)^n}{l_0! l_1! \ldots l_{m-1}!}$$
$$\times \sum_{i \in \mathcal{I}[l]} \prod_{k=0}^{m-1} c_{X_{\theta_0,1}}^{(l_k)} \mathbb{E}_{\theta_0}\Big[g(W^{(l,i)}(m)) - g(W^{(l,i^-)}(m))\Big].$$

For example, taking $X_{\theta,1}(k)$ to be exponentially distributed with mean $1/\theta$, we obtain in accordance with Example 4.1.4:

$$c_{X_{\theta_0,1}}^{(l_k)} = \frac{l_k!}{(\theta_0)^{l_k}}, \quad l = (l_0, \cdots, l_{m-1}) \in \mathcal{L}[0, m-1;n].$$

Inserting the above equality into the Taylor series for $\mathbb{E}_\theta[g(W(m))]$ yields

$$\mathbb{E}_\theta[g(W(m))] = \sum_{n=0}^{\infty} \sum_{l \in \mathcal{L}[0,m-1;n]} \frac{(\theta - \theta_0)^n}{(\theta_0)^n}$$
$$\times \sum_{i \in \mathcal{I}[l]} \mathbb{E}_{\theta_0}\Big[g(W^{(l,i)}(m)) - g(W^{(l,i^-)}(m))\Big].$$

It is worth noting that the complexity of the resulting Taylor series is independent of g. This stems from the fact that the weak approach works essentially uniformly for a class of performance functions and results are independent of any particular choice of performance function. This improves the result in [5], where expansions for second order moments were given that have considerably higher complexity than the expansions for first moments.

5.1.3 Variability Expansion

In this section we discuss an approach to performance evaluation of finite horizon performance indicators of stochastic max-plus linear systems, introduced in [63], called *variability expansion*. For applications of this technique to model predictive control of max-plus systems see [99, 100]. The basic setup for variability expansion is as follows. Let $\{A(k)\}$ be an i.i.d. sequence of square matrices over the max-plus algebra and consider the max-plus recurrence relation

$$x(k+1) = A(k) \otimes x(k), \quad k \geq 0, \tag{5.6}$$

with $x(0) = x_0$. Our goal is to evaluate $\mathbb{E}[g(x(m))]$ for fixed m and given performance indicator $g \in C_p$. To this end, we introduce a parameter θ and replace with probability $1 - \theta$ the random matrix $A(k)$ in the above recurrence relation by its mean. Parameter θ allows controlling the level of randomness in the system: letting θ go from 0 to 1 increases the level of stochasticity in the system. For example, $\theta = 0$ represents a completely deterministic system, whereas $\theta = 1$ represents the (fully) stochastic system (that is, the original one). Denote by $\{x_\theta(k)\}$ the θ-version of $\{x(k)\}$, for $\theta \in [0,1]$. For $\theta = 1$, it holds that $\mathbb{E}_1[g(x(m))] = \mathbb{E}[g(x(m))]$. In order to evaluate $\mathbb{E}_1[g(x(m))]$, we consider the Taylor series for $\mathbb{E}_\theta[g(x(m))]$ at $\theta = 0$. For the sake of sake implicity, we illustrate our approach with the waiting time in the $G/G/1$ queue.

Consider a $G/G/1$ queue with i.i.d. interarrival times $\{\sigma_0(k)\}$ and i.i.d. service times $\{\sigma_1(k)\}$. Denote by σ_0 the mean interarrival time and by σ_1 the mean service time, and assume that $\rho \stackrel{\text{def}}{=} \sigma_1/\sigma_0 < 1$. The system is initially empty and the waiting time of the kth customer, denoted by $W(k)$, follows:

$$W(k+1) = \sigma_1(k) \otimes (-\sigma_0(k+1)) \otimes W(k) \oplus 0$$
$$= \max(\sigma_1(k) - \sigma_0(k+1) + W(k), 0), \quad k \geq 0,$$

with $W(0) = 0$ and $\sigma_1(0) = 0$, see Example 1.5.4. We write the above equation as a homogeneous equation, like (5.6). To this end, we set, for $k \geq 1$:

$$A(k) = \begin{pmatrix} \sigma_1(k) - \sigma_0(k+1) & 0 \\ \varepsilon & 0 \end{pmatrix}.$$

Remark 5.1.2 *There are numerous ways of arriving at a homogeneous representation for $W(k)$. For example, let $x(k)$ model the kth beginning of service at the station, then*

$$x(k+1) = \sigma_1(k) \otimes x(k) \oplus \tau(k+1), \quad k \geq 0,$$

with $x(0) = 0$ and $\sigma_1(0) = 0$, where $\tau(k)$ denotes the arrival epoch of the kth customer. Including the source into the state-vector, we arrive at the equation

$$\begin{pmatrix} x_0(k+1) \\ x_1(k+1) \end{pmatrix} = \begin{pmatrix} \sigma_0(k+1) & \varepsilon \\ \sigma_0(k+1) & \sigma_1(k) \end{pmatrix} \otimes \begin{pmatrix} x_0(k) \\ x_1(k) \end{pmatrix},$$

where $x_1(k)$ is the time of the kth beginning of service at the station and $x_0(k)$ the time of the kth arrival of a customer. As in the previous section, the waiting time of the kth customer equals $x_1(k) - x_0(k)$ and it holds that $W(k) = x_1(k) - x_0(k)$, $k \geq 1$, see (5.4) on page 183.

Let $w(0) = (0, 0)$ and set

$$w(k+1) = A(k) \otimes w(k), \quad k \geq 0,$$

then $w(k) = (W(k), 0)$. In words, the first component of $w(k)$ is the actual waiting time of the kth customer. Set $g_W(w(k)) = g(W(k))$, then

$$g(W(k)) = g_W(w(k)) = g_W \left(\bigotimes_{j=0}^{k-1} A(j) \otimes w(0) \right), \quad k \geq 1.$$

The deterministic variant of the system is obtained by replacing the random entries of $A(k)$ by their means, that is, by considering the transition matrix

$$\mathbf{A} = \begin{pmatrix} \sigma_1 - \sigma_0 & 0 \\ \varepsilon & 0 \end{pmatrix}.$$

In order to construct a version of $\{W(k)\}$ that combines deterministic transitions according to $\mathbf{A}$ with random ones according to $\{A(k)\}$, proceed as follows. Let $A(k)$ have distribution μ and recall that the Dirac measure in x is denoted by δ_x. Let $D_\theta(k) \in \mathbb{R}^{2 \times 2}_{\max}$, for $k \in \mathbb{N}$, be an i.i.d. sequence with distribution $\theta\mu + (1 - \theta)\delta_{\mathbf{A}}$. In words, with probability θ, $D_\theta(k)$ behaves like $A(k)$, whereas, with probability $(1 - \theta)$, $D_\theta(k)$ is equal to $\mathbf{A}$. For $\theta \in [0, 1]$, set

$$w_\theta(k + 1) = D_\theta(k) \otimes w_\theta(k), \quad k \geq 0,$$

with $w_\theta(0) = (0, 0)$. We call the transition from $w_\theta(k)$ to $w_\theta(k+1)$ *deterministic* if $D_\theta(k) = \mathbf{A}$ and *stochastic* otherwise. We write $\mathbb{E}_\theta[g_W(w(m))]$ to indicate that the θ-version is considered. For fixed $m > 0$, the performance characteristic of the transient waiting time of the θ-version is thus given by $\mathbb{E}_\theta[g_W(w(m))]$, where $\mathbb{E}[g_W(w(m))] = \mathbb{E}_1[g_W(w(m))]$ and

$$\mathbb{E}_0[g_W(w(m))] = g_W(\mathbf{A}^m \otimes w(0)) .$$

Let $p \geq 0$ be such that, for any $g \in C_p$ and $k < m$, $\mathbb{E}[g(A(k))]$ and $g(\mathbf{A})$ are finite. Then, for any $k < m$,

$$\mathbb{E}[g(D_\theta(k))] = \theta\mathbb{E}[g(A(k))] - (1 - \theta)\, g(\mathbf{A}) ,$$

which implies C_p-analyticity of $D_\theta(k)$ on $[0, 1]$. In particular, for any $k < m$, $D'_\theta(k) = (1, A(k), \mathbf{A})$ and all higher order C_p-derivatives of $D_\theta(k)$ are not significant, in symbols: $s(D_\theta(k)) = 1$.

Applying Corollary 5.1.1 with $B(k) = (\varepsilon, \cdots, \varepsilon)$ yields that $\bigotimes_{k=0}^{m-1} D_\theta(k)$ is C_p-analytical. Recall that we have assumed that $g \in C_p$. Following the train of thought put forward in the previous section, this implies $g_W \in C_p$, see (5.5). Hence, for any $\theta \in [0, 1]$, the Taylor series for $\mathbb{E}_\theta[g_W(w(m))]$ at θ has domain of convergence $[0, 1]$. The nth derivative at a boundary point has to be understood as a sided limit; specifically, set

$$\lim_{\theta \downarrow 0} \frac{d^n}{d\theta^n} \mathbb{E}_\theta[g_W(w(m))] = \frac{d^n}{d\theta^n} \mathbb{E}_0[g_W(w(m))]$$

and

$$\lim_{\theta \uparrow 1} \frac{d^n}{d\theta^n} \mathbb{E}_\theta[g_W(w(m))] = \frac{d^n}{d\theta^n} \mathbb{E}_1[g_W(w(m))] ,$$

then $\mathbb{E}[g(W(m))] = \mathbb{E}_1[g_W(w(m))]$, the 'true' expected performance characteristic of the mth waiting time, is given by

$$\mathbb{E}[g(W(m))] = \sum_{n=0}^{h} \frac{1}{n!} \frac{d^n}{d\theta^n} \mathbb{E}_0[g_W(w(m))] + R_{h+1}(m)$$

where, for $h < m$,

$$R_{h+1}(m) = \frac{1}{h!} \int_0^1 (1 - t)^h \left. \frac{d^{h+1}}{d\theta^{h+1}} \right|_{\theta=t} \mathbb{E}_\theta[g_W(w(m))]\, dt$$

and $R_{h+1}(m) = 0$ otherwise, see Theorem 5.1.1.

Let $n \leq m$. For $0 \leq l_1 < l_2 < \cdots < l_n \leq m - 1$, let $\eta(l_1, \ldots, l_n) \in \{0,1\}^m$ denote the vector with entry 1 at position l_k, $1 \leq k \leq n$, and zero otherwise. This leads to the following expression for the nth order derivative of $\mathbb{E}_\theta[g_W(w(m))]$

$$\frac{d^n}{d\theta^n}\mathbb{E}_\theta[g_W(w(m))]$$

$$= n! \sum_{l_1=0}^{m-n} \sum_{l_2=l_1+1}^{m-n+1} \cdots \sum_{l_n=l_{n-1}+1}^{m-1}$$

$$\sum_{i\in\mathcal{I}[\eta(l_1,\ldots,l_n)]} \mathbb{E}\left[g_W\left(\left(\bigotimes_{k=0}^{m-1} D_\theta(k)\right)^{(\eta(l_1,\ldots,l_n),i)} \otimes w(0)\right)\right]$$

$$- \sum_{i\in\mathcal{I}[\eta(l_1,\ldots,l_n)]} \mathbb{E}\left[g_W\left(\left(\bigotimes_{k=0}^{m-1} D_\theta(k)\right)^{(\eta(l_1,\ldots,l_n),i^-)} \otimes w(0)\right)\right],$$

whereas the nth derivative is zero for $n > m$.

Letting θ tend to zero, those $D_\theta(k)$ for which $l_k = 0$ converge in total variation to $\mathbf{A}$. In the following, explicit representations of the first three derivatives of $\mathbb{E}_\theta[g_W(w(m))]$ at $\theta = 0$ are given.

For $0 \leq j < m$, set

$$V_g(m; j) = \mathbb{E}\left[g_W\left(\mathbf{A}^{m-j-1} \otimes A(j) \otimes \mathbf{A}^j \otimes w(0)\right)\right]$$

and

$$V_g(m) = g_W\left(\mathbf{A}^m \otimes w(0)\right).$$

Then

$$\frac{d}{d\theta}\mathbb{E}_0[g_W(w(m))] = \sum_{j=0}^{m-1}\left(V_g(m; j) - V_g(m)\right). \tag{5.7}$$

In the same vein, set, for $0 \leq j_1 < j_2 < m$,

$$V_g(m; j_1, j_2) = \mathbb{E}\left[g_W\left(\mathbf{A}^{m-j_2-1} \otimes A(j_2) \otimes \mathbf{A}^{j_2-j_1-1} \otimes A(j_1) \otimes \mathbf{A}^{j_1} \otimes w(0)\right)\right],$$

then

$$\frac{d^2}{d\theta^2}\mathbb{E}_0[g_W(w(m))]$$

$$= 2\sum_{j_1=0}^{m-2}\sum_{j_2=j_1+1}^{m-1}\left(V_g(m; j_1, j_2) + V_g(m) - V_g(m; j_1) - V_g(m; j_2)\right).$$

For the third element set, for $0 \leq j_1 < j_2 < j_3 < m$,

$$V_g(m; j_1, j_2, j_3) = \mathbb{E}\Big[g_W\Big(\mathbf{A}^{m-j_3-1}\otimes A(j_3) \otimes \mathbf{A}^{j_3-j_2-1}$$

$$\otimes A(j_2) \otimes \mathbf{A}^{j_2-j_1-1} \otimes A(j_1) \otimes \mathbf{A}^{j_1} \otimes w(0)\Big)\Big]$$

and the third-order derivative is obtained from

$$\frac{d^3}{d\theta^3}\mathbb{E}_0[g_W(w(m))]$$

$$= 6 \sum_{j_1=0}^{m-3} \sum_{j_2=j_1+1}^{m-2} \sum_{j_3=j_2+1}^{m-1} \Big(V_g(m;j_1,j_2,j_3) + V_g(m;j_1) + V_g(m;j_2) + V_g(m;j_3)$$

$$- V_g(m;j_1,j_2) - V_g(m;j_2,j_3) - V_g(m;j_1,j_3) - V_g(m) \Big).$$

The derivatives can be verbally described as follows. The factor is $n!$ when the nth order derivative is evaluated. The outer summation ranges over all possible combinations of marking n out of m transitions. The inner sum ranges over all possible combinations of letting the n marked transitions be either stochastic or not. The sign of an element in the inner sum is given by -1 to the power of the number of deterministic substitutions among the n marked transitions.

The resulting Taylor series approximation of degree $h = 3$ is given by

$$\mathbb{E}[g(W(m))] \approx \frac{d}{d\theta}\mathbb{E}_0[g_W(w(m))]$$

$$+ \frac{1}{2}\frac{d^2}{d\theta^2}\mathbb{E}_0[g_W(w(m))] + \frac{1}{6}\frac{d^3}{d\theta^3}\mathbb{E}_0[g_W(w(m))].$$

Proceeding as above, we can define factors $V_g(m; j_1, \ldots, j_k)$, for $1 \leq k \leq m$. The nth order derivative of $\mathbb{E}_0[g_W(w(m))]$ is then given through $V_g(m; j_1, \ldots, j_k)$, for $1 \leq k \leq n$. Let

$$V_g(m,k) = \sum_{j_1=0}^{m-k} \sum_{j_2=j_1+1}^{m-k+1} \cdots \sum_{j_k=j_{k-1}+1}^{m-1} V_g(m;j_1,\ldots,j_k),$$

for $k \leq m$, and $V_g(m,0) = V_g(m)$. The term $V_g(m,l)$ yields the total effect of making l out of m transitions stochastic. For the nth derivative of $\mathbb{E}[g_W(w(m))]$ we mark in total n transitions out of which l are stochastic. Hence, there are

$$\binom{m-l}{n-l}$$

possibilities of reaching at $(m - l)$ deterministic transitions provided that there l stochastic ones, and we obtain

$$\frac{d^n}{d\theta^n}\mathbb{E}_0[g_W(w(m))]$$

$$= n! \sum_{l=0}^{n} \binom{m-l}{n-l} (-1)^{n-l} V_g(m,l).$$

Inserting the above expression into the Taylor series and rearranging terms gives

$$\mathbb{E}[g(W(m))] \approx \sum_{n=0}^{h} \sum_{l=0}^{n} \binom{m-l}{n-l} (-1)^{n-l} V_g(m,l)$$

$$= \sum_{l=0}^{h} \sum_{n=l}^{h} \binom{m-l}{n-l} (-1)^{n-l} V_g(m,l)$$

$$= \sum_{l=0}^{h} C(h,m,l) V_g(m,l), \quad h \leq m,$$

with

$$C(h,m,l) = \frac{1}{(h-l)!} \prod_{j=l+1}^{h} (j-m),$$

where we set the product to one for $l = h$. For a proof that

$$C(h,m,l) = \sum_{n=l}^{h} \binom{m-l}{n-l} (-1)^{n-l}$$

see, for example, formula (18) on page 57 in [76].

5.1.3.1 Computation of the Taylor Series Expansion

The coefficients of the Taylor series enjoy a recursive structure which can be exploited when calculating the series. In the following we will discuss this in more detail where the key observation is that a stochastic transition only contributes to the overall derivative if the waiting time introduced by that stochastic transition doesn't die out before the following stochastic transition occurs.

For $0 \leq i_1 < i_2 < \cdots < i_h < m$, let $W(m; i_1, i_2, \ldots, i_h)$ denote the mth waiting time in the system with deterministic transitions except for transitions $i_1, i_2, \cdots, i_h$. Let $\mathbf{W}(\cdot)$ be the projection onto the first component of the vector $w(k)$ and introduce the variables

$$W[m;i] = \mathbf{W}(\mathbf{A}^{m-i-1} \otimes A(i) \otimes \mathbf{A}^i \otimes w(0)), \quad 0 \leq i \leq m,$$

$$W[m;i_1,i_2] = \mathbf{W}(\mathbf{A}^{m-i_2-1} \otimes A(i_2) \otimes \mathbf{A}^{i_2-i_1-1} \otimes A(i_1) \otimes \mathbf{A}^{i_1} \otimes w(0)),$$

for $0 \leq i_1 < i_2 < m$ and

$$W[m;i_1,i_2,i_3] = \mathbf{W}(\mathbf{A}^{m-i_3-1} \otimes A(i_3) \otimes \mathbf{A}^{i_3-i_2-1} \otimes A(i_2)$$
$$\otimes \mathbf{A}^{i_2-i_1-1} \otimes A(i_1) \otimes \mathbf{A}^{i_1} \otimes w(0)),$$

for $0 \leq i_1 < i_2 < i_3 < m$. In addition to that, set

$$W[m] = \mathbf{W}(\mathbf{A}^m \otimes w(0)).$$

Consider the G/G/1 queue with mean interarrival time σ_0 and mean service time σ_1, and set $c = \sigma_1 - \sigma_0$. Then, c is the so-called *drift* of the random walk $\{W(k)\}$ and because we have assumed that the system is stable, i.e., $\rho \stackrel{\text{def}}{=} \sigma_1/\sigma_0 < 1$, the drift is negative. This can be phrased by saying that a deterministic transition decreases the amount of work present at the server by c. Denote the density of the interarrival times by f^A and the density of the service times by f^S and assume that f^A and f^S have support $(0, \infty)$.

We now turn to the computation of the derivative of $\mathbb{E}[g(W(m))]$ with respect to θ, see (5.7). First, notice that

$$V_g(m) = g(\mathbf{W}(\mathbf{A}^m \otimes w(0))).$$

For $m > 0$, it is easily checked that

$$V_g(m; i) = g(0) P(W[m; i] = 0) + \mathbb{E}[1_{W[m;i]>0} g(W[m; i])]$$

$$= \int_0^\infty \int_{a-(m-i-1)c}^\infty g(s - a + (m - i - 1)c) f^S(ds) f^A(da)$$
$$+ g(0) P(W[m; i] = 0),$$

where

$$P(W[m; i] = 0) = \int_0^\infty \int_0^{a-(m-i-1)c} f^S(ds) f^A(da)$$

and $1_{W[m;i]>0}$ denotes the indicator mapping for the event $\{W[m; i] > 0\}$, that is, $1_{W[m;i]>0} = 1$ if $W[m; i] > 0$ and otherwise zero.

We now turn to the second order derivative. For $0 \le i_1 < i_2 < m$, $W[i_2; i_1] > 0$ describes the event that a stochastic transition at i_1 generated a workload at the server that (possibly) hasn't been completely worked away until transition i_2. With the help of this event we can compute as follows

$$V_g(m; i_1, i_2) = \mathbb{E}\big[g(W[m; i_1, i_2])\big]$$
$$= \mathbb{E}\big[1_{W[i_2;i_1]>0} 1_{W[m;i_1,i_2]>0}\, g(W[m; i_1, i_2])\big]$$
$$+ \mathbb{E}\big[1_{W[i_2;i_1]=0} 1_{W[m;i_1,i_2]>0}\, g(W[m; i_1, i_2])\big]$$
$$+ \mathbb{E}\big[1_{W[i_2;i_1]>0} 1_{W[m;i_1,i_2]=0}\, g(W[m; i_1, i_2])\big]$$
$$+ \mathbb{E}\big[1_{W[i_2;i_1]=0} 1_{W[m;i_1,i_2]=0}\, g(W[m; i_1, i_2])\big].$$

On the event $\{W[i_2; i_1] = 0\}$ the effect of the first stochastic transition dies out before transition i_2. By independence,

$$\mathbb{E}\big[1_{W[i_2;i_1]=0} 1_{W[m;i_1,i_2]>0}\, g(W[m; i_1, i_2])\big]$$
$$= \mathbb{E}\big[1_{W[i_2;i_1]=0} 1_{W[m;i_2]>0}\, g(W[m; i_2])\big]$$
$$= P(W[i_2; i_1] = 0)\mathbb{E}\big[1_{W[m;i_2]>0}\, g(W[m; i_2])\big]$$

and

$$\mathbb{E}\big[1_{W[i_2;i_1]=0} 1_{W[m;i_1,i_2]=0}\, g(W[m; i_1, i_2])\big]$$
$$= g(0)P(W[i_2; i_1] = 0)\, P(W[m; i_2] = 0).$$

Moreover, it is easily checked that

$$\mathbb{E}\big[1_{W[i_2;i_1]>0}1_{W[m;i_1,i_2]=0}\,g(W[m;i_1,i_2])\big]$$
$$= g(0)\,P\big(W[i_2;i_1]>0 \wedge W[m;i_2]=0\big).$$

We obtain $V_g(m;i_1,i_1)$ as follows:

$$V_g(m;i_1,i_2) = \mathbb{E}\big[1_{W[m;i_1,i_2]>0}\,1_{W[i_2;i_1]>0}\,g(W[m;i_1,i_2])\big]$$
$$+ \mathbb{E}\big[1_{W[m;i_2]>0}\,g(W[m;i_2])\big]\,P\big(W[i_2;i_1]=0\big)$$
$$+ g(0)\,\Big(P\big(W[i_2;i_1]=0\big)\,P\big(W[m;i_2]=0\big)$$
$$+ P\big(W[i_2;i_1]>0 \wedge W[m;i_1,i_2]=0\big)\Big),$$

where noticeably some of the expressions in the product on the right-hand side in the above formula have already been calculated in the process of computing the first order derivative. Specifically, in order to compute the second order derivative only $m(m+1)/2$ terms have to be computed, namely $\mathbb{E}\big[1_{W[m;i_1,i_2]>0}\,1_{W[i_2;i_1]>0}\,g(W[m;i_1,i_2])\big]$ for $0 \le i_1 < i_2 < m$. These terms can be computed as follows:

$$\mathbb{E}\big[1_{W[i_2;i_1]>0}\,1_{W[m;i_1,i_2]>0}\,g(W[m;i_1,i_2])\big]$$
$$= \int_0^\infty \int_0^\infty \int_{a_1+a_2-(m-i_1-2)c}^\infty \int_0^\infty g(s_1+s_2-a_1-a_2+(m-i_1-2)c)$$
$$\times f^S(s_2)f^S(s_1)f^A(a_2)f^A(a_1)\,ds_2\,ds_1\,da_2\,da_1$$
$$+ \int_0^\infty \int_0^\infty \int_{a_1-(i_2-i_1-1)c}^{a_1+a_2-(m-i_1-2)c}$$
$$\int_{a_1+a_2-s_1-(m-i_1-2)c}^\infty g(s_1+s_2-a_1-a_2+(m-i_1-2)c)$$
$$\times f^S(s_2)f^S(s_1)f^A(a_2)f^A(a_1)\,ds_2\,ds_1\,da_2\,da_1\,.$$

Setting $g = 1$ and adjusting the boundaries of the integrals, we can compute from the above equations the probability of the event $\{W[i_2;i_1] > 0 \wedge W[i_3;i_1,i_2] = 0\}$, as well.

For the third order derivative the computations become more cumbersome. To abbreviate the notation, we set

$$h_i(s_1,s_2,s_3,a_1,a_2,a_3)$$
$$= g(s_1+s_2+s_3-a_1-a_2-a_3+(m-i-3)c)$$
$$\times f^S(s_3)f^S(s_2)f^S(s_1)f^A(a_3)f^A(a_2)f^A(a_1)$$

and we obtain

$$\mathbb{E}\big[\,1_{W[i_2;i_1]>0}1_{W[i_3;i_1,i_2]>0}1_{W[m;i_1,i_2,i_3]>0}\,g(W[m;i_1,i_2,i_3])\,\big]$$

$$=\int_0^\infty\int_0^\infty\int_0^\infty\int_{a_1+a_2+a_3-(m-i_1-3)c}^\infty$$
$$\int_0^\infty$$
$$\int_0^\infty h_{i_1}(s_1,s_2,s_3,a_1,a_2,a_3)\,ds_3\,ds_2\,ds_1\,da_3\,da_2\,da_1$$

$$+\int_0^\infty\int_0^\infty\int_0^\infty\int_{a_1-(i_2-i_1-1)c}^{a_1+a_2-(i_3-i_1-2)c}$$
$$\int_{a_1+a_2+a_3-(m-i_1-3)c-s_1}^\infty$$
$$\int_0^\infty h_{i_1}(s_1,s_2,s_3,a_1,a_2,a_3)\,ds_3\,ds_2\,ds_1\,da_3\,da_2\,da_1$$

$$+\int_0^\infty\int_0^\infty\int_0^\infty\int_{a_1-(i_2-i_1-1)c}^{a_1+a_2-(i_3-i_1-2)c}$$
$$\int_{a_1+a_2-(i_3-i_1-2)c-s_1}^{a_1+a_2+a_3-(m-i_1-3)c-s_1}$$
$$\int_{a_1+a_2+a_3-s_1-s_2-(m-i_1-3)c}^\infty$$
$$\times h_{i_1}(s_1,s_2,s_3,a_1,a_2,a_3)\,ds_3\,ds_2\,ds_1\,da_3\,da_2\,da_1$$

$$+\int_0^\infty\int_0^\infty\int_0^\infty\int_{a_1+a_2-(i_3-i_1-2)c}^{a_1+a_2+a_3-(m-i_1-3)c}$$
$$\int_{a_1+a_2+a_3-(m-i_1-3)c-s_1}^\infty$$
$$\int_0^\infty h_{i_1}(s_1,s_2,s_3,a_1,a_2,a_3)\,ds_3\,ds_2\,ds_1\,da_3\,da_2\,da_1$$

$$+\int_0^\infty\int_0^\infty\int_0^\infty\int_{a_1+a_2-(i_3-i_1-2)c}^{a_1+a_2+a_3-(m-i_1-3)c}$$
$$\int_0^{a_1+a_2+a_3-(m-i_1-3)c-s_1}$$
$$\int_{a_1+a_2+a_3-(m-i_1-3)c-s_1-s_2}^\infty$$
$$\times h_{i_1}(s_1,s_2,s_3,a_1,a_2,a_3)\,ds_3\,ds_2\,ds_1\,da_3\,da_2\,da_1\,.$$

Following the line of argument used for the second derivative, the third order derivative can now be expressed as a combination of the above variables together with the variables already computed while calculating the first and second order derivatives. The precise formula is:

$$V_g(m; i_1, i_2, i_3) = \mathbb{E}\left[1_{W[i_2;i_1]>0} 1_{W[i_3;i_1,i_2]>0} 1_{W[m;i_1,i_2,i_3]>0}\, g(W[m; i_1, i_2, i_3]) \right]$$

$$+ \mathbb{E}\left[1_{W[i_3;i_2]>0} 1_{W[m;i_2,i_3]>0}\, g(W[m; i_2, i_3]) \right] P\big(W[i_2; i_1] = 0\big)$$

$$+ \mathbb{E}\left[1_{W[m;i_3]>0}\, g(W[m; i_3]) \right]$$

$$\times \Big(P\big(W[i_2; i_1] > 0 \wedge W[i_3; i_1, i_2] = 0\big)$$

$$+ P\big(W[i_2; i_1] = 0\big) P\big(W[i_3; i_2] = 0\big) \Big)$$

$$+ g(0)\, P\big(W[m; i_1, i_2, i_3] = 0\big) ,$$

where

$$P\big(W[m; i_1, i_2, i_3] = 0\big) = P\big(W[i_2; i_1] = 0\big) P\big(W[i_3; i_2] > 0 \wedge W[m; i_3, i_2] = 0\big)$$

$$+ P\big(W[i_2; i_1] > 0 \wedge W[i_3; i_1, i_2] > 0 \wedge W[m; i_3, i_2] = 0\big)$$

$$+ P\big(W[i_2; i_1] = 0\big) P\big(W[i_3; i_2] = 0\big) P\big(W[m; i_3] = 0\big)$$

$$+ P\big(W[i_2; i_1] > 0 \wedge W[i_3; i_1, i_2] = 0\big) P\big(W[m; i_3] = 0\big) .$$

5.1.3.2 Numerical Examples

Consider $g = \mathrm{id}$, that is, $g(W(m)) = W(m)$, $m \geq 0$. Note that $\rho < 1$ implies that $V_g(m) = g_W\big(\mathbf{A}^m \otimes w(0)\big) = 0$. Direct computation of $\mathbb{E}[W(m)]$ involves performing an m fold integration over a complex polytope. In contrast to this, the proposed variability expansion allows to build an approximation of $\mathbb{E}[W(m)]$ out of terms that involve h fold integration with $h < m$ (below we have taken $h = 2, 3$). This reduces the complexity of evaluating $\mathbb{E}[W(m)]$ considerably. To illustrate the performance of the variability expansion, we applied our approximation scheme to the transient waiting time in a stable (that is, $\rho < 1$) M/M/1 queue and D/M/1 queue, respectively.

The M/M/1 Queue

Figure 5.1 illustrates the relative error of the Taylor polynomial of degree $h = 2$ for various traffic loads. For $h = 2$, we are performing two stochastic transitions and a naive approximation of $\mathbb{E}[W(m)]$ is given through $V_{id}(2; 0, 1) = \mathbb{E}[W(2)]$ and the numerical results are depicted in Figure 5.2. The exact values used to construct the figures in this section are provided in Section H in the Appendix.

To illustrate the influence of h, we also evaluated the Taylor polynomial of degree $h = 3$. See Figure 5.3 for numerical results. Here, the naive approximation is given by $V_{id}(3; 0, 1, 2) = \mathbb{E}[W(3)]$ and the corresponding results are depicted in Figure 5.4.

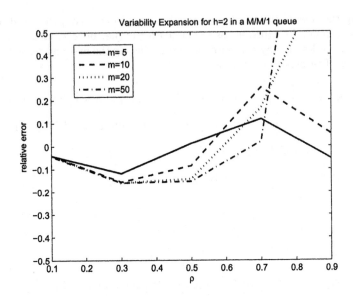

Figure 5.1: Relative error for the M/M/1 queue for $h = 2$.

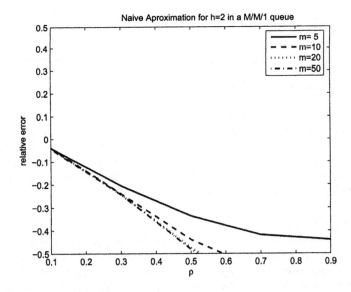

Figure 5.2: Relative error for the naive approximation for the M/M/1 queue for $h = 2$.

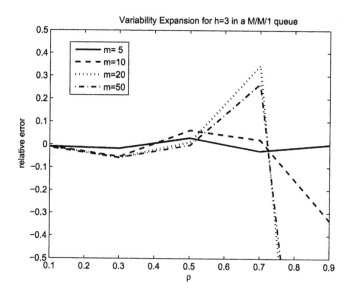

Figure 5.3: Relative error for the M/M/1 queue for $h = 3$.

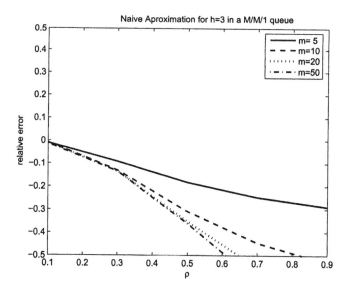

Figure 5.4: Relative error for the naive approximation of the M/M/1 queue for $h = 3$.

It turns out that for $\rho < 0.5$ the Taylor polynomial of degree 3 provides a good approximation for the transient waiting time. However, the quality of the approximation decreases with increasing time horizon. For $\rho \geq 0.5$, the approximation works well only for relatively small time horizons ($m \leq 10$). It is worth noting that in heavy traffic ($\rho = 0.9$) the quality of the approximation decreases when the third order derivative is taken into account. The erratic behavior of the approximation for large values of ρ is best illustrated by the kink at $\rho = 0.7$ for $m = 20$ and $m = 50$. However, for $m = 5$, the approximation still works well. In addition, the results illustrate that variability expansion outperforms the naive approach. To summarize, the quality of the approximation decreases with growing traffic intensity when the time horizon increases.

Comparing the figures, one notes that the outcome of the Taylor series approximation can be independent of the time horizon m. For example, at $\rho = 0.1$, the values of the Taylor polynomial do not vary in m. This stems from the fact that for such a small ρ the dependence of the mth waiting time on waiting times $W(m-k)$, $k \geq 5$, is negligible. Hence, allowing transitions $m - k$, $k \geq 5$, to be stochastic doesn't contribute to the outcome of $\mathbb{E}[W(m)]$, which is reflected by the true values as well.

In heavy traffic, the quality of the approximation decreases for growing h. This stems from the fact that convergence of the Taylor series is forced by the fact that the nth derivative of $\mathbb{E}_\theta[W(m)]$ jumps to zero at $n = m$. As discussed in Section G.4 in the Appendix, in such a situation, the quality of the approximation provided by the Taylor polynomial may worsen through increasing h as long as $h \leq m$.

The numerical values were computed with the help of a computer algebra program. The calculations were performed on a Laptop with Intel Pentium III processor and the computation times are listed in Table 5.1.

Table 5.1: CPU time (in seconds) for computing Taylor polynomials of degree h for time horizon m in a M/M/1 queue.

m	h=2	h=3
5	1.8	3.9
10	1.8	4.4
20	1.9	6.1
50	2.2	36.7

Note that the computational effort is independent of the traffic rate and only influenced by the time horizon. The table illustrates that the computational effort for computing the first two elements of the Taylor polynomial grows very slowly in m, whereas the computational effort for computing the first three elements of the Taylor series increases rapidly in m. This indicates that computing higher-degree Taylor polynomials will suffer from high computational costs.

The D/M/1 Queue

Figure 5.5 plots the relative error of the Taylor polynomial of degree $h = 2$ for various traffic loads. For the naive approximation, the values $V_{id}(2; 0, 1)$ are used to predict the waiting times and Figure 5.6 presents the numerical values. The exact values used to construct the figures are provided in Section H in the Appendix.

Figure 5.7 plots the relative error of the Taylor polynomial of degree $h = 3$

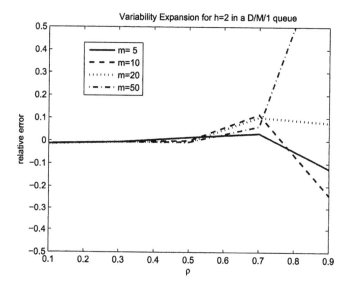

Figure 5.5: Relative error for the D/M/1 queue for $h = 2$.

for various traffic loads. The naive approximation is given by $V_{id}(3; 0, 1, 2)$ and Figure 5.8 depicts the numerical results.

Figure 5.5 up to Figure 5.8 show the same behavior of variability expansion as already observed for the M/M/1 queue. Like for the M/M/1 queue, the quality of the approximation decreases with growing traffic intensity when the time horizon increases. It is worth noting that variability expansion outperforms the naive approach.

The numerical values were computed with the help of a computer algebra program. The calculations were performed on a Laptop with Intel Pentium III processor and the computation times are listed in Table 5.2. Due to the fact that the interarrival times are deterministic, calculating the elements of the variability expansions for the D/M/1 queue requires less computation time than for the M/M/1 queue, see Table 5.1.

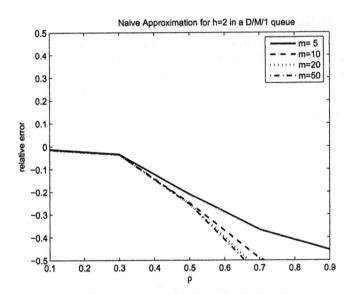

Figure 5.6: Relative error for the naive approximation of the D/M/1 queue for $h = 2$.

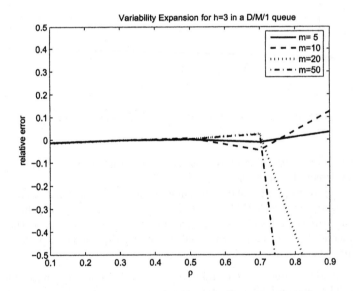

Figure 5.7: Relative error for the D/M/1 queue for $h = 3$.

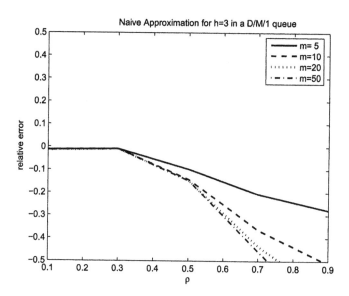

Figure 5.8: Relative error for the naive approximation of the D/M/1 queue for $h = 3$.

Table 5.2: CPU time (in seconds) for computing Taylor polynomials of degree h for time horizon m in a D/M/1 queue.

m	h=2	h=3
5	0.8	1.8
10	0.9	2.0
20	1.4	4.7
50	1.9	47.3

5.2 Random Horizon Experiments

Analytic expansions of n-fold products in the max-plus semiring were given in the previous section. This section extends these results to random horizon products, that is, we consider the case when n is random. For a motivation, revisit the multi-server system with server breakdowns in Example 1.5.5. Suppose that we are interested in the point in time when the server breaks down twice in a row. The time of the kth beginning of service at the multi server station is given by $x_3(k)$. The event that the second of two consecutive breakdowns occurs at the kth transition is given by $\{A(k-1) = D_1 = A(k)\}$ and the time at which this event occurs is given by $x_3(k)$. Set

$$\tau_{(D_1,D_1)} = \tau_{(D_1,D_1)}(\theta) = \inf\{k \geq 1 : A(k) = D_1 = A(k-1)\} .$$

Then $\mathbb{E}_\theta[x_3(\tau_{(D_1,D_1)})]$ yields the expected time of the occurrence of the second breakdown in row. Our goal is to compute $\mathbb{E}_\theta[x_3(\tau_{(D_1,D_1)})]$ via a Taylor series.

The general setup is as follows. Let $\{A_\theta(k)\}$ have (discrete) state space $\mathcal{A}$. For $\tilde{a}_i \in \mathcal{A}$, $1 \leq i \leq M$, set $\tilde{a} = (\tilde{a}_1, \ldots, \tilde{a}_M)$ and denote by

$$\tau_{\tilde{a}}(\theta) = \inf\{k \geq M - 1 : A_\theta(k - M + 1) = \tilde{a}_1, \ldots$$
$$\ldots, A_\theta(k-1) = \tilde{a}_{M-1}, A_\theta(k) = \tilde{a}_M\} \quad (5.8)$$

the time at which the sequence $\tilde{a}$ occurs for the first time in $\{A_\theta(k) : k \geq 0\}$. This section addresses the following problem.

The Random Horizon Problem: Let $\theta \in \Theta$ be a real-valued parameter, Θ being an interval. We shall take θ to be a variational parameter of an i.i.d. sequence $\{A_\theta(k)\}$ of square matrices in $\mathbb{R}_{\max}^{J \times J}$ with discrete state space $\mathcal{A}$ and study sequences $\{x_\theta(k)\}$ following

$$x_\theta(k+1) = A_\theta(k) \otimes x_\theta(k), \quad k \geq 0,$$

with $x_\theta(0) = x_0$ for all θ. Let $\tau_{\tilde{a}}(\theta)$ be defined as in (5.8). For a given performance function $g : \mathbb{R}_{\max}^J \to \mathbb{R}$ compute the Taylor series for the expected performance of the random horizon experiment, given by

$$\mathbb{E}_\theta[g(x(\tau_{\tilde{a}}))] . \quad (5.9)$$

5.2.1   The 'Halted' Max-Plus System

In Section 4.5, sufficient conditions for the analyticity of $E_\theta[\bigotimes_{k=0}^m A(k) \otimes x_0]$ were given, for fixed $m \in \mathbb{N}$. Unfortunately, the situation we are faced with here is more complicated, since $\tau_{\tilde{a}}$ is random and depends on θ. To deal with the situation, we borrow an idea from the theory of Markov chains. There, the expectation over a random number of transitions of a Markov chain is analyzed by introducing an absorbing state. More precisely, a new Markov kernel is defined such that, once the chain reaches a specified criterion (like entering a certain set), the chain is forced to jump to the absorbing state and to remain there forever. Following a similar train of thought, we introduce in this section an operator, denoted by $[\cdot]_{\tilde{a}}$, that yields a 'halted' version of $A(k)$, denoted by $[A(k)]_{\tilde{a}}$, where $[A(k)]_{\tilde{a}}$ will be constructed in such a way that it equals $A(k)$ as long as the sequence $\tilde{a}$ has not occurred in the sequence $A(0), A(1), \ldots, A(k)$. Once $\tilde{a}$ has occurred, the operator $[\cdot]_{\tilde{a}}$ sets $A(k)$ to E, the identity matrix. In other words, $[\cdot]_{\tilde{a}}$ 'halts' the evolution of the system dynamics as soon as the sequence $\tilde{a}$ occurs and we denote the halted version of $\{A(k)\}$ by $\{[A(k)]_{\tilde{a}}\}$.

In the following we explain our approach with the multi-server example, with $\tilde{a} = (D_1, D_1)$. Suppose that we observe the sequence:

$$(A(k) : k \geq 0) = (D_1, D_2, D_1, D_2, D_2, D_2, D_1, D_1, D_1, D_2, \ldots) . \quad (5.10)$$

Hence, $\tau_{(D_1, D_1)} = 7$ and the nominal observation is based on the cycle

$$(A(0), A(1), \ldots, A(\tau_{(D_1,D_1)})) = (D_1, D_2, D_1, D_2, D_2, D_2, D_1, D_1).$$

We call

$$(D_1, D_2, D_1, D_2, D_2, D_2, D_1, D_1)$$

the *initial segment* and

$$A(\tau_{(D_1,D_1)} + 1), A(\tau_{(D_1,D_1)} + 2), \ldots$$

the *tail* of $\{A(k)\}$:

$$\underbrace{D_1, D_2, D_1, D_2, D_2, D_2, D_1, D_1,}_{\text{inital segement of } \{A(k)\}} \ \underbrace{D_1, D_2, \ldots}_{\text{tail of } \{A(k)\}} .$$

The halted version of $\{A(k)\}$, denoted by $\{[A(k)]_{(D_1,D_1)}\}$, is obtained from $\{A(k)\}$ through replacing the tail segment by the sequence $E, E, \ldots$, in formula:

$$\underbrace{D_1, D_2, D_1, D_2, D_2, D_2, D_1, D_1,}_{\text{inital segement of } \{[A(k)]_{(D_1,D_1)}\}} \quad \underbrace{E, E, \ldots}_{\text{tail of } \{[A(k)]_{(D_1,D_1)}\}}$$

which implies that

$$g\left(\bigotimes_{k=0}^{m} [A(k)]_{(D_1,D_1)} \otimes x_0 \right) = g\left(\bigotimes_{k=0}^{\min(m, \tau_{(D_1,D_1)})} A(k) \otimes x_0 \right),$$

for any g and any initial value x_0. Moreover, letting m tend to ∞ in the above equation yields:

$$g\left(\bigotimes_{k=0}^{\infty} [A(k)]_{(D_1,D_1)} \otimes x_0 \right) = g\left(\bigotimes_{k=0}^{\tau_{(D_1,D_1)}} A(k) \otimes x_0 \right). \qquad (5.11)$$

This reflects the fact that $[A(k)]_{(D_1,D_1)}$ behaves just like $A(k)$ until (D_1, D_1) occurs. Provided that $\tau_{(D_1,D_1)} < \infty$ a.s., the limit in (5.11) holds a.s. without g, resp. g^τ, being continuous. Once (D_1, D_1) occurs, $[A(k)]_{(D_1,D_1)}$ is set to E, the neutral elements of $\otimes$ matrix product.

By equation (5.11), differentiating

$$\bigotimes_{k=0}^{\infty} [A(k)]_{(D_1,D_1)}$$

is equivalent to differentiating

$$\bigotimes_{k=0}^{\tau_{(D_1,D_1)}} A(k).$$

Of course we would like to apply the differentiation techniques developed in Chapter 3 and 4 to $\{[A(k)]_{(D_1,D_1)}\}$. Unfortunately, we cannot apply our theory straightforwardly because $[A(k)]_{(D_1,D_1)}$ fails to be an i.i.d. sequence. Indeed, the distribution of $[A(k)]_{(D_1,D_1)}$ depends on whether the string (D_1, D_1) has occurred prior to k or not.

The trick that allows us to apply our theory of $\mathcal{D}$-differentiation to $\{[A(k)]_{(D_1,D_1)}\}$ is to show that the order in which the differential operator and the operator $[\cdot]_{(D_1,D_1)}$ are applied can be interchanged. If we knew that we are allowed to interchange differentiation and application of the $[\cdot]_{(D_1,D_1)}$ operator, we could boldly compute as follows:

$$
\left(\bigotimes_{k=0}^{\min(m,\tau_{(D_1,D_1)})} A(k) \right)' \equiv \left(\bigotimes_{k=0}^{m} [A(k)]_{(D_1,D_1)} \right)'
$$

$$
\equiv \left[\left(\bigotimes_{k=0}^{m} A(k) \right)' \right]_{(D_1,D_1)}
$$

$$
\equiv \sum_{j=0}^{m} \left[\bigotimes_{k=j+1}^{m} A(k) \otimes \left(A(j) \right)' \otimes \bigotimes_{k=0}^{j-1} A(k) \right]_{(D_1,D_1)} .
$$

Notice that for the motivating example of the multi server model we have $A(k)' = (1, D_1, D_2)$. For example, let $m = 9$ and take $j = 6$. Then the above formula transforms the realization of $\{A(k)\}$ given in (5.10) as follows:

$$
(A(0), \ldots, A(5), A^+(6), A(7), A(8), A(9))
$$
$$
= (D_1, D_2, D_1, D_2, D_2, D_2, \boldsymbol{D_1}, D_1, D_1, D_2)
$$

and

$$
(A(0), \ldots, A(5), A^-(6), A(7), A(8), A(9))
$$
$$
= (D_1, D_2, D_1, D_2, D_2, D_2, \boldsymbol{D_2}, D_1, D_1, D_2) ,
$$

where the bold faced elements of the realization are those effected by the derivative. Applying the $[\cdot]_{(D_1,D_1)}$ operator yields

$$
[(A(0), \ldots, A(5), A^+(6), A(7), A(8), A(9))]_{(D_1,D_1)}
$$
$$
= (D_1, D_2, D_1, D_2, D_2, D_2, \boldsymbol{D_1}, D_1, E, E)
$$

and

$$
[(A(0), \ldots, A(5), A^-(6), A(7), A(8), A(9))]_{(D_1,D_1)}
$$
$$
= (D_1, D_2, D_1, D_2, D_2, D_2, \boldsymbol{D_2}, D_1, D_1, E) .
$$

Notice that the lengths of the cycles differ from the nominal ones.

Now, consider $j = 9$ and notice that for this value of j it holds that $j > \tau_{(D_1, D_1)}$. Then,

$$(A(0), \ldots, A(8), A^+(9))$$
$$= (D_1, D_2, D_1, D_2, D_2, D_2, D_1, D_1, D_1, \boldsymbol{D_1})$$

and

$$(A(0), \ldots, A(8), A^-(9))$$
$$= (D_1, D_2, D_1, D_2, D_2, D_2, D_1, D_1, D_1, \boldsymbol{D_2}),$$

which implies

$$[(A(0), \ldots, A(8), A^+(9))]_{(D_1, D_1)} = [(A(0), \ldots, A(8), A^-(9))]_{(D_1, D_1)}$$
$$= (D_1, D_2, D_1, D_2, D_2, D_2, D_1, D_1, E, \boldsymbol{E}).$$

If the positive part and the negative part of a derivative are equal, then the derivative doesn't contribute to the overall derivative, which stems from the fact that for any mapping $g \in \mathbb{R}_{\max}^{4 \times 4} \mapsto \mathbb{R}$ it holds that

$$g^\tau ([A'(9) \otimes A(8) \otimes \cdots \otimes A(0)]_{(D_1, D_1)})$$
$$= g([A^+(9) \otimes A(8) \otimes \cdots \otimes A(0)]_{(D_1, D_1)})$$
$$\quad - g([A^-(9) \otimes A(8) \otimes \cdots \otimes A(0)]_{(D_1, D_1)})$$
$$= 0. \tag{5.12}$$

In words, for $j > \tau_{(D_1, D_1)}$, the derivatives of $A(k)$ do not contribute to the overall derivative. Hence,

$$\left[\left(\bigotimes_{k=0}^m A(k) \right)' \right]_{(D_1, D_1)} \equiv \sum_{j=0}^{\min(m, \tau_{(D_1, D_1)})} \left[\bigotimes_{k=j+1}^m A(k) \otimes A(j)' \otimes \bigotimes_{k=0}^{j-1} A(k) \right]_{(D_1, D_1)}.$$

In the following we show that interchanging the differentiation operator and the $[\cdot]_{\tilde{a}}$ operator is indeed justified. Let μ_i, $0 \le i \le m$, be probability measures on a discrete state space $\mathcal{A} \subset \mathbb{R}_{\max}^{J \times J}$, and let $E \in \mathcal{A}$. Let $\tilde{a} = (\tilde{a}_1, \ldots, \tilde{a}_M)$ be a sequence of elements out of $\mathcal{A}$. For fixed $m > 0$, denote by $\mathbf{A}_m(j)$ the set of sequences $(a_0, \ldots, a_m) \in \mathcal{A}^{m+1}$ such that the first occurrence of $\tilde{a}$ is completed at the entry with label j, for $0 \le j < m$. More formally, for $M - 1 \le j < m$, set

$$\mathbf{A}_m(j) \stackrel{\text{def}}{=} \Big\{ (a_0, a_1, \ldots, a_m) \in \mathcal{A}^{m+1} :$$

$$j = \min\{k \ge M - 1 : a_{k-M+1} = \tilde{a}_1, \ldots, a_{k-1} = \tilde{a}_{M-1}, a_k = \tilde{a}_M\} \Big\}.$$

The set $\mathbf{A}_m(m)$ is defined as follows:

$$\mathbf{A}_m(m) \stackrel{\text{def}}{=} \mathcal{A}^m \setminus \bigcup_{j=0}^{m-1} \mathbf{A}_m(j).$$

Moreover, we set $\mathbf{A}_m(j) = \emptyset$ for any other combination of m and any j. We denote the (independent) product measure of the μ_i's by $\prod_{i=0}^{m} \mu_i$. In order to construct the halted version, we introduce the measure-theoretic version of the operator $[\cdot]_{\tilde{a}}$ as follows: for $0 \leq j \leq m$ and $a \in \mathcal{A}_m(j)$, we set

$$
\left[\prod_{i=0}^{m} \mu_i \right]_{\tilde{a}} (a_0, \ldots, a_m) = \left(\prod_{i=0}^{j} \mu_i \right) (a_0, \ldots, a_j) \times \left(\prod_{i=j+1}^{m} \delta_E \right) (a_{j+1}, \ldots, a_m)
$$

$$
= \prod_{i=0}^{j} \mu_i(a_i) \times \prod_{i=j+1}^{m} \delta_E(a_i) , \tag{5.13}
$$

where δ_E denotes the Dirac measure in E and we disregard $\prod_{i=j+1}^{m} \delta_E$ for $j = m$.

Theorem 5.2.1 *Let $\mathcal{A} \subset \mathbb{R}_{\max}^{J \times J}$, with $E \in \mathcal{A}$, and let μ_i, for $0 \leq i \leq m$, be a sequence of n times $C_p(\mathcal{A})$-differentiable probability measures on $\mathcal{A}$, for $p \in \mathbb{N}$. Then, the product measure $\prod_{i=0}^{m} \mu_i$ is n times $C_p(\mathcal{A}^{m+1})$-differentiable and it holds that*

$$
\left(\left[\prod_{i=0}^{m} \mu_i \right]_{\tilde{a}} \right)^{(n)} = \left[\left(\prod_{i=0}^{m} \mu_i \right)^{(n)} \right]_{\tilde{a}} .
$$

Proof: For any $g \in C_p(\mathcal{A}^{m+1})$,

$$
\sum_{(a_0, \ldots, a_m) \in \mathcal{A}^{m+1}} g(a_0, \ldots, a_m) \left[\prod_{i=0}^{m} \mu_i \right]_{\tilde{a}} (a_0, \ldots, a_m)
$$

$$
= \sum_{j=0}^{m} \sum_{(a_0, \ldots, a_m) \in \mathbf{A}_m(j)} g(a_0, \ldots, a_m)
$$

$$
\times \left(\prod_{i=0}^{j} \mu_i \right) (a_0, \ldots, a_j) \times \left(\prod_{i=j+1}^{m} \delta_E \right) (a_{j+1}, \ldots, a_m) ,
$$

which implies

$$
\frac{d^n}{d\theta^n} \sum_{(a_0, \ldots, a_m) \in \mathcal{A}^m} g(a_0, \ldots, a_m) \left[\prod_{i=0}^{m} \mu_i \right]_{\tilde{a}} (a_0, \ldots, a_m)
$$

$$
= \sum_{j=0}^{m} \frac{d^n}{d\theta^n} \sum_{(a_0, \ldots, a_m) \in \mathbf{A}_m(j)} g(a_0 \ldots, a_m)
$$

$$
\times \left(\prod_{i=0}^{j} \mu_i \right) (a_0, \ldots, a_j) \times \left(\prod_{i=j+1}^{m} \delta_E \right) (a_{j+1}, \ldots, a_m)
$$

$$= \sum_{j=0}^{m} \frac{d^n}{d\theta^n} \sum_{(a_0,\ldots,a_m)\in\mathcal{A}^m} 1_{\mathbf{A}_m(j)}(a_0 \ldots, a_m)\, g(a_0 \ldots, a_m)$$

$$\times \left(\prod_{i=0}^{j} \mu_i\right)(a_0,\ldots,a_j) \times \left(\prod_{i=j+1}^{m} \delta_E\right)(a_{j+1},\ldots,a_m).$$

The sets $\mathbf{A}_m(j)$ are measurable subsets of $\mathcal{A}^{m+1}$ and independent of θ. Notice that for $g \in C_p(\mathcal{A}^{m+1})$ it follows that $1_{\mathbf{A}_m(j)}g \in C_p(\mathcal{A}^{m+1})$. Adapting Lemma 4.2.1 to the situation of the independent product of non-identical probability measures is straightforward. The above derivatives is thus equal to

$$\sum_{j=0}^{m} \sum_{(a_0,\ldots,a_m)\in\mathcal{A}^m} 1_{\mathbf{A}_m(j)}(a_0 \ldots, a_m)\, g(a_0 \ldots, a_m)$$

$$\times \left(\prod_{i=0}^{j} \mu_i\right)^{(n)}(a_0,\ldots,a_j) \times \left(\prod_{i=j+1}^{m} \delta_E\right)(a_{j+1},\ldots,a_m)$$

$$= \sum_{j=0}^{m} \sum_{(a_0,\ldots,a_m)\in\mathbf{A}_m(j)} g(a_0,\ldots,a_m)$$

$$\times \left(\prod_{i=0}^{j} \mu_i\right)^{(n)}(a_0,\ldots,a_j) \times \left(\prod_{i=j+1}^{m} \delta_E\right)(a_{j+1},\ldots,a_m)$$

and invoking (5.13)

$$= \sum_{(a_0,\ldots,a_m)\in\mathcal{A}^{m+1}} g(a_0,\ldots,a_m) \left[\left(\prod_{i=0}^{m} \mu_i\right)^{(n)}\right]_{\tilde{a}}(a_0,\ldots,a_m),$$

which concludes the proof of the theorem. $\square$

For $l \in \mathcal{L}[0,m;n]$ and $i \in \mathcal{I}[l]$, let $(A^{(l,i)}(k) : 0 \le k \le m)$ be distributed according to $\prod_{k=0}^{m} \mu_k^{(l_k,i_k)}$ and let $(A^{(l,i^-)}(k) : 0 \le k \le m)$ be distributed according to $\prod_{k=0}^{m} \mu_k^{(l_k,i_k^-)}$. Furthermore, for $l \in \mathcal{L}[0,m;n]$ and $i \in \mathcal{I}[l]$, let $([A^{(l,i)}(k)]_{\tilde{a}} : 0 \le k \le m)$ be distributed according to $\left[\prod_{k=0}^{m} \mu_k^{(l_k,i_k)}\right]_{\tilde{a}}$ and $([A^{(l,i^-)}(k)]_{\tilde{a}} : 0 \le k \le m)$ be distributed according to $\left[\prod_{k=0}^{m} \mu_k^{(l_k,i_k^-)}\right]_{\tilde{a}}$. Assume that $(A^{(l,i)}(k) : 0 \le k \le m)$, $(A^{(l,i^-)}(k) : 0 \le k \le m)$, $([A^{(l,i)}(k)]_{\tilde{a}} : 0 \le k \le m)$ and $([A^{(l,i^-)}(k)]_{\tilde{a}} : 0 \le k \le m)$ are constructed on a common probability space.

For $l \in \mathcal{L}[0,m;n]$ and $i \in \mathcal{I}[l]$ let $\tau_{\tilde{a}}^{(l,i)}$ denote the position of the first occurrence of $\tilde{a}$ in $(A^{(l,i)}(k) : 0 \le k \le m)$ and let $\tau_{\tilde{a}}^{(l,i^-)}$ denote the position of the first occurrence of $\tilde{a}$ in $(A^{(l,i^-)}(k) : 0 \le k \le m)$. The $[\cdot]_{\tilde{a}}$ operator translates to random sequences as follows. Applying $[\cdot]_{\tilde{a}}$ to $(A^{(l,i)}(k) : 0 \le k \le m)$, resp. to

$(A^{(l,i)}(k) : 0 \leq k \leq m)$, yields

$$[A^{(l,i)}(k)]_{\tilde{a}} = \begin{cases} A^{(l_k,i_k)}(k) & \text{for } k \leq \tau_{\tilde{a}}^{(l,i)}, \\ E & \text{for } k > \tau_{\tilde{a}}^{(l,i)}, \end{cases} \tag{5.14}$$

$$[A^{(l,i^-)}(k)]_{\tilde{a}} = \begin{cases} A^{(l_k,i_k^-)}(k) & \text{for } k \leq \tau_{\tilde{a}}^{(l,i^-)}, \\ E & \text{for } k > \tau_{\tilde{a}}^{(l,i^-)}, \end{cases} \tag{5.15}$$

and

$$c_{[A(k)]_{\tilde{a}}}^{(l,i)} = \begin{cases} c_A^{(l_k)} & \text{for } 1 \leq k \leq \max(\tau_{\tilde{a}}^{(l,i)}, \tau_{\tilde{a}}^{(l,i^-)}), \\ 1 & \text{for } k > \max(\tau_{\tilde{a}}^{(l,i)}, \tau_{\tilde{a}}^{(l,i^-)}) . \end{cases}$$

The statement in Theorem 5.2.1 can now be phrased as follows:

$$\frac{d^n}{d\theta^n} E\left[g\left(\bigotimes_{k=0}^m [A(k)]_{\tilde{a}} \otimes x_0 \right) \right]$$

$$= \sum_{l \in \mathcal{L}([0,m;n])} \sum_{i \in \mathcal{I}(l)} \frac{n!}{l_0! l_1! \ldots l_m!}$$

$$\times \left(E\left[\prod_{k=0}^m c_{[A(k)]_{\tilde{a}}}^{(l,i)} g\left(\bigotimes_{k=0}^m [A^{(l,i)}(k)]_{\tilde{a}} \otimes x_0 \right) \right] \right.$$

$$\left. - E\left[\prod_{k=0}^m c_{[A(k)]_{\tilde{a}}}^{(l,i)} g\left(\bigotimes_{k=0}^m [A^{(l,i^-)}(k)]_{\tilde{a}} \otimes x_0 \right) \right] \right) .$$

We summarize the above analysis in the following theorem.

Theorem 5.2.2 *Let $(A(k) : 0 \leq k \leq m)$ be an i.i.d. sequence of n times C_p-differentiable matrices in $\mathbb{R}_{\max}^{J \times J}$, then it holds that*

$$\left(\bigotimes_{k=0}^m [A(k)]_{\tilde{a}} \right)^{(n)} \equiv \left[\left(\bigotimes_{k=0}^m A(k) \right)^{(n)} \right]_{\tilde{a}}$$

$$\equiv \sum_{l \in \mathcal{L}[0,m;n]} \sum_{i \in \mathcal{I}(l)} \frac{n!}{l_0! \ldots l_m!} \left(\prod_{k=0}^m c_{[A(k)]_{\tilde{a}}}^{(l,i)}, \bigotimes_{k=0}^m [A^{(l,i)}(k)]_{\tilde{a}}, \bigotimes_{k=0}^m [A^{(l,i^-)}(k)]_{\tilde{a}} \right) .$$

Remark 5.2.1 *If $A \in \mathbb{R}_{\max}^{J \times J}$ is n times C_p-differentiable and $x_0 \in \mathbb{R}_{\max}^J$ is independent of θ, then $(A \otimes x_0)^{(n)} \equiv A^{(n)} \otimes x_0$ and*

$$\left(\bigotimes_{k=0}^m [A(k)]_{\tilde{a}} \otimes x_0 \right)^{(n)} \equiv \left(\bigotimes_{k=0}^m [A(k)]_{\tilde{a}} \right)^{(n)} \otimes x_0 .$$

The intuitive explanation for the above formula is that, since x_0 does not depend on θ, all (higher-order) C_p-derivatives of x_0 are 'zero.'

We now turn to pathwise derivatives of random horizon products. Let $\tau_{\tilde{a}}^{(n)}$ denote the index of the $(n+1)^{st}$ occurrence of $\tilde{a}$ in $\{A(k)\}$, with $\tau_{\tilde{a}}^{(0)} = \tau_{\tilde{a}}$. Let $l \in \mathcal{L}[0, m; n]$ and $i \in \mathcal{I}[l]$. Suppose that l has only one element different from zero and that this perturbation falls into the tail $\{A^{(l,i)}(k) : 0 \le k \le m\}$. As we have already explained for first order derivatives, see (5.12), applying the operator $[\cdot]_{\tilde{a}}$ has the effect that this perturbation doesn't contribute to the overall derivative, in formula:

$$0 = g\left(\left[\bigotimes_{k=0}^{m} A^{(l,i)}(k)\right]_{\tilde{a}}\right) - g\left(\left[\bigotimes_{k=0}^{m} A^{(l,i^-)}(k)\right]_{\tilde{a}}\right).$$

For higher-order derivatives a similar rule applies: If l has least one element different from zero that falls into the tail of $\{A^{(l,i)}(k) : 0 \le k \le m\}$, then this perturbation doesn't contribute to the overall derivative. This is a direct consequence of Theorem 5.2.1 which allows to interchange the order of differentiation and application of the operator $[\cdot]_{\tilde{a}}$. The following example will illustrate this.

Example 5.2.1 *Consider our motivating example of the multi-server model again. Here, it holds that $A(k)' = (1, D_1, D_2)$ and all higher-order derivatives of $A(k)$ are not significant. For example, let $m = 9$, $j = 6$ and take $\tilde{a} = (D_1, D_1)$. Consider $l = (0, 0, 1, 0, 0, 0, 0, 0, 1, 0)$, then $\mathcal{I}(l) = \{i_1, i_2\}$, with $i_1 = (0, 0, +1, 0, 0, 0, 0, 0, +1, 0)$, $i_1^- = (0, 0, +1, 0, 0, 0, 0, 0, -1, 0)$, $i_2 = (0, 0, -1, 0, 0, 0, 0, 0, -1, 0)$ and $i_2^- = (0, 0, -1, 0, 0, 0, 0, 0, +1, 0)\}$. Let the realization of $(A(k) : k \ge 0)$ be given as in (5.10). Recall that $\tau_{(D_1, D_1)} = 7$. Hence, l places one perturbation before $\tau_{(D_1, D_1)}$ and the second perturbation after $\tau_{(D_1, D_1)}$. Then,*

$$(A^{(l,i_1)}(k) : 0 \le m) = \underbrace{(D_1, D_2, \boldsymbol{D_1}, D_2, D_2, D_2, D_1, D_1, \boldsymbol{D_1}, D_2)}_{\text{inital segement of } \{A(k)\}}$$

and

$$(A^{(l,i_1^-)}(k) : 0 \le m) = \underbrace{(D_1, D_2, \boldsymbol{D_1}, D_2, D_2, D_2, D_1, D_1, \boldsymbol{D_2}, D_2)}_{\text{inital segement of } \{A(k)\}},$$

where the bold faced elements of the realization are those effected by l. Applying the operator $[\cdot]_{\tilde{a}}$ for $\tilde{a} = (D_1, D_1)$ yields

$$[(A^{(l,i_1)}(k) : 0 \le m)]_{(D_1,D_1)} = (D_1, D_2, \boldsymbol{D_1}, D_2, D_2, D_2, D_1, D_1, \boldsymbol{E}, E)$$
$$= [(A^{(l,i_1^-)}(k) : 0 \le m)]_{(D_1,D_1)}.$$

Hence,

$$0 = g\left(\left[\bigotimes_{k=0}^{m} A^{(l,i_1)}(k)\right]_{(D_1,D_1)}\right) - g\left(\left[\bigotimes_{k=0}^{m} A^{(l,i_1^-)}(k)\right]_{(D_1,D_1)}\right).$$

It is easily checked that it also holds that

$$0 = g\left(\left[\bigotimes_{k=0}^{m} A^{(l,i_2)}(k)\right]_{(D_1,D_1)}\right) - g\left(\left[\bigotimes_{k=0}^{m} A^{(l,i_2^-)}(k)\right]_{(D_1,D_1)}\right).$$

The above examples illustrates that any vector l that places at least one perturbation into the tail of $(A(k) : k \geq 0)$ does not contribute to the derivative.

The building principle for higher-order derivatives, as illustrated in Example 5.2.1, has the following implication. Any $l \in \mathcal{L}[0, m; n]$ has at most n entries different from zero. Hence, in order to obtain an initial segment of $\{A^{(l,i)}(k) : 0 \leq k \leq m\}$ that is of maximal length, l has to be such that it places a perturbation on the first n occurrences of $\tilde{a}$ in $\{A(k)\}$. In other words, the initial segment of $\{A^{(l,i)}(k) : 0 \leq k \leq m\}$ is at most of length $\tau_{\tilde{a}}^{(n)}$. In formula:

$$\lim_{m \to \infty} \sum_{l \in \mathcal{L}[0,m;n]} \sum_{i \in \mathcal{I}(l)} g\left(\left[\bigotimes_{k=0}^{m} A^{(l,i)}(k)\right]_{\tilde{a}}\right)$$

$$= \sum_{l \in \mathcal{L}[0,\tau_{\tilde{a}}^{(n-1)};n]} \sum_{i \in \mathcal{I}(l)} \lim_{m \to \infty} g\left(\left[\bigotimes_{k=0}^{m} A^{(l,i)}(k)\right]_{\tilde{a}}\right).$$

Following the same line of argument,

$$\lim_{m \to \infty} g\left(\left[\bigotimes_{k=0}^{m} A^{(l,i)}(k)\right]_{\tilde{a}}\right) = g\left(\left[\bigotimes_{k=0}^{\tau_{\tilde{a}}^{(n)}} A^{(l,i)}(k)\right]_{\tilde{a}}\right).$$

Indeed, the initial segment of $(A^{(l,i)}(k) : k \geq 0)$ cannot be longer than $\tau_{\tilde{a}}^{(n)}$, i.e., the point in time when sequence $\tilde{a}$ occurs for the $(n+1)^{st}$ time in $\{A(k) : k \geq 0\}$. For (l, i^-) we argue the same way. The nth order C_p-derivative thus satisfies

$$\lim_{m \to \infty} g^\tau\left(\left[\left(\bigotimes_{k=0}^{m} A(k)\right)^{(n)}\right]_{\tilde{a}}\right) = g^\tau\left(\left(\bigotimes_{k=0}^{\tau_{\tilde{a}}} [A(k)]_{\tilde{a}}\right)^{(n)}\right),$$

with

$$\left(\bigotimes_{k=0}^{\tau_{\tilde{a}}^{(0)}} [A(k)]_{\tilde{a}}\right)^{(n)} \tag{5.16}$$

$$\stackrel{\text{def}}{=} \sum_{l \in \mathcal{L}[0,\tau_{\tilde{a}}^{(n-1)};n]} \frac{n!}{l_0! l_1! \ldots l_{\tau_{\tilde{a}}^{(n-1)}}!} \sum_{i \in \mathcal{I}[l]}$$

$$\times \left(\prod_{k=0}^{\tau_{\tilde{a}}^{(n,i)}} c_{[A(k)]_{\tilde{a}}}^{(l,i)}, \bigotimes_{k=0}^{\tau_{\tilde{a}}^{(n)}} [A^{(l,i)}(k)]_{\tilde{a}}, \bigotimes_{k=0}^{\tau_{\tilde{a}}^{(n)}} [A^{(l,i^-)}(k)]_{\tilde{a}}\right).$$

As the following lemma shows, the expression defined in (5.16) yields an unbiased estimator for the nth order derivative of $\mathbb{E}_\theta[g(\bigotimes_{k=0}^{\tau_{\tilde{a}}} A(k) \otimes x_0)]$ (which justifies the notation). In Lemma 4.4.2 we have shown that $B_{g,m,\{A(k)\}}(n,p)$ is an upper bound for

$$\left| g^\tau \left(\left(\bigotimes_{k=0}^{m} A(k) \otimes x_0 \right)^{(n)} \right) \right|$$

for any $g \in C_p$, and since $\|E\|_\oplus = 0$ this implies that

$$\left| g^\tau \left(\left(\bigotimes_{k=0}^{m} [A(k)]_{\tilde{a}} \otimes x_0 \right)^{(n)} \right) \right| \leq B_{g,m,\{A(k)\}}(n,p) \tag{5.17}$$

as well. Moreover, note that, for any $g \in C_p$,

$$\left| g^\tau \left(\left(\bigotimes_{k=0}^{m} [A(k)]_{\tilde{a}} \otimes x_0 \right)^{(n)} \right) \right| = \left| g^\tau \left(\left(\bigotimes_{k=0}^{m} [A(k)]_{\tilde{a}} \right)^{(n)} \otimes x_0 \right) \right| ,$$

see Remark 5.2.1.

When we replace m by $\tau_{\tilde{a}}^{(0)}$, we have to take into account the fact that the horizon of the product depends on the order of the derivative. To this end, we set

$$B_{g,\tau_{\tilde{a}},\{A(k)\}}(n,p)$$

$$\overset{\text{def}}{=} \sum_{l \in \mathcal{L}[0,\tau_{\tilde{a}}^{(n-1)};n]} \frac{n!}{l_0! l_1! \dots l_{\tau_{\tilde{a}}^{(n-1)}}!} \sum_{i \in \mathcal{I}[l]} \prod_{k=0}^{\tau_{\tilde{a}}^{(n-1)}} c_{[A(k)]_{\tilde{a}}}^{(l,i)}$$

$$\times \left(2a_g + b_g \left(\sum_{k=0}^{\tau_{\tilde{a}}^{(n)}} \left\| [A^{(l,i)}(k)]_{\tilde{a}} \right\|_\oplus + \|x_0\|_\oplus \right)^p \right.$$

$$\left. + b_g \left(\sum_{k=0}^{\tau_{\tilde{a}}^{(n)}} \left\| [A^{(l,i^-)}(k)]_{\tilde{a}} \right\|_\oplus + \|x_0\|_\oplus \right)^p \right) \tag{5.18}$$

and, in particular,

$$B_{g,\tau_{\tilde{a}},\{A(k)\}}(0,p) \overset{\text{def}}{=} a_g + b_g \left(\sum_{k=0}^{\tau_{\tilde{a}}} \|A(k)\|_\oplus + \|x_0\|_\oplus \right)^p .$$

Following the line of argument in the proof of Lemma 4.4.2, we deduce from (5.17) that for any $g \in C_p$:

$$\left| g^\tau \left(\left(\bigotimes_{k=0}^{\tau_{\tilde{a}}} A(k) \otimes x_0 \right)^{(n)} \right) \right| \leq B_{g,\tau_{\tilde{a}},\{A(k)\}}(n,p), \quad n \geq 0, \tag{5.19}$$

where, for $n = 0$, we set in accordance with (4.8):

$$g \left(\bigotimes_{k=0}^{\tau_{\tilde{a}}} A(k) \otimes x_0 \right) = g \left(\left(\bigotimes_{k=0}^{\tau_{\tilde{a}}} A(k) \otimes x_0 \right)^{(0)} \right).$$

We obtain the following result.

Lemma 5.2.1 *For $n \geq 1$, let $A(k)$ ($k \geq 0$) be mutually stochastically indepen-dent and $(n+1)$ times C_p-differentiable matrices in $\mathbb{R}_{\max}^{J \times J}$. If, for $0 \leq m \leq n+1$,*

$$\sup_{\theta \in \Theta} \mathbb{E}_\theta \left[B_{g, \tau_{\tilde{a}}, \{A(k)\}}(m, p) \right] < \infty,$$

then

$$\frac{d^n}{d\theta^n} \mathbb{E}_\theta \left[g \left(\bigotimes_{k=0}^{\tau_{\tilde{a}}} A(k) \otimes x_0 \right) \right] = \mathbb{E}_\theta \left[g^\tau \left(\left(\bigotimes_{k=0}^{\tau_{\tilde{a}}} [A(k)]_{\tilde{a}} \right)^{(n)} \otimes x_0 \right) \right],$$

where the expression on the right-hand side is defined in (5.16).

Proof: We prove the lemma by induction. For $i = 0, 1, 2$, it holds for any For $g \in C_p$ and $m \in \mathbb{N}$ that

$$\left| g^\tau \left(\left(\bigotimes_{k=0}^{m} [A(k)]_{\tilde{a}} \right)^{(i)} \otimes x_0 \right) \right| \leq B_{g, \tau_{\tilde{a}}, \{A(k)\}}(i, p). \qquad (5.20)$$

In particular, it holds that

$$\left| g^\tau \left(\bigotimes_{k=0}^{m} [A(k)]_{\tilde{a}} \otimes x_0 \right) \right| \leq B_{g, \tau_{\tilde{a}}, \{A(k)\}}(0, p). \qquad (5.21)$$

By definition,

$$\lim_{m \to \infty} g \left(\bigotimes_{k=0}^{m} [A(k)]_{\tilde{a}} \otimes x_0 \right) = g \left(\bigotimes_{k=0}^{\tau_{\tilde{a}}} [A(k)]_{\tilde{a}} \otimes x_0 \right),$$

with probability one. We have assumed that $\mathbb{E}_\theta[B_{g, \tau_{\tilde{a}}, \{A(k)\}}(0, p)] < \infty$. This together with inequality (5.21) justifies applying the dominated convergence theorem, which yields

$$\lim_{m \to \infty} \mathbb{E}_\theta \left[g \left(\bigotimes_{k=0}^{m} [A(k)]_{\tilde{a}} \otimes x_0 \right) \right] = \mathbb{E}_\theta \left[g \left(\bigotimes_{k=0}^{\infty} [A(k)]_{\tilde{a}} \otimes x_0 \right) \right]$$

$$= \mathbb{E}_\theta \left[g \left(\bigotimes_{k=0}^{\tau_{\tilde{a}}} [A(k)]_{\tilde{a}} \otimes x_0 \right) \right].$$

Theorem 5.2.2 (the Leibnitz rule for the $[\cdot]_{\tilde{a}}$ operator) implies that, for any m and $i = 1, 2$,

$$\frac{d^i}{d\theta^i} \mathbb{E}_\theta \left[g \left(\bigotimes_{k=0}^m [A(k)]_{\tilde{a}} \otimes x_0 \right) \right] = \mathbb{E}_\theta \left[g^\tau \left(\left(\bigotimes_{k=0}^m [A(k)]_{\tilde{a}} \right)^{(i)} \otimes x_0 \right) \right].$$

Hence, following the line of argument in the proof of Lemma 4.4.2, we obtain for any m and $i \in \{1, 2\}$:

$$\sup_{\theta \in \Theta} \frac{d^i}{d\theta^i} \mathbb{E}_\theta \left[g \left(\bigotimes_{k=0}^m [A(k)]_{\tilde{a}} \otimes x_0 \right) \right] \leq \sup_{\theta \in \Theta} \mathbb{E}_\theta [B_{g,\tau_{\tilde{a}},\{A(k)\}}(i,p)],$$

where we make use of (5.20). We have assumed that $\sup_{\theta \in \Theta} \mathbb{E}_\theta [B_{g,\tau_{\tilde{a}},\{A(k)\}}(i,p)] < \infty$ for $i = 1, 2$. The proof of the statement of the theorem for $n = 1$ thus follows from Theorem G.3.1 in the Appendix.

The proof of the lemma now follows by finite induction. □

The key condition for unbiasedness in the above lemma is that $\sup_{\theta \in \Theta} \mathbb{E}_\theta [B_{g,\tau_{\tilde{a}},\{A(k)\}}(n,p)]$ is finite and we provide an explicit upper bound for $\sup_{\theta \in \Theta} \mathbb{E}_\theta [B_{g,\tau_{\tilde{a}},\{A(k)\}}(n,p)]$ in the next lemma. For the definitions of $\|A\|_\oplus$ and $c_{A(0)}$, see Lemma 4.4.2.

Lemma 5.2.2 *Let $\{A(k)\}$ be an i.i.d. sequence of n times C_p-differentiable matrices in $\mathbb{R}_{\max}^{J \times J}$ with state space $\mathcal{A}$. Let $p_{\tilde{a}}(\theta)$ denote the probability that sequence $\tilde{a}$ occurs in $\{A(k)\}$ and let $\tilde{a}$ be of length M. Provided that $\|A\|_\oplus$ is finite and that $x_0 = e$, it holds that*

$$\mathbb{E}_\theta [B_{g,\tau_{\tilde{a}},\{A(k)\}}(n,p)] \leq 2^n (c_{A(0)})^n \left(a_g + b_g (\|A\|_\oplus)^p a(p_{\tilde{a}}(\theta), M, n, p) \right),$$

where

$$a(p_{\tilde{a}}(\theta), M, n, p) = \begin{cases} n! \, M^{n+p} (n+1)^p & \text{for } p_{\tilde{a}}(\theta) = 1, \, p \geq 1, \\ (n+1)! \, M^{n+1} (n+1) \frac{2^{n-1}}{(p_{\tilde{a}}(\theta))^{n+1}} & \text{for } p_{\tilde{a}}(\theta) \in (0,1), \, p = 1. \end{cases}$$

Proof: For $l \in \mathcal{L}[0, \tau_{\tilde{a}}^{(n-1)}; n]$, let

$$V \left(g, l, \tau_{\tilde{a}}^{(n)} \right) = \sum_{i \in \mathcal{I}[l]} g^\tau \left(\left(\prod_{k=0}^{\tau_{\tilde{a}}^{(n)}} c_{A(0)}^{(l_k, i_k)}, \bigotimes_{k=0}^{\tau_{\tilde{a}}^{(n)}} [A^{(l,i)}(k)]_{\tilde{a}}, \bigotimes_{k=0}^{\tau_{\tilde{a}}^{(n)}} [A^{(l,i^-)}(k)]_{\tilde{a}} \right) \right).$$

Following the line of argument in the proof of Lemma 4.4.2, we show that, for any $g \in C_p$:

$$V\left(g, l, \tau_{\tilde{a}}^{(n)}\right)$$

$$\leq 2^{n-1}\left(\mathbf{c}_{A(0)}\right)^n$$

$$\times\left(2a_g + b_g\left(\sum_{k=0}^{\tau_{\tilde{a}}^{(n)}}\left\|\left[A^{(l,i)}(k)\right]_{\tilde{a}}\right\|_{\oplus}\right)^p + b_g\left(\sum_{k=0}^{\tau_{\tilde{a}}^{(n)}}\left\|\left[A^{(l,i^-)}(k)\right]_{\tilde{a}}\right\|_{\oplus}\right)^p\right)$$

$$\leq 2^n(\mathbf{c}_{A(0)})^n\left(a_g + b_g\left(\tau_{\tilde{a}}^{(n)}\right)^p\left(\|\mathcal{A}\|_{\oplus}\right)^p\right).$$

It remains to be shown that

$$\mathbb{E}_\theta\left[\sum_{l \in \mathcal{L}[0, \tau_{\tilde{a}}^{(n-1)}; n]} \frac{n!}{l_0! l_1! \dots l_{\tau_{\tilde{a}}^{(n-1)}}!} 1_{V(g, l, \tau_{\tilde{a}}^{(n)}) \neq 0}\left(\tau_{\tilde{a}}^{(n)}\right)^p\right] \leq a(p_{\tilde{a}}(\theta), M, n, p).$$

Let $\beta_{\tilde{a}}(k)$ denote the number of transitions in $\{A(m) : m \geq k + 1\}$ until $\tilde{a}$ has occurred for the first time. The key observation for the proof is that $l \in \mathcal{L}[0, \tau_{\tilde{a}}^{(n-1)}; n]$ only contributes if $l_0 \leq \tau_{\tilde{a}}$ and if the following condition holds:

$$l_{k-1} < l_k \leq l_{k-1} + \beta_{\tilde{a}}(l_{k-1}), \quad 1 \leq k \leq n, \tag{5.22}$$

with $l_{-1} = 0$. In words, a perturbation l_k at transition k may not occur after the sequence $\tilde{a}$ has occurred; see Example 5.2.1 for details. Let $\mathcal{L}(n)$ denote the set of vectors $l \in \mathcal{L}[0, \tau_{\tilde{a}}^{(n-1)}; n]$ that satisfy condition (5.22), that is, the set of perturbation vectors l that possibly contribute to the nth derivative, then

$$\mathbb{E}_\theta\left[\sum_{l \in \mathcal{L}[0, \tau_{\tilde{a}}^{(n-1)}; n]} \frac{n!}{l_0! l_1! \dots l_{\tau_{\tilde{a}}^{(n-1)}}!} 1_{V(g, l, \tau_{\tilde{a}}^{(n)}) \neq 0}\left(\tau_{\tilde{a}}^{(n)}\right)^p\right]$$

$$\leq n! \, \mathbb{E}_\theta\left[\sum_{l \in \mathcal{L}[0, \tau_{\tilde{a}}^{(n-1)}; n]} 1_{l \in \mathcal{L}(n)}\left(\tau_{\tilde{a}}^{(n)}\right)^p\right].$$

If $p_{\tilde{a}}(\theta) = 1$, then $\tau_{\tilde{a}} = M - 1$ and $\beta_{\tilde{a}}(k) = M$ for any k and $\tau_{\tilde{a}}^{(n)} \leq (n+1)M$. This yields

$$\mathbb{E}_\theta\left[\sum_{l \in \mathcal{L}[0, \tau_{\tilde{a}}^{(n-1)}; n]} 1_{l \in \mathcal{L}(n)}\left(\tau_{\tilde{a}}^{(n)}\right)^p\right]$$

$$\leq (n+1)^p M^p \, \mathbb{E}_\theta\left[\sum_{l \in \mathcal{L}[0, \tau_{\tilde{a}}^{(n-1)}; n]} 1_{l \in \mathcal{L}(n)}\right].$$

A necessary condition for $l \in \mathcal{L}(n)$ is that

$$\sum_{k=0}^{\tau_{\tilde{a}}^{(n-1)}} l_k = n \quad \wedge \quad l_{k+1} - l_k \leq M , \quad |l_0| \leq M ,$$

which implies:

$$\mathbb{E}_\theta \left[\sum_{l \in \mathcal{L}[0,\tau_{\tilde{a}}^{(n-1)};n]} 1_{l \in \mathcal{L}(n)} \right] \leq M^n$$

and, combining the above results, we obtain

$$\mathbb{E}_\theta \left[\sum_{l \in \mathcal{L}[0,\tau_{\tilde{a}}^{(n-1)};n]} \frac{n!}{l_0! l_1! \ldots l_{\tau_{\tilde{a}}^{(n-1)}}!} 1_{V(g,l,\tau_{\tilde{a}}^{(n)}) \neq 0} \left(\tau_{\tilde{a}}^{(n)} \right)^p \right]$$
$$\leq n! \, M^{n+p} \, (n+1)^p ,$$

which proves the first part of the lemma.

We now turn to the proof of the second part of the lemma whereby we assume that $0 < p_{\tilde{a}}(\theta) < 1$. For this part of the proof we work with the assumption $p = 1$, that is, we consider $g \in C_1$. We divide $\{A(k) : k \geq 0\}$ into blocks of length M. Let $\tau_{\tilde{a}}(1) = \tau_{\tilde{a}}$ denote the number of blocks until the first M-block equals $\tilde{a}$, i.e., for $p(\theta) = 1$ we have $\tau_{\tilde{a}}(1) = M$ and $\tau_{\tilde{a}}(1) \geq M$ for $p(\theta) < 1$, and let $\tau_{\tilde{a}}(k)$ denote the number of blocks between the $(k-1)^{st}$ and k^{th} occurrence of a M-block that equals $\tilde{a}$ (including the $\tilde{a}$ block itself). Consider $l \in \mathcal{L}[0, \tau_{\tilde{a}}^{(n-1)}; n]$. Recall that $k(l)$ denotes the position of the highest non-zero element in l. Let $k(l)$ fall into segment m, that is,

$$\sum_{k=1}^{m-1} M\tau_{\tilde{a}}(m-1) < k(l) \leq \sum_{k=1}^{m} M\tau_{\tilde{a}}(m) ,$$

where $\tau_{\tilde{a}}(0) = 0$. Such l doesn't contribute to the derivative if one of the first $(m-1)$ M-blocks equals $\tilde{a}$. In other words, we have to place at least one perturbation in each M-block (in order to destroy $\tilde{a}$). If we place at least one perturbation in each segment, then the nth derivative can at most effect the first n M-blocks. We now introduce the set

$$\mathcal{H}(n) = \left\{ h \in \{0, \ldots, n\}^n \mid \sum_{k=1}^{n} h_k = n \wedge h_k = 0 \Rightarrow h_m = 0 \text{ for } m > k \right\} .$$

We now split up $\mathcal{L}[0, \tau_{\tilde{a}}^{(n-1)}; n]$ in the following way: we first decide how many perturbations we place in the kth segment (given by h_k) and then we consider all possible combinations of distributing h_k perturbations over the $M\tau_{\tilde{a}}(k)$ places of this segment:

$$\mathbb{E}_\theta \left[\sum_{l \in \mathcal{L}[0,\tau_{\tilde{a}}^{(n-1)};n]} \frac{n!}{l_0! l_1! \dots l_{\tau_{\tilde{a}}^{(n-1)}}!} 1_{V(g,l,\tau_{\tilde{a}}^{(n)}) \neq 0} \tau_{\tilde{a}}^{(n)} \right]$$

$$\leq \mathbb{E}_\theta \left[\sum_{h \in \mathcal{H}(n)} \prod_{k=1}^{n} \sum_{\substack{l \in \{0,\dots,h_k\}^{M\tau_{\tilde{a}}(k)} \\ \sum l_m = h_k}} \frac{h_k!}{l_1! \dots l_{M\tau_{\tilde{a}}(k)}!} \sum_{j=1}^{n+1} M\tau_{\tilde{a}}(j) \right] .$$

Observe that

$$\sum_{\substack{l \in \{0,\dots,h_k\}^{M\tau_{\tilde{a}}(k)} \\ \sum l_m = h_k}} \frac{h_k!}{l_1! \dots l_{M\tau_{\tilde{a}}(k)}!} = (M\tau_{\tilde{a}}(k))^{h_k} ,$$

see Section G.5 in the Appendix. Hence,

$$\mathbb{E}_\theta \left[\sum_{l \in \mathcal{L}[0,\tau_{\tilde{a}}^{(n-1)};n]} \frac{n!}{l_0! l_1! \dots l_{\tau_{\tilde{a}}^{(n-1)}}!} 1_{V(g,l,\tau_{\tilde{a}}^{(n)}) \neq 0} \tau_{\tilde{a}}^{(n)} \right]$$

$$\leq \mathbb{E}_\theta \left[\sum_{h \in \mathcal{H}(n)} \prod_{k=1}^{n} (M\tau_{\tilde{a}}(k))^{h_k} \sum_{j=1}^{n+1} M\tau_{\tilde{a}}(j) \right]$$

$$\leq M^{n+1} \mathbb{E}_\theta \left[\sum_{h \in \mathcal{H}(n)} \prod_{k=1}^{n} (\tau_{\tilde{a}}(k))^{h_k} \sum_{j=1}^{n+1} \tau_{\tilde{a}}(j) \right] .$$

Because $\tau_{\tilde{a}}(1)$ is geometrically distributed with probability of success $p_{\tilde{a}}(\theta)$, it holds that

$$\mathbb{E}_\theta \left[(\tau_{\tilde{a}}(1))^m \right] \leq \frac{m!}{(p_{\tilde{a}}(\theta))^m} , \quad m \geq 1 ,$$

see Section C in the Appendix. Using the fact that $\{\tau_{\tilde{a}}(k)\}$ is an i.i.d. sequence, we obtain

$$
\mathbb{E}_\theta \left[\sum_{h \in \mathcal{H}(n)} \prod_{k=1}^{n} (\tau_{\tilde{a}}(k))^{h_k} \sum_{j=1}^{n+1} \tau_{\tilde{a}}(j) \right]
$$

$$
= \mathbb{E}_\theta \left[\sum_{h \in \mathcal{H}(n)} \left(\sum_{j=1}^{n} \left(\prod_{k=1, k\neq j}^{n} (\tau_{\tilde{a}}(k))^{h_k} \right) (\tau_{\tilde{a}}(j))^{h_j+1} + \tau_{\tilde{a}}(n+1) \prod_{k=1}^{n} (\tau_{\tilde{a}}(k))^{h_k} \right) \right]
$$

$$
= \sum_{h \in \mathcal{H}(n)} \sum_{j=1}^{n} \left(\prod_{k=1, k\neq j}^{n} \mathbb{E}_\theta \left[(\tau_{\tilde{a}}(k))^{h_k} \right] \mathbb{E}_\theta \left[(\tau_{\tilde{a}}(j))^{h_j+1} \right] \right)
$$

$$
+ \sum_{h \in \mathcal{H}(n)} \mathbb{E}_\theta \left[\tau_{\tilde{a}}(n+1) \right] \prod_{k=1}^{n} \mathbb{E}_\theta \left[(\tau_{\tilde{a}}(k))^{h_k} \right]
$$

$$
= \sum_{h \in \mathcal{H}(n)} \sum_{j=1}^{n} \left(\prod_{k=1, k\neq j}^{n} \mathbb{E}_\theta \left[(\tau_{\tilde{a}}(1))^{h_k} \right] \mathbb{E}_\theta \left[(\tau_{\tilde{a}}(1))^{h_j+1} \right] \right)
$$

$$
+ \sum_{h \in \mathcal{H}(n)} \mathbb{E}_\theta \left[\tau_{\tilde{a}}(1) \right] \prod_{k=1}^{n} \mathbb{E}_\theta \left[(\tau_{\tilde{a}}(1))^{h_k} \right]
$$

$$
= \sum_{h \in \mathcal{H}(n)} \left(\sum_{j=1}^{n} \left(\prod_{k=1, k\neq j}^{n} \frac{h_k!}{(p_{\tilde{a}}(\theta))^{h_k}} \frac{(h_j+1)!}{(p_{\tilde{a}}(\theta))^{h_j+1}} \right) + \frac{1}{p_{\tilde{a}}(\theta)} \prod_{k=1}^{n} \frac{h_k!}{(p_{\tilde{a}}(\theta))^{h_k}} \right)
$$

$$
\leq \sum_{h \in \mathcal{H}(n)} (n+1) \frac{(n+1)!}{(p_{\tilde{a}}(\theta))^{n+1}} \, ,
$$

where the last inequality follows from the fact that $(n_1 + n_2)! \geq (n_1)! \, (n_2)!$ for $n_1, n_2 \in \mathbb{N}$. It is easily seen that $\mathcal{H}(n)$ has 2^{n-1} elements, for a proof see Section G.5 in the Appendix. Hence,

$$
\mathbb{E}_\theta \left[\sum_{l \in \mathcal{L}[0, \tau_{\tilde{a}}^{(n-1)}; n]} \frac{n!}{l_0! l_1! \ldots l_{\tau_{\tilde{a}}^{(n-1)}}!} 1_{V(g, l, \tau_{\tilde{a}}^{(n)}) \neq 0} (\tau_{\tilde{a}}^{(n)})^p \right]
$$

$$
\leq (n+1) \, M^{n+1} \, 2^{n-1} \, \frac{(n+1)!}{(p_{\tilde{a}}(\theta))^{n+1}} \, ,
$$

which concludes the proof of the lemma. □

We will use upper bounds for $\mathbb{E}_\theta[B_{g, \tau_{\tilde{a}}, \{A(k)\}}(n, p)]$, like the one in Lemma 5.2.2, for two purposes: (a) to calculate an upper bound for the remainder term of the Taylor polynomial, and (b) to compute a lower bound for the radius of convergence of the Taylor series. The following lemma gives an alternative upper bound for $\mathbb{E}_\theta[B_{g, \tau_{\tilde{a}}, \{A(k)\}}(n, p)]$ at $p = 1$. The main difference between the two upper bounds is that the bound in the following lemma turns out to perform numerically better than that in Lemma 5.2.2. However, this superiority of the new bound comes at the cost that this bound will be only implicitly given.

Lemma 5.2.3 *Let $\{A(k)\}$ be an i.i.d. sequence of n times C_1-differentiable matrices in $\mathbb{R}_{\max}^{J \times J}$ with state space $\mathcal{A}$ such that $||A||_\oplus$ is finite. Let $p_{\tilde{a}}(\theta)$ denote the probability that sequence $\tilde{a}$ occurs for θ and let $\tilde{a}$ be of length M. Provided that $x_0 = \mathbf{e}$, it then holds for $0 < p_{\tilde{a}}(\theta) \leq 1$ that*

$$\mathbb{E}_\theta[B_{g,\tau_{\tilde{a}},\{A(k)\}}(n,1)] \leq 2^n \left(\mathbf{c}_{A(0)}\right)^n \left(a_g + b_g \, ||A||_\oplus \, b(p_{\tilde{a}}(\theta), M, n)\right),$$

where

$$b(q, M, n) \stackrel{\text{def}}{=} M^{n+1} \sum_{j=0}^{n} (-1)^j \frac{q^{n+1-j}}{(1-q)^{n+1-j}} \frac{d^j}{dq^j} \left(\frac{(1-q)^{n+1}}{q^{n+1-j}}\right),$$

for $q < 1$ and for $q = 1$

$$b(1, M, n) \stackrel{\text{def}}{=} M^{n+1} \sum_{j=0}^{n} (n+1-j)^{j+1}.$$

Proof: Let $V(g, l, \tau_{\tilde{a}}^{(n)})$ be defined as in the proof of Lemma 5.2.2. We argue as for the proof of Lemma 5.2.2, however, we will provide an alternative upper bound for

$$\mathbb{E}\left[\sum_{l \in \mathcal{L}[0, \tau_{\tilde{a}}^{(n-1)}; n]} \frac{n!}{l_0! l_1! \ldots l_{\tau_{\tilde{a}}^{(n-1)}}!} \mathbf{1}_{V(g, l, \tau_{\tilde{a}}^{(n)}) \neq 0} \tau_{\tilde{a}}^{(n)}\right].$$

Recall that M denotes the length of $\tilde{a}$. From the definition of the stopping time $\tau_{\tilde{a}}^{(n)}$ it follows that

$$\tau_{\tilde{a}}^{(n)} \leq \tau_{\tilde{a}}^{(n+1)}, \quad n \geq 0.$$

We divide the sequence $\{A(k)\}$ into blocks of length M. The probability that a block is equal to $\tilde{a}$ is $q \stackrel{\text{def}}{=} p_{\tilde{a}}(\theta)$. Assume that $0 < q < 1$. Let $\beta_1 \in \mathbb{N}$ be distributed as follows

$$P(\beta_1 = k) = (1-q)^{k-1} q, \quad k \geq 0.$$

In words, $\{\beta_1 = k\}$ is the event that the kth M-block in $\{A(k)\}$ is the first M-block that equals $\tilde{a}$. Since $\tau_{\tilde{a}}^{(0)}$ denotes the first occurrence of an $\tilde{a}$-block, this implies

$$\tau_{\tilde{a}}^{(0)} \leq M \beta_1 \quad \text{a.s.}$$

Let $B(k, q; \cdot)$ denote the binomial distribution with parameter q, that is, $B(k, q; n)$ is the probability of observing n successes in k independent trials, where the probability of success per trial is q, more formally

$$B(k, q; n) = \binom{k}{n} (1-q)^{k-n} q^n, \quad k \geq n \geq 0.$$

For technical convenience we set $B(k, q; n) = 0$ for $k < n$. The event $\{\beta_{n+1} = k+1\}$ is constituted as follows: there are n blocks among the first k blocks that are equal to $\tilde{a}$ and the $(k+1)^{st}$ block (which is the last block) is equal to $\tilde{a}$. Hence, for $n \geq 0$, the distribution of β_{n+1} is given by

$$P(\beta_{n+1} = k+1) = B(k, q; n)\, q$$
$$= \binom{k}{n} (1-q)^{k-n} q^{n+1},$$

which is the negative binomial distribution shifted by n. An upper bound for the moments of β_{n+1} is computed in Section D the Appendix.

The stopping time $\tau_{\tilde{a}}^{(n)}$ is of maximal length if we place a perturbation on each of the first n occurrences of $\tilde{a}$; see Example 5.2.1 for details. In this case, $\tau_{\tilde{a}}^{(n)} \leq M\beta_{n+1}$ and there are M^n possibilities of destroying the first n occurrences of $\tilde{a}$ through placing perturbations. Hence,

$$\mathbb{E}\left[\sum_{\substack{l \in \mathcal{L}[0, \tau_{\tilde{a}}^{(n-1)}; n] \\ \text{all pert. fall on the first } n \text{ strings } \tilde{a}}} \frac{n!}{l_0! l_1! \ldots l_{\tau_{\tilde{a}}^{(n-1)}}!} 1_{V(g, l, \tau_{\tilde{a}}^{(n)}) \neq 0} \tau_{\tilde{a}}^{(n)} \right]$$
$$\leq M^{n+1} \mathbb{E}_\theta[\beta_{n+1}].$$

If we place $n-1$ perturbations on the first $n-1$ occurrences of $\tilde{a}$ (there are M^{n-1} possibilities of doing this), then $\tau_{\tilde{a}}^{(n)}$ is at most $M\beta_n$. Moreover, there is one perturbation we are free to place on any of the $M\beta_n$ places. Hence,

$$\mathbb{E}\left[\sum_{\substack{l \in \mathcal{L}[0, \tau_{\tilde{a}}^{(n-1)}; n] \\ \text{all but one pert. fall on the first } n-1 \text{ strings } \tilde{a}}} \frac{n!}{l_0! l_1! \ldots l_{\tau_{\tilde{a}}^{(n-1)}}!} 1_{V(g, l, \tau_{\tilde{a}}^{(n)}) \neq 0} \tau_{\tilde{a}}^{(n)} \right]$$
$$\leq M^{n-1} \mathbb{E}_\theta[(M\beta_n)^2]$$
$$= M^{n+1} \mathbb{E}_\theta[\beta_n^2].$$

In general, for $0 \leq j \leq n$,

$$\mathbb{E}\left[\sum_{\substack{l \in \mathcal{L}[0, \tau_{\tilde{a}}^{(n-1)}; n] \\ (n-j) \text{ pert. fall on the first } j \text{ strings } \tilde{a}}} \frac{n!}{l_0! l_1! \ldots l_{\tau_{\tilde{a}}^{(n-1)}}!} 1_{V(g, l, \tau_{\tilde{a}}^{(n)}) \neq 0} \tau_{\tilde{a}}^{(n)} \right]$$
$$\leq M^{n+1} \mathbb{E}_\theta[\beta_{n+1-j}^{j+1}],$$

with the understanding that for $j = n$ the sum is w.r.t. the case that no perturbation is placed on the first occurrence of $\tilde{a}$, which gives

$$
\mathbb{E}\left[\sum_{l\in\mathcal{L}[0,\tau_{\tilde{a}}^{(n-1)};n]} \frac{n!}{l_0!l_1!\dots l_{\tau_{\tilde{a}}^{(n-1)}}!} 1_{V(g,l,\tau_{\tilde{a}}^{(n)})\neq 0} \tau_{\tilde{a}}^{(n)}\right]
$$

$$
\leq M^{n+1}\sum_{j=0}^{n}\mathbb{E}_\theta\left[\beta_{n+1-j}^{j+1}\right] \tag{5.23}
$$

$$
\leq M^{n+1}\sum_{j=0}^{n}(-1)^{j+1}\frac{q^{n+1-j}}{(1-q)^{n+1-j}}\frac{d^{j+1}}{dq^{j+1}}\frac{(1-q)^{n+2}}{q^{n+1-j}}\ .
$$

Inserting $p_{\tilde{a}}(\theta)$ for q concludes the proof of the lemma for the case $p_{\tilde{a}}(\theta) > 1$.

For $p_{\tilde{a}}(\theta) = 1$, it holds that $\beta_n = n$, for $n \geq 1$. Inserting this equality into (5.23) yields the second part of the lemma. $\square$

We summarize our analysis in the following theorem.

Theorem 5.2.3 *Let $\{A(k)\}$ be an i.i.d. sequence of n times C_1-differentiable matrices with state space $\mathcal{A}\subset\mathbb{R}_{\max}^{J\times J}$ such that $||\mathcal{A}||_\oplus$ is finite. Let $\tilde{a}$ be of length M and let $\tilde{a}$ occur with probability $p_{\tilde{a}}(\theta)$ and assume that*

$$
\mathbf{C}_{A(0)} \stackrel{\text{def}}{=} \sup_{\theta\in[\theta_0,\theta_0+\Delta]}\max_{0\leq m\leq n} c_{A(0)}^{(m)} < \infty\ .
$$

Provided that $x_0 = \mathbf{e}$ it holds, for any $g \in C_p$:

$$
\mathbb{E}_{\theta_0+\Delta}\left[g\left(\bigotimes_{k=0}^{\tau_{\tilde{a}}}A(k)\otimes x_0\right)\right]
$$

$$
=\sum_{m=0}^{h}\frac{\Delta^m}{m!}\mathbb{E}_{\theta_0}\left[g^\tau\left(\left(\bigotimes_{k=0}^{\tau_{\tilde{a}}}[A(k)]_{\tilde{a}}\right)^{(n)}\otimes x_0\right)\right] + r_{h+1}^{(\tilde{a},g)}(\theta_0,\Delta)\ ,
$$

whereby

$$
r_{h+1}^{(\tilde{a},g)}(\theta_0,\Delta) \leq R_{h+1}^{(\tilde{a},g)}(\theta_0,\Delta)
$$

$$
\stackrel{\text{def}}{=} \frac{\Delta^{h+1}2^{h+1}}{(h+1)!}\mathbf{C}_{A(0)}^{h+1}a_g
$$

$$
+\frac{2^{h+1}}{h!}\mathbf{C}_{A(0)}^{h+1}b_g\,||\mathcal{A}||_\oplus\int_{\theta_0}^{\theta_0+\Delta}\Big((\theta_0+\Delta-t)^h f(p_{\tilde{a}}(t),M,h+1)\Big)dt\ ,
$$

with $f(q,M,h+1)$ either equal to $b(p_{\tilde{a}}(\theta),M,h+1)$ as defined in Lemma 5.2.3, or equal to $a(q,M,h+1,1)$ as defined in Lemma 5.2.2.

Proof: We only prove the statement for $f(q,M,h+1) = b(p_{\tilde{a}}(\theta),M,h+1)$ since for $f(q,M,h+1) = a(p_{\tilde{a}}(\theta),M,h+1,1)$ the proof follows from the same line of argument.

Note that for $t \in (\theta_0, \theta_0 + \Delta)$ it holds that

$$
\left| \frac{d^{h+1}}{d\theta^{h+1}} \right|_{\theta=t} \mathbb{E}_\theta \left[g \left(\bigotimes_{k=0}^{\tau_{\tilde{a}}} A(k) \otimes x_0 \right) \right] \right|
$$

$$
= \left| \mathbb{E}_t \left[g^\tau \left(\left(\bigotimes_{k=0}^{\tau_{\tilde{a}}} [A(k)]_{\tilde{a}} \right)^{(h+1)} \otimes x_0 \right) \right] \right|
$$

$$
\overset{(5.19)}{\leq} \mathbb{E}_t \left[B_{g,\tau_{\tilde{a}},\{A(k)\}}(h+1,1) \right]
$$

$$
\leq 2^{h+1} \, \mathbf{C}_{A(0)}^{h+1} \left(a_b + b_g \, ||\mathcal{A}||_\oplus \, b(p_{\tilde{a}}(t), M, h+1) \right),
$$

where the last inequality follows from Lemma 5.2.3. Hence, the remainder term is bounded by

$$
\frac{1}{h!} 2^{h+1} \, \mathbf{C}_{A(0)}^{h+1} \int_{\theta_0}^{\theta_0 + \Delta} (\theta_0 + \Delta - t)^h \left(a_b + b_g \, ||\mathcal{A}||_\oplus \, b(p_{\tilde{a}}(t), M, h+1) \right) dt \, ,
$$

see equation (G.2) on page 294 in the Appendix. Rearranging terms, the upper bound for the remainder equals

$$
\frac{1}{h!} 2^{h+1} \, \mathbf{C}_{A(0)}^{h+1} \, a_g \int_{\theta_0}^{\theta_0 + \Delta} (\theta_0 + \Delta - t)^h dt
$$

$$
+ \frac{2^{h+1} \, \mathbf{C}_{A(0)}^{h+1} \, b_g \, ||\mathcal{A}||_\oplus}{h!} \int_{\theta_0}^{\theta_0 + \Delta} \left((\theta_0 + \Delta - t)^h b(p_{\tilde{a}}(t), M, h+1) \right) dt
$$

$$
= \frac{\Delta^{h+1} 2^{h+1} \, \mathbf{C}_{A(0)}^{h+1} \, a_g}{(h+1)!}
$$

$$
+ \frac{2^{h+1} \, \mathbf{C}_{A(0)}^{h+1} \, b_g \, ||\mathcal{A}||_\oplus}{h!} \int_{\theta_0}^{\theta_0 + \Delta} \left((\theta_0 + \Delta - t)^h b(p_{\tilde{a}}(t), M, h+1) \right) dt \, ,
$$

which concludes the proof of the theorem. □

We conclude this section by showing that C_p-analyticity of halted sequences is preserved under the $\otimes$-operation.

Theorem 5.2.4 Let $\{A(k)\}$ be an i.i.d. sequence in $\mathbb{R}_{max}^{J \times J}$. If $A(k)$ is C_p-analytic on Θ, then $A_{\tilde{a}}(k)$ is C_p-analytical on Θ for any k, and $[A_{\tilde{a}}(k+1)]_{\tilde{a}} \otimes [A(k)]_{\tilde{a}}$ is C_p-analytical on Θ for $k \geq 0$. Moreover, if, for $\theta_0 \in \Theta$, the Taylor series for $A(k)$ has domain of convergence $U_{\theta_0}^A$, then the domain of convergence of the Taylor series for $[A(k)]_{\tilde{a}}$ is $U_{\theta_0}^A$ for any k. Moreover, the domain of convergence of the Taylor series for $[A(k+1)]_{\tilde{a}} \otimes [A(k)]_{\tilde{a}}$ is $U_{\theta_0}^A$, for any k.

Proof: Observe that all arguments used in the proof of Corollary 4.6.2 remain valid when we integrate over a measurable subset of the state space. Hence, if we split up the state space in disjunct sets representing the possible outcomes of $[A(k+1)]_{\tilde{a}} \otimes [A(k)]_{\tilde{a}}$, then the proof follows from the same line of argument as the proof of Theorem 4.2.1 and Theorem 4.1.1, respectively. □

Remark 5.2.2 *The framework of this section can be extended to the more general case of halting the evaluation of the sequence $x(k)$ whenever $\underline{x(k)}$ hits a certain set. For example, let $B \subset \mathbb{PR}_{\max}^J$ and halt the system when $\overline{x(k)} \in B$. Take*

$$\tau_{x_0,B}(\theta) = \inf \left\{ m : \overline{\bigotimes_{n=0}^{m} A_\theta(n) \otimes x_0} \in B \right\} \tag{5.24}$$

as stopping time, and let $\mathbf{A}_{m,x_0,B}(j)$ in the definition of the halted version, see (5.13) on 206, be defined as

$$\left\{ (a_0, a_1, \ldots, a_m) \in \mathcal{A}^{m+1} : j = \min \left\{ k : \overline{\bigotimes_{n=0}^{k} a_n \otimes x_0} = B \right\} \right\} .$$

Then, the results in this section readily extend to (higher-order) C_p-differentiability and C_p-analyticity, respectively, of

$$\bigotimes_{k=0}^{\tau_{x_0,B}(\theta)} A_\theta(k) \otimes x_0 .$$

We illustrate the above setup with the following example. Consider an open queuing system with J stations, see Example 1.5.2. Denote by B the set of state-vectors x such that $x_J - x_0$ is greater than a threshold value h. Assume that the system is initially empty and model the evolution of the system via a homogeneous recursion, see Section 1.4.3. Consequently, when $\overline{x(k)}$ enters B, then the total sojourn time of the kth customer exceeds h. Hence, $x_J(\tau_{x_0,B}(\theta))$ yields the time at which the first customer that leaves the system violates the sojourn time restriction.

5.2.2   The Time Until Two Successive Breakdowns

Let $\{A(k)\}$ be a sequence of i.i.d. Bernoulli-(θ)-distributed matrices with state space $\mathcal{A} = \{D_1, D_2\} \subset \mathbb{R}_{\max}^{J \times I}$, as defined in Example 1.5.5. Take $\tilde{a} = (D_1, D_1)$ the event that two successive breakdowns occur, then the probability of observing the sequence is $p_{(D_1,D_1)}(\theta) = \theta^2$. Only the first-order C_p-derivative of $A(k)$ is significant with $A^{(1,+1)}(k) = D_1$, $A^{(1,-1)}(k) = D_2$ and $c^{(1)} = 1$.

We are interested in the expected time at which (for the first time) the second breakdown of two consecutive breakdown occurs. As already explained at the beginning of Section 5.2, for $x \in \mathbb{R}_{\max}^4$, we take $g(x) = (x)_3$ and the quantity of interest is $\mathbb{E}_\theta[g(x(\tau_{(D_1,D_1)}))]$. In particular, $|g(x)| \leq \|x\|_\oplus$, and we may take $a_g = 0, b_g = 1$.

For $\theta \in (0,1]$, $\tau_{\tilde{a}} = \tau_{(D_1,D_1)}$ is a.s. finite and the C_p-derivative of $x(\tau_{(D_1,D_1)})$

at θ reads

$$\left(\bigotimes_{k=0}^{\tau_{(D_1,D_1)}} [A(k)]_{(D_1,D_1)}\right)^{(1)} \otimes x_0$$

$$= \sum_{l\in\mathcal{L}[0,\tau_{(D_1,D_1)};1]} \sum_{i\in\mathcal{I}[l]}$$

$$\left(1, \bigotimes_{k=0}^{\tau_{(D_1,D_1)}^{(1)}} [A^{(l,i)}(k)]_{(D_1,D_1)} \otimes x_0, \bigotimes_{k=0}^{\tau_{(D_1,D_1)}^{(1)}} [A^{(l,i^-)}(k)]_{(D_1,D_1)} \otimes x_0\right).$$

In particular, at $\theta_0 = 1$ it holds that

$$(A(k) : k \geq 0) = (D_1, D_1, D_1, \dots),$$

which shows that $\tau_{(D_1,D_1)} = \tau_{(D_1,D_1)}^{(0)} = 1$ and $\tau_{(D_1,D_1)}^{(1)} = 3$. This gives

$$\left(\bigotimes_{k=0}^{\tau_{(D_1,D_1)}} [A(k)]_{(D_1,D_1)}\right)^{(1)} \otimes x_0$$

$$\equiv \sum_{l\in\mathcal{L}[0,1;1]} \sum_{i\in\mathcal{I}[l]} \left(1, \bigotimes_{k=0}^{3} [A^{(l,i)}(k)]_{(D_1,D_1)}, \bigotimes_{k=0}^{3} [A^{(l,i^-)}(k)]_{(D_1,D_1)}\right) \otimes x_0.$$

Moreover, taking the C_p-derivative at $\theta_0 = 1$, we obtain

$$\mathcal{L}[0,1;1] = \{(0,1),(1,0)\}$$

and

$$\mathcal{I}[(0,1)] = \{(0,+1)\} \quad \text{and} \quad \mathcal{I}[(1,0)] = \{(+1,0)\}.$$

To illustrate the construction of the first order C_p-derivative, take, for example, $l = (1,0)$ and $i = (+1,0)$. The positive part of the C_p-derivative of $A(k)$ is D_1, which gives

$$(A^{((1,0),(+1,0))}(k) : k \geq 0) = (A(k) : k \geq 0) = (D_1, D_1, D_1, D_1, \dots).$$

The operator $[\cdot]_{(D_1,D_1)}$ sets $A^{((1,0),(+1,0))}(k)$ to E after the first occurrence of the sequence (D_1, D_1) and applying the $[\cdot]_{(D_1,D_1)}$ operator gives

$$([A^{((1,0),(+1,0))}(k)]_{(D1,D_1)} : k \geq 0) = (D_1, D_1, E, E, \dots).$$

The second occurrence of (D_1, D_1) in $\{A(k)\}$ happens for $\theta_0 = 1$ at $\tau_{(D_1,D_1)}^{(1)} = 3$ and for any $l \in \mathcal{L}[0,\tau_{(D_1,D_1)};1]$ and $i \in \mathcal{I}[l]$ the initial segment of $\{[A^{((1,0),(+1,0))}(k)]_{(D1,D_1)} : k \geq 0\}$ will be at most of length 3. We thus obtain

$$\bigotimes_{k=0}^{3} [A^{((1,0),(+1,0))}(k)]_{(D_1,D_1)} \otimes x_0 \equiv E \otimes E \otimes D_1 \otimes D_1 \otimes x_0$$

$$\equiv D_1 \otimes D_1 \otimes x_0.$$

The negative part of the C_p-derivative of $A(k)$ is D_2 and we obtain for $l = (1,0)$ and $i = (-1,0)$:

$$(A^{((1,0),(-1,0))}(k) : k \geq 0) = (D_2, D_1, D_1, D_1, D_1 \ldots)$$

and applying the $[\cdot]_{(D_1,D_1)}$ operator gives

$$([A^{((1,0),(-1,0))}(k)]_{(D_1,D_1)} : k \geq 0) = (D_2, D_1, D_1, E, E, \ldots).$$

Hence, the first occurrence of the sequence (D_1, D_1) is at $k = 2$, which yields

$$\bigotimes_{k=0}^{3} [A^{((1,0),(-1,0))}(k)]_{(D_1,D_1)} \otimes x_0 \equiv E \otimes D_1 \otimes D_1 \otimes D_2 \otimes x_0$$

$$\equiv D_1 \otimes D_1 \otimes D_2 \otimes x_0.$$

In the same vein, we obtain for $l = (0,1)$ and $i = (0,+1)$,

$$\bigotimes_{k=0}^{3} [A^{((0,1),(0,+1))}(k)]_{(D_1,D_1)} \otimes x_0 \equiv D_1 \otimes D_1 \otimes x_0$$

and for $l = (0,1)$ and $i = (0,-1)$,

$$\bigotimes_{k=0}^{3} [A^{((0,1),(0,-1))}(k)]_{(D_1,D_1)} \otimes x_0 \equiv D_1 \otimes D_1 \otimes D_2 \otimes D_1 \otimes x_0.$$

For any $g \in C_p$, it therefore holds that

$$\frac{d}{d\theta} \mathbb{E}_1[g(x(\tau_{(D_1,D_1)}))] = 2\, g(D_1 \otimes D_1 \otimes x_0)$$
$$- g(D_1 \otimes D_1 \otimes D_2 \otimes x_0)$$
$$- g(D_1 \otimes D_1 \otimes D_2 \otimes D_1 \otimes x_0).$$

Following the above line of argument, the second order C_p-derivative reads in explicit from

$$\frac{d^2}{d\theta^2} \mathbb{E}_1[g(x(\tau_{(D_1,D_1)}))] = g(D_1 \otimes D_1 \otimes x_0)$$
$$+ g(D_1 \otimes D_1 \otimes D_2 \otimes D_2 \otimes x_0)$$
$$+ g(D_1 \otimes D_1 \otimes D_2 \otimes D_1 \otimes D_2 \otimes x_0)$$
$$+ g(D_1 \otimes D_1 \otimes D_2 \otimes D_2 \otimes D_1 \otimes x_0)$$
$$+ g(D_1 \otimes D_1 \otimes D_2 \otimes D_1 \otimes D_2 \otimes D_1 \otimes x_0)$$
$$- 2g(D_1 \otimes D_1 \otimes D_2 \otimes x_0)$$
$$- 3\, g(D_1 \otimes D_1 \otimes D_2 \otimes D_1 \otimes x_0),$$

for any $g \in C_p$. Explicit expressions for higher-order C_p-derivatives can be obtained just as easy.

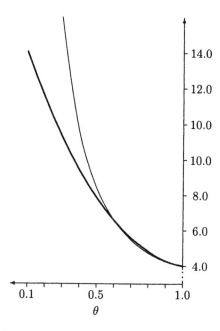

Figure 5.9: Taylor polynomial of degree $h = 3$ at $\theta_0 = 1$ for the expected time until two consecutive breakdowns.

Figure 5.9 shows the Taylor polynomial for degree $h = 3$, where we let $\sigma = 1$ and $\sigma' = 2$. The thin line indicates the true value of $\mathbb{E}_\theta[x_3(\tau_{(D_1,D_1)})]$ and the thick line shows the Taylor polynomial. The figure shows that the approximation is fairly accurate for values of Δ up to 0.4. To illustrate the influence of the order of the Taylor polynomial, we plot in Figure 5.10 the Taylor polynomial at $\theta_0 = 1$ of degree 5, where we again take $\sigma = 1$ and $\sigma' = 2$. The thick line shows the Taylor polynomial and the thin line gives the true value. Figure 5.11 plots the actual error for $h = 3$, where the actual error is obtained by taking the difference between $\mathbb{E}_\theta[x_3(\tau_{(D_1,D_1)})]$ and a Taylor polynomial of degree $h = 3$. Figure 5.12 shows the error for predicting $\mathbb{E}_\theta[x_3(\tau_{(D_1,D_1)})]$ by a Taylor polynomial of degree $h = 5$.

We now discuss the quality of our bound for the remainder term. Table 5.3 lists the bound for the remainder term for $h = 3$ and $h = 5$, respectively, for various Δ's where we evaluate the remainder term by $b(\theta_0^2, h + 1, M)$ given in Lemma 5.2.3. Comparing the values in Table 5.3 with the true error as shown in Figure 5.9 (Figure 5.11) and Figure 5.10 (Figure 5.12), respectively, we conclude that our upper bound for the remainder term is of only poor quality. As a last point of discussion, we turn to the upper bound for the remainder term as obtained by the mapping $a(\theta_0^2, h+1, M, 1)$ given in Lemma 5.2.2. Table 5.4 shows the numerical values for the (upper bound of the) remainder term. Comparing the values in Table 5.3 with those in Table 5.4 shows that the upper bound by

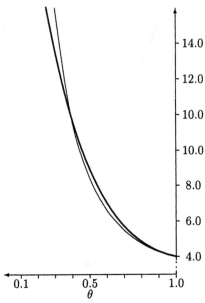

Figure 5.10: Taylor polynomial of degree $h = 5$ at $\theta_0 = 1$ for the expected time until two consecutive breakdowns.

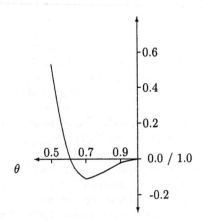

Figure 5.11: Error for the Taylor polynomial of degree $h = 3$ at $\theta_0 = 1$ (see Figure 5.9).

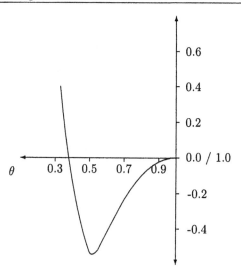

Figure 5.12: Error for the Taylor polynomial of degree $h = 5$ at θ_0 (see Figure 5.10).

Table 5.3: Bound for the remainder term using $b(\theta_0^2, h+1, M)$ ($\theta_0 = 1$, $h = 3, 5$ and $M = 2$).

Δ	$R_4^{((D_1,D_1),(\cdot)_3)}(1,\Delta)$	$R_6^{((D_1,D_1),(\cdot)_3)}(1,\Delta)$
0.05	4.0779×10^{-1}	3.2365×10^{-2}
0.1	8.3073	3.2675
0.15	55.3554	58.3219

Table 5.4: Bound for the remainder term using $a(\theta_0^2, h+1, M, 1)$ ($\theta_0 = 1$, $h = 3, 5$ and $M = 2$).

Δ	$R_4^{((D_1,D_1),(\cdot)_3)}(1,\Delta)$	$R_6^{((D_1,D_1),(\cdot)_3)}(1,\Delta)$
0.05	1.9327	7.4427×10^{-1}
0.1	47.9138	91.6786
0.15	386.7890	2094.8819

$b(\theta_0^2, h+1, M)$ out-performs the one by $a(\theta_0^2, h+1, M, 1)$.

We now turn to our lower bound for the radius of convergence of the Taylor series. According to the formula of Cauchy-Hadamard, see (G.4) in Section G.4 in the Appendix, the radius of convergence of the Taylor series, denoted by r, is given by

$$r = \left(\limsup \left(\frac{1}{n!} \left| \frac{d^n}{d\theta^n} \mathbb{E}_1[x_3(\tau_{(D_1,D_1)})] \right| \right)^{\frac{1}{n}} \right)^{-1}.$$

At $\theta_0 = 1$, the upper bound for the nth order C_1-derivative in Lemma 5.2.2 applies, and the expression on the right-hand side of the above formula is bounded by

$$r \geq \left(\limsup \left(\frac{2^{n+1}}{n!} a(1,2,n,1) \right)^{\frac{1}{n}} \right)^{-1}.$$

Inserting the explicit representation for $a(1,2,n,1)$, we obtain a lower bound for the radius of convergence through

$$r \geq \left(\limsup_n \left(2^{n+1} (n+1) 2^{n+1} \right)^{\frac{1}{n}} \right)^{-1} = \frac{1}{4}, \qquad (5.25)$$

where we use the fact that $\|\mathcal{A}\|_\oplus = 2$ and $a_{(\cdot)_3} = 0, b_{(\cdot)_3} = 1$. Hence, we obtain $1/4$ as lower bound for the radius of convergence.

As we have already noticed, the quality of the approximation increases with the order of the Taylor polynomial, see Figure 5.9 and Figure 5.10. Specifically, the Taylor series of degree 5 provides a feasible approximation for $\Delta \leq 0.6$, whereas that of degree 3 only yields good results for $\Delta \leq 0.4$. This illustrates that our lower bound for the radius of convergence of the Taylor series, which turns out to be 0.25 ($M = 2$), is a rather conservative lower bound. It is worth noting that for Δ large enough, our upper bound for the remainder term is increasing with respect to the degree of the Taylor polynomial. This effect already occurs while $\Delta < r$, which illustrates the imperfection of our bound because for these values of Δ convergence of the Taylor series implies that eventually the remainder term has to decrease when the degree of the Taylor polynomial is increased.

In the general case (that is, $\theta_0 < 1$, M and $\|\mathcal{A}\|_\oplus$ arbitrary), the lower bound for the radius of convergence of the Taylor series for $\mathbb{E}_\theta[x_3(\tau_{(D_1,D_1)})]$ at θ_0 reads

$$\frac{1}{r(\theta_0)} = \limsup_n \left(\frac{2^n \|\mathcal{A}\|_\oplus c_{A(0)}^n}{n!} \right.$$

$$\left. \times \left((n+1)! M^{n+1} (n+1) 2^{n-1} \frac{1}{\theta_0^{M(n+1)}} \right) \right)^{\frac{1}{n}},$$

which gives

$$r(\theta_0) = \frac{1}{4\, M\, c_{A(0)}} \theta_0^M$$

$$= \frac{1}{4\, M\, c_{A(0)}} p_{\tilde{a}}(\theta_0).$$

as lower bound for the radius of convergence. Hence, for $M = 2$ and $\mathbf{c}_{A(0)} = 1$,

$$r(\theta_0) = \frac{1}{8}\theta_0^2,$$

for $\theta_0 \in (0, 1]$. Observe that $r(1) = 1/8$ is smaller by a factor of 2 than the lower bound $r = 1/4$ in (5.25).

5.3 Taylor Series Expansions for the Lyapunov Exponent

In this section, we study sequences $\{A(k)\} = \{A_\theta(k)\}$ with $\theta \in \Theta$. We adjust conditions (C1) to (C3) in Section 2.5.1 (see page 100) accordingly:

(C1) For any $\theta \in \Theta$, the sequence $\{A_\theta(k)\}$ is i.i.d. with common countable state space $\mathcal{A}$.

(C2) Each $A \in \mathcal{A}$ is regular.

(C3) There is a set $\mathcal{C}$ of matrices such that each $C \in \mathcal{C}$ is primitive. Furthermore, each $C \in \mathcal{C}$ is a pattern of $\{A_\theta(k)\}$ for any $\theta \in \Theta$.

By assumption (C3), we may choose a pattern C and take $\tilde{a}$ as the $c(C)$-fold concatenation of C, where $c(C)$ denotes the coupling time of C. Under (C1) to (C3), the Lyapunov exponent of $\{A_\theta(k)\}$, denoted by $\lambda(\theta)$, exists, see Theorem 2.6.2. The goal of this section is to represent the Lyapunov exponent of $\{A(k)\}$ by a Taylor series.

In Section 2.6.2, we showed that the Lyapunov exponent of a max-plus linear system can be written as the difference between two products over a random number of matrices. In this section, we combine this representation with our results on Taylor series expansions over random horizon products as established in the previous section. We thereby will obtain Taylor series expansions for the Lyapunov exponent of max-plus linear systems. This approach has the following benefits:

- θ may influence either particular entries of the max-plus model or the distribution of the entire matrix.

- The Taylor series can computed at any point of analyticity, which is in contrast to the results known so far, where only Maclaurin series have been studied.

- Lower bounds for the radius of convergence of the Taylor series for the Lyapunov exponent are deduced from more elementary properties of the system, which allows us to establish lower bounds for the radius of convergence in a very simple manner.

- Upper bounds for the remainder term are obtained in explicit form.

We illustrate our approach with the Bernoulli scheme in Example 1.5.5.

The Lyapunov Problem: Let $\theta \in \Theta$ be a real-valued parameter, Θ being an interval. We shall take θ to be a variational parameter of an i.i.d. sequence $\{A_\theta(k)\}$ of square matrices in $\mathbb{R}_{\max}^{J \times J}$ and study sequences $\{x_\theta(k)\}$ following

$$x_\theta(k+1) = A_\theta(k) \otimes x_\theta(k), \quad k \geq 0,$$

with $x_\theta(0) = x_0$ for all θ. We assume that $\{A_\theta(k)\}$ satisfies (**C1**) to (**C3**), for $\theta \in \Theta$. The aim of this section is to write the Lyapunov exponent of $\{A_\theta(k)\}$, given by

$$\lambda(\theta) \otimes \mathbf{e} = \lim_{k \to \infty} \frac{1}{k} \mathbb{E}[(x_\theta(k))_j], \quad 1 \leq j \leq J, \tag{5.26}$$

as a Taylor series.

In Section 5.3.1 we will establish sufficient conditions for analyticity of the Lyapunov exponent. In Section 5.3.2, we apply these results to the Bernoulli scheme. Finally, Section 5.3.3 discusses the relation between our result and the Taylor series known in the literature.

5.3.1 Analytic Expansion of the Lyapunov Exponent

As explained in Section 2.6.2, under appropriate conditions, the Lyapunov exponent can be represented by the difference between two products of matrices, where the range of each product is given by stopping time η. The main difference between the setup of Section 2.6.2 and the current section is that in the setup in Section 2.6.2 time runs backwards, whereas in (5.26) time runs forward. Thus, in order to use results of the previous section for the current analysis we have to reverse time. When $\{A(k)\}$ is i.i.d., this can be done without any difficulty and we will freely use results from the previous sections in reversed time.

Following Section 2.6.2, analyticity of the Lyapunov exponent can be deduced from analyticity of the product $\mathbb{E}[\bigotimes_{k=-\eta}^{0} A(k) \otimes x_0]$, where η is the time of the first occurrence of $\tilde{a}$ going backward from time 0.

We write $\eta_{\tilde{a}}$ for the number of transitions in $\{A(k) : 0 \geq k\}$ until the first occurrence of $\tilde{a}$, that is, $\eta_{\tilde{a}}$ is the counterpart of $\tau_{\tilde{a}}$. Following Section 2.6.2 we let $\eta_{\tilde{a}} > 0$ and the actual time of the first occurrence of $\tilde{a}$ in $\{A(k) : 0 \geq k\}$ is thus given by $-\eta_{\tilde{a}}$. In the same vein we adjust the notation introduced in the previous section(s) to time running backwards. For $k \leq 0$, we define $[A(k)]_{\tilde{a}}$ such that $[A(k)]_{\tilde{a}} = A(k)$ as long as $\tilde{a}$ hasn't occurred in $\{A(k) : 0 \geq k\}$ and $[A(k)]_{\tilde{a}} = E$ otherwise, that is, $[A(k)]_{\tilde{a}} = A(k)$ for $0 \geq k \geq -\eta_{\tilde{a}}$ and $[A(k)]_{\tilde{a}} = E$ for $-\eta_{\tilde{a}} > k$.

Furthermore, denoting by $\eta_{\tilde{a}}^{(n)}$ the index k such that at k the $(n+1)^{st}$ occurrence of the pattern $\tilde{a}$ in $\{A(k) : 0 \geq k\}$ takes place, we adapt definition

(5.16) to time running backwards as follows:

$$\left(\bigotimes_{k=-\eta_{\tilde{a}}}^{0} [A(k)]_{\tilde{a}} \right)^{(n)} \tag{5.27}$$

$$\equiv \sum_{l \in \mathcal{L}[-\eta_{\tilde{a}}^{(n-1)}, 0; n]} \frac{n!}{l_{-\eta_{\tilde{a}}^{(n-1)}}! \ldots l_{-1}! l_0!} \sum_{i \in \mathcal{I}[l]}$$

$$\left(\prod_{k=-\eta_{\tilde{a}}^{(n)}}^{0} c_{[A(k)]_{\tilde{a}}}^{(l,i)} , \bigotimes_{k=-\eta_{\tilde{a}}^{(n)}}^{0} [A^{(l,i)}(k)]_{\tilde{a}} , \bigotimes_{k=-\eta_{\tilde{a}}^{(n)}}^{0} [A^{(l,i^-)}(k)]_{\tilde{a}} \right).$$

The bound $B_{g, \eta_{\tilde{a}}, \{A(k)\}}(n, p)$, defined in (5.18) reads in reserved time

$$B_{g, \eta_{\tilde{a}}, \{A(k)\}}^{0}(n, p)$$

$$\stackrel{\text{def}}{=} \sum_{l \in \mathcal{L}[-\eta_{\tilde{a}}^{(n-1)}, 0; n]} \frac{n!}{l_{-\eta_{\tilde{a}}^{(n-1)}}! \ldots l_{-1}! l_0!} \sum_{i \in \mathcal{I}[l]} \prod_{k=-\eta_{\tilde{a}}^{(n)}}^{0} c_{[A(k)]_{\tilde{a}}}^{(l,i)}$$

$$\times \left(2 a_g + b_g \left(\sum_{k=-\eta_{\tilde{a}}^{(n)}}^{0} \left\| A^{(l_k, i_k)}(k) \right\|_{\oplus} + \|x_0\|_{\oplus} \right)^p \right.$$

$$\left. + b_g \left(\sum_{k=-\eta_{\tilde{a}}^{(n)}}^{0} \left\| A^{(l_k, i_k^-)}(k) \right\|_{\oplus} + \|x_0\|_{\oplus} \right)^p \right),$$

for $n \geq 1$, and, for $n = 0$,

$$B_{g, \eta_{\tilde{a}}, \{A(k)\}}^{0}(0, p) \stackrel{\text{def}}{=} 2 a_g + b_g \left(\sum_{k=-\eta_{\tilde{a}}}^{0} \|A(k)\|_{\oplus} + \|x_0\|_{\oplus} \right)^p.$$

Following the line of the proof of Lemma 4.4.2 it follows for any $g \in C_p$ that

$$\left| g^\tau \left(\left(\bigotimes_{k=-\eta_{\tilde{a}}}^{0} [A(k)]_{\tilde{a}} \otimes x_0 \right)^{(n)} \right) \right| \leq B_{g, \eta_{\tilde{a}}, \{A(k)\}}^{0}(n, p),$$

for $n \geq 1$, and

$$\left| g^\tau \left(\left(\bigotimes_{k=-\eta_{\tilde{a}}}^{0} [A(k)]_{\tilde{a}} \otimes x_0 \right)^{(0)} \right) \right| \leq B_{g, \eta_{\tilde{a}}, \{A(k)\}}^{0}(0, p),$$

where we set

$$\bigotimes_{k=-\eta_{\tilde{a}}}^{0} A(k) \otimes x_0 = \left(\bigotimes_{k=-\eta_{\tilde{a}}}^{0} [A(k)]_{\tilde{a}} \otimes x_0 \right)^{(0)}.$$

The following lemma is a straightforward adaptation of Lemma 5.2.1.

Lemma 5.3.1 *For $n \geq 1$, let $A(k)$ $(0 \geq k)$ be mutually stochastically indepen-dent and $(n+1)$ times C_p-differentiable matrices in $\mathbb{R}_{\max}^{J \times J}$. If, for $0 \leq m \leq n+1$,*

$$\sup_{\theta \in \Theta} \mathbb{E}_\theta [B^0_{\{g, \eta_{\tilde{a}}, A(k)\}}(m, p)] < \infty,$$

then

$$\frac{d^n}{d\theta^n} \mathbb{E}_\theta \left[g \left(\bigotimes_{k=-\eta_{\tilde{a}}}^{0} A(k) \otimes x_0 \right) \right] = \mathbb{E}_\theta \left[g^\tau \left(\left(\bigotimes_{k=-\eta_{\tilde{a}}}^{0} [A(k)]_{\tilde{a}} \right)^{(n)} \otimes x_0 \right) \right],$$

for any $g \in C_p$.

Before we can state the main result of this section, we provide an upper bound for

$$\left(A(1) \otimes \bigotimes_{k=-\eta_{\tilde{a}}}^{0} [A(k)]_{\tilde{a}} \otimes x_0 \right)^{(n)}.$$

However, before we state the result we note that one has to distinguish

$$\left(A(1) \otimes \bigotimes_{k=-\eta_{\tilde{a}}}^{0} [A(k)]_{\tilde{a}} \otimes x_0 \right)^{(n)}$$

and

$$\left([A(1)]_{\tilde{a}} \otimes \bigotimes_{k=-\eta_{\tilde{a}}}^{0} [A(k)]_{\tilde{a}} \otimes x_0 \right)^{(n)}.$$

In the former expression $[\cdot]_{\tilde{a}}$ is applied to $\{A^{(l,i)}(k) : 0 \geq k\}$ whereas in the latter it is applied to $\{A^{(l,i)}(k) : 1 \geq k\}$, for $l \in \mathcal{L}[-\eta_{\tilde{a}}^{(n-1)}, \ldots, 1; n]$ and $i \in \mathcal{I}[l]$. To illustrate the difference, we consider the multi-server example. Let

$$(A(1), A(0), A(-1), A(-2), \ldots) = (D_1, D_1, D_1, D_2, D_1, \ldots)$$

and consider $l = (1, 0, 0, \ldots) \in \mathcal{L}[-\eta_{\tilde{a}}^{(n-1)}, \ldots, 1; n]$, $i = (1, 0, 0, 0, \ldots) \in \mathcal{I}[l]$. Then,

$$(A^{(l,i)}(k) : 1 \geq k) = (D_1, D_1, D_1, D_2, D_1, \ldots)$$

and

$$([A^{(l,i)}(k)]_{\tilde{a}} : 1 \geq k) = (D_1, D_1, E, E, E, \ldots),$$

whereas

$$(A^{(l_1,i_1)}(1), [A^{(l,i)}(k)]_{\tilde{a}} : 0 \geq k) = (D_1, D_1, D_1, E, E, \ldots).$$

The following lemma, which is a variant of the Lemma 4.4.2, provides the desired upper bound.

Lemma 5.3.2 *Let $\{A(k)\}$ be an i.i.d. sequence of n times C_p-differentiable matrices in $\mathbb{R}_{\max}^{J \times J}$. For any $g \in C_p$ it holds that*

$$\left| g^\tau \left(\left(A(1) \otimes \bigotimes_{k=-\eta_{\tilde{a}}}^{0} [A(k)]_{\tilde{a}} \otimes x_0 \right)^{(n)} \right) \right| \leq B^1_{g,\eta_{\tilde{a}},\{A(k)\}}(n,p) \, ,$$

where

$$B^1_{g,\eta_{\tilde{a}},\{A(k)\}}(n,p)$$

$$= \sum_{l \in \mathcal{L}[-\eta_{\tilde{a}}^{(n-1)},1;n]} \frac{n!}{l_{-\eta_{\tilde{a}}^{(n-1)}}! \dots l_0! l_1!} \sum_{i \in \mathcal{I}[l]} \prod_{k=-\eta_{\tilde{a}}^{(n)}}^{1} c_{[A(k)]_{\tilde{a}}}^{(l,i)}$$

$$\times \left(2a_g + b_g \left(\sum_{k=-\eta_{\tilde{a}}^{(n)}}^{1} \left\| [A^{(l,i)}(k)]_{\tilde{a}} \right\|_\oplus + \|x_0\|_\oplus \right)^p \right.$$

$$\left. + b_g \left(\sum_{k=-\eta_{\tilde{a}}^{(n)}}^{1} \left\| [A^{(l,i^-)}(k)]_{\tilde{a}} \right\|_\oplus + \|x_0\|_\oplus \right)^p \right) .$$

Proof: The proof follows from the same line of argument as the proof of Lemma 4.4.2 and is therefore omitted. $\square$

We now turn to the Lyapunov exponent. Note that in case of the Lyapunov exponent we take as performance function g the projection on any component of the state-vector; more formally, we take $g(x) = (x)_j$ for some $j \in \{1, \dots, J\}$.

Theorem 5.3.1 *Let assumptions* **(C1)** *to* **(C3)** *be satisfied. If $A(0)$ is C_1-analytic on Θ with domain of convergence $U(\theta_0)$, for $\theta_0 \in \Theta$, and if, for some $j \in \{1, \dots, J\}$, $\mathbb{E}_{\theta_0}[B^1_{(\cdot)_j,\eta_{\tilde{a}},\{A(k)\}}(n,1)]$ is finite for any n and*

$$\sum_{n=0}^{\infty} \frac{1}{n!} \sup_{\theta \in U(\theta_0)} \mathbb{E}_\theta \left[B^1_{(\cdot)_j,\eta_{\tilde{a}},\{A(k)\}}(n,1) \right] |\theta - \theta_0|^n < \infty \, ,$$

then

$$\lim_{k \to \infty} \mathbb{E}_\theta[x(k+1) - x(k)] = \lambda(\theta) \otimes \mathbf{e}$$

exists and is analytic on Θ. For $\theta_0 \in \Theta$, the domain of convergence is at least $U(\theta_0)$. Moreover, the nth derivative of the Lyapunov exponent is given by

$$\frac{d^n}{d\theta^n} \lambda(\theta) \otimes \mathbf{e} = \mathbb{E}_\theta \left[\left(A(1) \otimes \bigotimes_{k=-\eta_{\tilde{a}}}^{0} [A(k)]_{\tilde{a}} \right)^{(n)} \otimes x_0 \right] - \mathbb{E}_\theta \left[\left(\bigotimes_{k=-\eta_{\tilde{a}}}^{0} [A(k)]_{\tilde{a}} \right)^{(n)} \otimes x_0 \right] .$$

Proof: Theorem 2.6.2 implies

$$
\lambda(\theta) = \mathbb{E}_\theta \left[\left(A(1) \otimes \bigotimes_{k=-\eta_{\bar{a}}}^{0} A(k) \otimes x_0 \right)_j - \left(\bigotimes_{k=-\eta_{\bar{a}}}^{0} A(k) \otimes x_0 \right)_j \right]
$$

$$
= \mathbb{E}_\theta \left[\left(A(1) \otimes \bigotimes_{k=-\infty}^{0} [A(k)]_{\bar{a}} \otimes x_0 \right)_j - \left(\bigotimes_{k=-\infty}^{0} [A(k)]_{\bar{a}} \otimes x_0 \right)_j \right],
$$

for any component j. Hence, for the proof it suffices to show the analyticity of

$$
\mathbb{E}_\theta \left[\left(\bigotimes_{k=-\eta_{\bar{a}}}^{0} A(k) \otimes x_0 \right)_j \right] = \lim_{m \to \infty} \mathbb{E}_\theta \left[\left(\bigotimes_{k=-m}^{0} [A(k)]_{\bar{a}} \otimes x_0 \right)_j \right] \qquad (5.28)
$$

and

$$
\mathbb{E}_\theta \left[\left(A(1) \otimes \bigotimes_{k=-\eta_{\bar{a}}}^{0} A(k) \otimes x_0 \right)_j \right] = \lim_{m \to \infty} \mathbb{E}_\theta \left[\left(A(1) \otimes \bigotimes_{k=-m}^{0} [A(k)]_{\bar{a}} \otimes x_0 \right)_j \right]
$$
$$(5.29)$$

separately. Note that

$$
\mathbb{E}_{\theta_0}[B^0_{(\cdot)_j,\eta_{\bar{a}},\{(k)\}}(n,1)] \le \mathbb{E}_{\theta_0}[B^1_{(\cdot)_j,\eta_{\bar{a}},\{A(k)\}}(n,1)], \quad n \ge 0,
$$

for any component j, In accordance with Theorem 5.2.4, the finite products on the right-hand side of (5.28) and (5.29) are analytic and we obtain, for $i = 0, 1$,

$$
\mathbb{E}_\theta \left[\bigotimes_{k=-\eta_{\bar{a}}}^{i} A(k) \otimes x_0 \right] = \lim_{m \to \infty} \sum_{n=0}^{\infty} \frac{d^n}{d\theta^n} \Big|_{\theta=\theta_0} \mathbb{E}_\theta \left[\bigotimes_{k=-m}^{i} [A(k)]_{\bar{a}} \otimes x_0 \right] \frac{(\theta - \theta_0)^n}{n!}
$$

$$
= \lim_{m \to \infty} \sum_{n=0}^{\infty} \mathbb{E}_{\theta_0} \left[\left(\bigotimes_{k=-m}^{i} [A(k)]_{\bar{a}} \right)^{(n)} \otimes x_0 \right] \frac{(\theta - \theta_0)^n}{n!}. \qquad (5.30)
$$

We now show that we may interchange the order of limit and summation. In accordance with Lemma 4.4.2 and Lemma 5.3.2, for any m and any component j it holds that

$$
\sum_{n=0}^{\infty} \left| \frac{(\theta - \theta_0)^n}{n!} \mathbb{E}_{\theta_0} \left[\left(\left(\bigotimes_{k=-m}^{i} [A(k)]_{\bar{a}} \right)^{(n)} \otimes x_0 \right)_j \right] \right|
$$

$$
\le \sum_{n=0}^{\infty} \mathbb{E}_{\theta_0} \left[B^1_{(\cdot)_j,\eta_{\bar{a}},\{A(k)\}}(n,1) \right] \frac{|\theta - \theta_0|^n}{n!},
$$

which is finite by assumption, for any $\theta \in U_{\theta_0}$. Hence, by dominated convergence,

$$\lim_{m \to \infty} \sum_{n=0}^{\infty} \frac{(\theta - \theta_0)^n}{n!} \mathbb{E}_{\theta_0} \left[\left(\left(\bigotimes_{k=-m}^{i} [A(k)]_{\tilde{a}} \right)^{(n)} \otimes x_0 \right)_j \right]$$

$$= \sum_{n=0}^{\infty} \frac{(\theta - \theta_0)^n}{n!} \lim_{m \to \infty} \mathbb{E}_{\theta_0} \left[\left(\left(\bigotimes_{k=-m}^{i} [A(k)]_{\tilde{a}} \right)^{(n)} \otimes x_0 \right)_j \right],$$

for $1 \leq j \leq J$. Following the line of argument for the proof of Lemma 5.2.1, we now show that

$$\lim_{m \to \infty} \mathbb{E}_{\theta_0} \left[\left(\left(\bigotimes_{k=-m}^{i} [A(k)]_{\tilde{a}} \right)^{(n)} \otimes x_0 \right)_j \right] = \mathbb{E}_{\theta_0} \left[\left(\left(\bigotimes_{k=-\eta_{\tilde{a}}}^{i} [A(k)]_{\tilde{a}} \right)^{(n)} \right)_j \right],$$

for $1 \leq j \leq J$, which concludes the proof of the theorem. $\square$

By Theorem 5.3.1, we obtain an explicit representation for the $(h+1)^{st}$ derivative of λ and, thereby, an upper bound for the error term of a Taylor polynomial of degree h.

Theorem 5.3.2 *Under assumptions* **(C1)** *to* **(C3)**, *denote by $\tilde{a}$ the sequence of matrices constituting the pattern. Let $A(0)$ with state space A be $(h+1)$ times C_1-differentiable and let $p_{\tilde{a}}(\theta)$ be the probability that the sequence $\tilde{a}$ occurs. Assume that $\|A\|_{\oplus}$ and $\mathbf{C}_{A(0)}$ are finite. Provided that $x_0 = \mathbf{e}$, it then holds that*

$$\lambda(\theta_0 + \Delta) \otimes \mathbf{e} = \sum_{m=0}^{h} \frac{\Delta^m}{m!} \left\{ \mathbb{E}_{\theta_0} \left[\left(A(1) \otimes \bigotimes_{k=-\eta_{\tilde{a}}}^{0} [A(k)]_{\tilde{a}} \right)^{(n)} \otimes x_0 \right] \right.$$

$$\left. - \mathbb{E}_{\theta_0} \left[\left(\bigotimes_{k=-\eta_{\tilde{a}}}^{0} [A(k)]_{\tilde{a}} \right)^{(n)} \otimes x_0 \right] \right\}$$

$$+ r_{h+1}^{\lambda}(\theta_0, \Delta),$$

for $\theta_0, \theta_0 + \Delta \in \Theta$, with

$$|r_{h+1}^{\lambda}(\theta_0, \Delta)| \leq R_{h+1}^{\lambda}(\theta_0, \Delta)$$

$$\stackrel{\text{def}}{=} \frac{2^{h+2}}{h!} \mathbf{C}_{A(0)}^{h+1} \|A\|_{\oplus} \int_{\theta_0}^{\theta_0+\Delta} (\theta_0 + \Delta - t)^h (1 + f(p_{\tilde{a}}(t), c, h+1)) \, dt,$$

where c denotes the length of $\tilde{a}$ and $f(\cdot, c, h+1)$ is either equal to $b(\cdot, c, h+1)$ as defined in Lemma 5.2.3, or is equal to $a(\cdot, c, h+1, 1)$ as defined in Lemma 5.2.2.

Proof: We only prove the statement for $f(p_{\tilde{a}}(\theta), M, h+1) = b(p_{\tilde{a}}(\theta), M, h+1)$ since for $f(q, M, h+1) = a(p_{\tilde{a}}(\theta), M, h+1, 1)$ the proof follows from the same line of argument.

By Theorem 5.3.1, the $(h+1)^{st}$ order derivative of λ is given by

$$\frac{d^{h+1}}{d\theta^{h+1}}\bigg|_{\theta=\theta_0} \lambda(\theta) \otimes \mathbf{e}$$

$$= \mathbb{E}_{\theta_0}\left[\left(A(1) \otimes \bigotimes_{k=-\eta_{\tilde{a}}}^{0} [A(k)]_{\tilde{a}}\right)^{(h+1)} \otimes x_0\right] - \mathbb{E}_{\theta_0}\left[\left(\bigotimes_{k=-\eta_{\tilde{a}}}^{0} [A(k)]_{\tilde{a}}\right)^{(h+1)} \otimes x_0\right],$$

and in accordance with Lemma 5.3.2 and equation (5.19) on page 211 this implies

$$\left|\frac{d^{h+1}}{d\theta^{h+1}}\bigg|_{\theta=\theta_0} \lambda(\theta)\right| \leq \mathbb{E}_{\theta_0}\left[B^1_{(\cdot)_j,\eta_{\tilde{a}},\{A(k)\}}(h+1,1)\right] + \mathbb{E}_{\theta_0}\left[B^0_{(\cdot)_j,\eta_{\tilde{a}},\{A(k)\}}(h+1,1)\right]$$

$$\leq 2\mathbb{E}_{\theta_0}\left[B^1_{(\cdot)_j,\eta_{\tilde{a}},\{A(k)\}}(h+1,1)\right], \tag{5.31}$$

for any component j. Following the line of argument for the proof of Lemma 5.2.3, we show that

$$\mathbb{E}_{\theta}\left[B^1_{(\cdot)_j,\eta_{\tilde{a}},\{A(k)\}}(h+1,1)\right] \leq 2^{h+1}\mathbf{C}^{h+1}_{A(0)}\|A\|_{\oplus}\left(1+b(p_{\tilde{a}}(\theta),c,h+1)\right).$$

Following the line of argument in the proof of Theorem 5.2.3, we calculate with the help of the above inequality an upper bound for the remainder term, which concludes the proof of the theorem. □

An example illustrating the above theorem will be given in the following section.

5.3.2   The Bernoulli Scheme

Let $\{A(k)\}$ be a sequence of i.i.d. Bernoulli-(θ)-distributed matrices with state space $\mathcal{A} = \{D_1, D_2\} \subset \mathbb{R}^{J \times I}_{\max}$, as defined in Example 1.5.5. For the numerical examples, we set $\sigma = 1$ and $\sigma' = 2$. Assumptions (**C1**) to (**C3**) hold. More specifically, D_2 is a primitive matrix that may serve as pattern. Since D_2 is already an element of $\mathcal{A}$, we have $N = c(D_2)$ in Definition 2.5.1, where $c(D_2)$ denotes the coupling time of D_2. We now take $\tilde{a}$ as the $c(D_2)$ fold concatenation of D_2:

$$\tilde{a} \stackrel{\text{def}}{=} \underbrace{(D_2, \ldots, D_2)}_{c(D_2) \text{ times}},$$

that is, $c \stackrel{\text{def}}{=} c(D_2)$ is the length of $\tilde{a}$. and the probability of observing $\tilde{a}$ equals $(1-\theta)^c$.

We calculate the first-order derivative of $\lambda(\theta)$ at $\theta = 0$. This implies that $A(k) = D_2$ for all k. Furthermore, the coupling time of D_2 equals c and since at $\theta = 0$ the sequence $\{A(k)\}$ is deterministic: $\eta = c - 1$. The first-order C_p-derivative of $A(k)$ is $(1, D_1, D_2)$ and all higher-order C_p-derivatives are not significant, see Section 5.2.2.

In the remainder of this section, we denote for $x, y \in \mathbb{R}^J$ the conventional component-wise difference of x and y by $x - y$ and their conventional component-wise addition by $x + y$. The symbol $\sum$ has to interpreted accordingly. In accordance with Theorem 5.3.1, we obtain

$$\frac{d}{d\theta}\bigg|_{\theta=0} \lambda(\theta) \otimes \mathbf{e} = \sum_{j=0}^{c} D_2^{c-j} \otimes D_1 \otimes D_2^c \otimes x_0 \ - \ \sum_{j=0}^{c} D_2^{c+1} \otimes x_0$$

$$- \sum_{j=0}^{c-1} D_2^{c-1-j} \otimes D_1 \otimes D_2^c \otimes x_0 \ + \ \sum_{j=0}^{c-1} D_2^c \otimes x_0 \,,$$

compare Section 5.2.2. We set $X_0 = D_2^c \otimes x_0$ and, since c is the coupling time of D_2, it follows that X_0 is an eigenvector of D_2. In accordance with (2.48) on page 113, we obtain

$$\frac{d}{d\theta}\bigg|_{\theta=0} \lambda(\theta) \otimes \mathbf{e} = \sum_{j=0}^{c} \left(D_2^{c-j} \otimes D_1 \otimes X_0 \ - \ D_2 \otimes X_0 \right)$$

$$- \sum_{j=0}^{c-1} \left(D_2^{c-1-j} \otimes X_0 \ - \ X_0 \right).$$

Higher-order C_1-derivatives are obtained from the same line of thought. The Taylor polynomial of degree $h = 3$ is shown in Figure 5.13 and Taylor polynomial for $h = 5$ is shown in Figure 5.14, where the thin line represents the true value, see Example 2.4.1.

Next we compute our upper bound for the remainder term. Note that $C_{A(0)} = 1$ and that $\|A\|_\oplus = \|D_1\|_\oplus \oplus \|D_2\|_\oplus$. We thus obtain for the remainder term of the Taylor polynomial of degree h:

$$R_{h+1}^\lambda(\theta_0, \Delta) = \frac{2^{h+2}}{h!} \left(\|D_1\|_\oplus \oplus \|D_2\|_\oplus \right)$$

$$\times \int_{\theta_0}^{\theta_0+\Delta} (\theta_0 + \Delta - t)^h \left(1 + b((1-t)^c, c, h+1) \right) dt \,,$$

with $\theta_0 \in [0, 1)$ and $\theta_0 < \theta_0 + \Delta < 1$. In the following, we address the actual quality of the Taylor polynomial approximation. At $\theta_0 = 0$, $\lambda(0)$ is just the Lyapunov exponent of D_1, and we obtain $\lambda(0) = 1$. From

$$\lambda(0 + \Delta) \leq \lambda(0) + R_1^\lambda(0, \Delta)$$

we obtain an immediate (that is, without estimating/calculating any derivatives) upper bound for $\lambda(\Delta)$. For example, elaborating on the numerical values in Table 5.5 (left), it follows $\lambda(0.01) \leq 3.0130$. Unfortunately, this is a rather useless bound because, for $\sigma = 1$ and $\sigma' = 2$, the Lyapunov exponent is at most 2 and thus $1 \leq \lambda(0.1) \leq 2$.

Table 5.5 shows the error term for the Taylor polynomial at $\theta_0 = 0$ and $\theta_0 = 0.1$ for $\Delta = 0.01$ and for various values of h. Comparing the results in

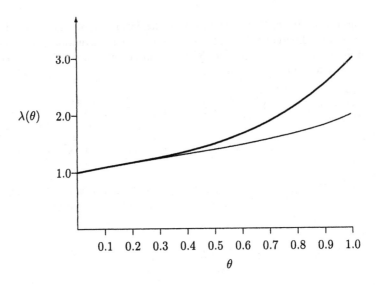

Figure 5.13: Taylor polynomial of degree $h = 3$ at $\theta_0 = 0$ for the Lyapunov exponent.

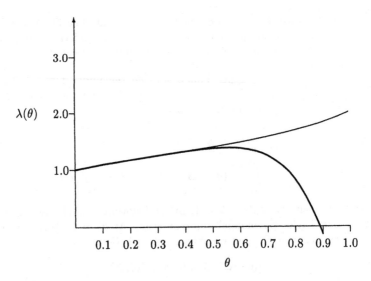

Figure 5.14: Taylor polynomial of degree $h = 5$ at $\theta_0 = 0$ for the Lyapunov exponent.

Table 5.5 (right) with the results in Table 5.5 (left), one observes (i) that the error terms at $\theta_0 = 0.1$ are larger than those at $\theta_0 = 0$ and (ii) that the error decreases at a slower pace at $\theta_0 = 0.1$ than at $\theta_0 = 0$. This comes as no surprise,

Table 5.5: $\theta_0 = 0, \Delta = 0.01$ (left side); $\theta_0 = 0.1, \Delta = 0.01$ (right side)

h	$R^\lambda_{h+1}(0.0, 0.01)$
0	13.0096
1	2.1376
2	2.8680×10^{-1}
3	3.4577×10^{-2}
4	3.8653×10^{-3}
5	4.1227×10^{-4}
10	3.7668×10^{-8}

h	$R^\lambda_{h+1}(0.1, 0.01)$
0	20.6007
1	4.2526
2	7.4748×10^{-1}
3	1.2096×10^{-1}
4	1.8505×10^{-2}
5	2.7206×10^{-3}
10	1.3633×10^{-7}

since the system at $\theta_0 = 0$ is deterministic whereas at $\theta_0 = 0.1$ we observe a stochastic system.

The most erratic behavior of the system will occur at $\theta_0 = 0.5$ and Table 5.6 presents numerical results for this case. According to (5.32) we have to choose $\Delta \leq 0.00390 (= 0.5^4/16)$.

Table 5.6: $\theta_0 = 0.5, \Delta = 0.003$ (left side); $\theta_0 = 0.5$, $h = 5$ (right side)

h	$R^\lambda_{h+1}(0.5, 0.003)$
5	6.6768
10	9.5109×10^{-2}
15	1.1311×10^{-3}

Δ	$R^\lambda_6(0.5, \Delta)$
10^{-2}	9699.6700
10^{-3}	9.0143×10^{-3}
10^{-4}	8.9506×10^{-9}
10^{-5}	8.9442×10^{-15}

Inspecting the numerical values, one concludes that the error term decreases at too slow a pace for a Taylor approximation for $\lambda(0.503)$ at $\theta_0 = 0.5$ at to be of any use. Finally, we illustrate in Table 5.6 the influence of Δ and h on the remainder term at $\theta_0 = 0.5$. Specifically, Table 5.6 (right) illustrates that $\Delta = 10^{-3}$ is a reasonable choice, when we assume that one is willing to evaluate the first five derivatives of λ with respect to θ at 0.5. However, the numerical values presented in the above tables are only upper bounds for the true remainder term, which stems from the fact that we only work with a (crude) upper bound given by $b(q, c, h + 1)$.

In the remainder of this section, we discuss our bound for the radius of convergence for the Taylor series. Denote the radius of convergence of the Taylor series at θ by $r(\theta)$. According to the formula of Cauchy-Hadamard, see (G.4) in Section G.4 in the Appendix, a lower bound for radius of convergence of the Taylor series at $\theta = 0$ is obtained from (5.31) together with Lemma 5.2.2 as follows

$$r(0) \geq \left(\limsup \left(\frac{1}{n!} 2^{n+1} \left(\|D_1\|_\oplus \oplus \|D_2\|_\oplus \right) \left(1 + a(1, c, n, 1) \right) \right)^{\frac{1}{n}} \right)^{-1}.$$

Hence, a lower bound for $r(0)$ is given by

$$\left(\limsup\left(\frac{2^{n+1}}{n!}\left(||D_1||_\oplus \oplus ||D_2||_\oplus\right)\left(1 + n!\, c^{n+1}(n+1)\right)\right)^{\frac{1}{n}}\right)^{-1}.$$

For example, let $\sigma = 1$, $\sigma' = 2$, then $c(D_1) = 4$, see Example 2.1.1, and

$$||D_1||_\oplus \oplus ||D_2||_\oplus = \max(\sigma, \sigma') = 2\,,$$

which implies

$$r(0) \geq \left(\limsup\left(\frac{1}{n!}2^{n+2}\left(1 + n!\, 4^{n+1}(n+1)\right)\right)^{\frac{1}{n}}\right)^{-1}$$

for the lower bound for radius of convergence. Hence,

$$r(0) \geq \frac{1}{8}\,,$$

which recovers the result in [7]. The above results were improved in [8] using a contraction argument for Hilbert's projective metric inspired by [88]. Elaborating on the 'memory loss' property implied by the occurrence of $\tilde{a}$, Gaubert and Hong improve in [48] the lower bound for the domain of convergence of the Taylor series at $\theta_0 = 0$ in [7, 8].

In the general case (that is, $\theta_0 > 0$, c and $||\mathcal{A}||_\oplus$ arbitrary), we obtain

$$\frac{1}{r(\theta_0)} \leq \limsup\left(\frac{2^{n+1}||\mathcal{A}||_\oplus \mathbf{C}^n_{A(0)}}{n!}\right.$$
$$\left. \times \left(1 + (n+1)!\, c^{n+1}(n+1)2^{n-1}\frac{1}{p_{\tilde{a}}(\theta_0)^{n+1}}\right)\right)^{\frac{1}{n}},$$

which gives

$$r(\theta_0) \geq \frac{1}{4\, c\, \mathbf{C}_{A(0)}}p_{\tilde{a}}(\theta_0) \qquad\qquad (5.32)$$

as lower bound for the radius of convergence.

5.3.3   A Note on the Elements of the Taylor Series for the Bernoulli System

The coefficients of the Taylor series are rather complex and can be represented in various ways; see for example the representations in [7]. Our analysis leads to yet another way of representing the coefficients of the Taylor series and in what follows we illustrate for the first-order derivative of the Lyapunov exponent of the Bernoulli system that the expression in Theorem 5.3.1 can indeed be algebraically manipulated in order to resemble the coefficients in Theorem 1 in [7].

We have already shown that

$$\frac{d}{d\theta}\bigg|_{\theta=0} \lambda(\theta) \otimes \mathbf{e} = \sum_{j=0}^{c} \left(D_2^{c-j} \otimes D_1 \otimes X_0 - D_2 \otimes X_0 \right)$$

$$- \sum_{j=0}^{c-1} \left(D_2^{c-1-j} \otimes D_1 \otimes X_0 - X_0 \right), \qquad (5.33)$$

where, like in the previous section, we denote for $x, y \in \mathbb{R}^J$ the conventional component-wise difference of x and y by $x - y$, their conventional component-wise addition by $x + y$ and interpret the symbol $\sum$ accordingly. Recall that

$$\lambda(0) \otimes \mathbf{e} = D_2 \otimes X_0 - X_0,$$

which gives

$$\frac{d}{d\theta}\bigg|_{\theta=0} \lambda(\theta) \otimes \mathbf{e} = -c\,\lambda(0) \otimes \mathbf{e} - D_2 \otimes X_0$$

$$+ \sum_{j=0}^{c} D_2^{c-j} \otimes D_1 \otimes X_0 - \sum_{j=0}^{c-1} D_2^{c-1-j} \otimes D_1 \otimes X_0 \,. (5.34)$$

It is easily checked that

$$D_2^c \otimes D_1 \otimes X_0 = \sum_{j=0}^{c} D_2^{c-j} \otimes D_1 \otimes X_0 - \sum_{j=0}^{c-1} D_2^{c-1-j} \otimes D_1 \otimes X_0 \,.$$

Inserting the above equality into (5.34) we obtain

$$\frac{d}{d\theta}\bigg|_{\theta=0} \lambda(\theta) \otimes \mathbf{e} = D_2^c \otimes D_1 \otimes X_0 - D_2 \otimes X_0 - c\,\lambda(0) \otimes \mathbf{e}\,.$$

Using the fact that $D_2 \otimes X_0 = \lambda(0) \otimes X_0$, which can be written as $D_2 \otimes X_0 = \lambda(0) \otimes \mathbf{e} + X_0$, we obtain

$$\frac{d}{d\theta}\bigg|_{\theta=0} \lambda(\theta) \otimes \mathbf{e} = D_2^c \otimes D_1 \otimes X_0 - X_0 - (c+1)\lambda(0) \otimes \mathbf{e}\,, \qquad (5.35)$$

which is the explicit form of the first-order derivative of the Lyapunov exponent at $\theta_0 = 0$ as given in [7].

For example, let $\sigma = 1$ and $\sigma' = 2$. The matrix

$$D_2 = \begin{pmatrix} 1 & \varepsilon & 2 & \varepsilon \\ 1 & \varepsilon & \varepsilon & \varepsilon \\ \varepsilon & \varepsilon & \varepsilon & e \\ \varepsilon & \varepsilon & 2 & \varepsilon \end{pmatrix}$$

has eigenvalue $\lambda(D_2) = 1$, and coupling time $c(D_2) = 4$, see Example 2.1.1. It is easily computed that

$$(D_2)^4 = \begin{pmatrix} 4 & 4 & 5 & 4 \\ 4 & 4 & 5 & 4 \\ 3 & 3 & 4 & 3 \\ 4 & 4 & 5 & 4 \end{pmatrix}.$$

The eigenspace of D_2 is given by

$$V(D_2) = \left\{ \begin{pmatrix} x_1 \\ x_2 \\ x_3 \\ x_4 \end{pmatrix} \in \mathbb{R}^4_{\max} \,\middle|\, \exists a \in \mathbb{R} : \begin{pmatrix} x_1 \\ x_2 \\ x_3 \\ x_4 \end{pmatrix} = a \otimes \begin{pmatrix} 1 \\ 1 \\ 0 \\ 1 \end{pmatrix} \right\},$$

see Theorem 2.1.2. Hence, Equation (5.35) reads

$$\left. \frac{d}{d\theta} \right|_{\theta=0} \lambda(\theta) \otimes \mathbf{e} = \begin{pmatrix} 4 & 4 & 5 & 4 \\ 4 & 4 & 5 & 4 \\ 3 & 3 & 4 & 3 \\ 4 & 4 & 5 & 4 \end{pmatrix} \otimes \begin{pmatrix} 1 & \varepsilon & 2 & \varepsilon \\ 1 & \varepsilon & \varepsilon & \varepsilon \\ \varepsilon & e & 2 & \varepsilon \\ \varepsilon & \varepsilon & 2 & \varepsilon \end{pmatrix} \otimes \begin{pmatrix} 1 \\ 1 \\ 0 \\ 1 \end{pmatrix} - \begin{pmatrix} 1 \\ 1 \\ 0 \\ 1 \end{pmatrix} - 5 \otimes \mathbf{e}$$

$$= \begin{pmatrix} 7 \\ 7 \\ 6 \\ 7 \end{pmatrix} - \begin{pmatrix} 1 \\ 1 \\ 0 \\ 1 \end{pmatrix} - 5 \otimes \mathbf{e}$$

$$= 1 \otimes \mathbf{e},$$

which implies that

$$\left. \frac{d}{d\theta} \right|_{\theta=0} \lambda(\theta) = 1.$$

Remark 5.3.1 *The coupling time of D_2 is of key importance for the above expressions for the derivative of $\lambda(\theta)$ at $\theta = 0$. Unfortunately, there are no efficient algorithms for evaluating the coupling time of a matrix. In particular, determining the coupling time of large matrices poses a serious problem. However, inspecting the above formulae for the derivative of $\lambda(\theta)$, one observes that the explicit knowledge of the coupling time can be avoided. We will explain this in the following. Starting point is the representation in (5.35). Notice that*

$$(c+1)\lambda(0) \otimes \mathbf{e} = D_2^{c+1} \otimes X_0 - X_0,$$

which implies

$$(c+1)\lambda(0) \otimes \mathbf{e} + X_0 = D_2^{c+1} \otimes X_0.$$

Inserting the above into (5.35) yields

$$\left. \frac{d}{d\theta} \right|_{\theta=0} \lambda(\theta) \otimes \mathbf{e} = D_2^c \otimes D_1 \otimes X_0 - D_2^{c+1} \otimes X_0.$$

For $j \geq 0$, let

$$X(j) = D_2^j \otimes D_1 \otimes X_0 \quad and \quad Y(j) = D_2^{j+1} \otimes X_0 \, .$$

Under the general assumptions, the eigenspace of D_2 reduces to the single point $\overline{X}_0$ in $\mathbb{PR}_{\max}^{J \times J}$ and we denote the number of transitions until $\overline{X(j)} = \overline{Y(j)} = \overline{X}_0$ by τ, or, more formally,

$$\tau = \inf \left\{ j \geq 0 : \overline{X(j)} = \overline{Y(j)} = \overline{X}_0 \right\} \, .$$

Note that $\tau \leq c$. Provided that $\overline{X(j)} = \overline{Y(j)} = \overline{X}_0$, it holds that

$$D_2^k \otimes X(j) - D_2^k \otimes Y(j) = X(j) - Y(j) \, , \quad k \geq 0 \, .$$

This implies

$$D_2^c \otimes D_1 \otimes X_0 - D_2^{c+1} \otimes X_0 = D_2^\tau \otimes D_1 \otimes X_0 - D_2^\tau \otimes D_2 \otimes X_0 \, .$$

Hence,

$$\left. \frac{d}{d\theta} \right|_{\theta=0} \lambda(\theta) = D_2^\tau \otimes D_1 \otimes X_0 - D_2^\tau \otimes D_2 \otimes X_0$$

and we obtain a representation of $d\lambda/d\theta$ that is independent of the coupling time. Moreover, the above representation can be implemented in a computer program in order to compute the derivative with a sequential algorithm. To see this, recall that efficient algorithms exists for computing an eigenvector of a max-plus matrix (see Section 2.1), and an eigenvector is the only input required for computing τ. Following the above line of argument, representations for higher-order derivatives avoiding the explicit knowledge of the coupling time can be obtained as well.

5.4 Stationary Waiting Times

In this section we turn to the analysis of stationary waiting times. In particular, we will provide a light-traffic approximation for stationary waiting times in open queuing networks with Poisson-λ-arrival stream. By 'light-traffic approximation' we mean a Taylor series expansion with respect to λ at $\lambda = 0$. Note that $\lambda = 0$ refers to the situation where no external customers arrive at the system.

Here 'λ' stands for the intensity of a Poisson process, which is in contrast to the previous sections where λ denoted the Lyapunov exponent. Both notations are classical and we have chosen to honor the notational traditions and speak of a Poisson-λ-process instead of a Poisson-θ-process, which would be the more logical notation in the context of this monograph. Specifically, since 'λ' is the parameter of interest in this section we will discuss derivatives with respect to λ rather than with respect to θ.

We consider the following situation: An open queuing network with J stations is given such that the vector of beginning of service times at the stations, denoted by $x(k)$, follows the recursion

$$x(k+1) = A(k) \otimes x(k) \oplus \tau(k+1) \otimes B(k) \, , \tag{5.36}$$

with $x_0 = \mathbf{e}$, where τ_k denotes the time of the kth arrival to the system; see equation (1.15) in Section 1.4.2.2 and equation (1.28) in Example 1.5.3, respectively. As usually, we denote by $\sigma_0(k)$ the kth interarrival time, which implies

$$\tau(k) = \sum_{i=1}^{k} \sigma_0(i), \quad k \geq 1,$$

with $\tau(0) = 0$. Then, $W_j(k) = x_j(k) - \tau(k)$ denotes the time the kth customer arriving to the system spends in the system until beginning of her/his service at server j. The vector of kth waiting times, denoted by $W(k) = (W_1(k), \ldots, W_J(k))$, follows the recursion

$$W(k+1) = A(k) \otimes C(\sigma_0(k+1)) \otimes W(k) \oplus B(k), \quad k \geq 0, \tag{5.37}$$

with $W(0) = x_0$ (we assume that the queues are initially empty), where $C(h)$ denotes a diagonal matrix with $-h$ on the diagonal and ε elsewhere, see Section 1.4.4 for details. Alternatively, $x_j(k)$ in (2.30) may model the times of the kth departure from station j. With this interpretation of $x(k)$, $W_j(k)$ defined above represents the time spend by the kth customer arriving to the system until departing from station j.

We assume that the arrival stream is a Poisson-λ-process for some $\lambda > 0$. In other words, the interarrival times are exponentially distributed with mean $1/\lambda$ and $\{\tau_\lambda(k)\}$ is a Poisson-λ-process, or, more formally, $\tau_\lambda(0) = 0$ and

$$\tau_\lambda(k) = \sum_{i=1}^{k} \sigma_0(i), \quad k \geq 1,$$

with $\{\sigma_0(k)\}$ an i.i.d. sequence of exponentially distributed random variables with mean $1/\lambda$.

Throughout this section, we assume that (**W1**) and (**W2**) are in force, see page 87. Moreover we assume that

(**W3**)' The sequence $\{(A(k), B^\alpha(k))\}$ is i.i.d. and independent of $\{\tau_\lambda(k)\}$.

See (**W3**) for a definition of $B^\alpha(k)$. Whenever, $W(k) = B(k-1)$, the kth customer arriving to the system receives immediate service at all stations on her/his way through the network. Suppose that $W(m) = B(m-1)$. From (5.37) together with (**W3**)' it follows that $\{W(k) : k < m\}$ and $W(k) : k \geq m\}$ are stochastically independent. The first time that $W(k)$ starts anew independent of the past is given by

$$\gamma_\lambda = \inf\{ k > 1 : W(k) = B(k-1) \}$$

and we call $\{W(k) : 1 \leq k < \gamma_\lambda\}$ a *cycle*. Condition (**W1**), (**W2**) and (**W3**)' imply that $\{W(k)\}$ is a regenerative process, see Section E.9 in the Appendix for basic definitions.

5.4.1 Cycles of Waiting Times

Let ϕ be some performance characteristics of the waiting times that can be computed from one cycle of the regenerative process $\{W(k)\}$. A cycle contains at least one customer arriving at time $\tau_\lambda(1)$ and we call this customer *initial customer*. This customer experiences time vector $B(0)$, that is, it takes the initial customer $B_j(0)$ time units until her/his beginning of service at station j and $B_j(0) = \varepsilon$ if the customer doesn't visit station j at all. This property is not obvious and we refer to Lemma 1.4.3 for a proof. A cycle may contain more than just the initial customer and these customers are called *additional customers*. The number of additional customers in the first cycle, denoted by β_λ, equals $\beta_\lambda = \gamma_\lambda - 1$. In words, on the event $\{\beta_\lambda = m\}$, the cycle contains one initial customer and m additional customers. The $(m+2)^{nd}$ customer experiences thus no waiting on her/his way through the network and she/he is the initial customer of a new cycle. Observe that for any max-plus linear queuing system, β_λ is measurable with respect to the σ-field generated by $\{(\tau_\lambda(k+1), A(k), B(k))\}$.

By conditions $(\mathbf{W1}) - (\mathbf{W3})'$, it holds with probability one that

$$\forall \lambda : \quad 0 < \lambda < \lambda_0 \quad \Longrightarrow \quad \beta_\lambda(k) \leq \beta_{\lambda_0}(k), \quad k \geq 0. \tag{5.38}$$

The reason for this is that, for $\lambda < \lambda_0$, the 'λ' system is visited by less customers than the 'λ_0' system and waiting times are thus smaller. For a rigorous proof of this statement use the fact that any finite element of $A(k)$ and $B(k)$ is positive and that $W(k)$ is thus monotone in $\tau_\lambda(k)$, see (5.37), and that $\tau_\lambda(k)$ is monotone decreasing in λ.

The fact that ϕ only depends on one cycle can be expressed as follows:

$$\phi_g(W(1), \ldots, W(\beta_\lambda + 1))$$
$$= \sum_{k=1}^{\beta_\lambda+1} g(W(k))$$
$$= \phi_g(\{\tau_\lambda(k) : 1 \leq k \leq \beta_\lambda + 1\}, \{(A(k), B(k)) : 0 \leq k \leq \beta_\lambda\})$$
$$\stackrel{\text{def}}{=} \phi_g(\{\tau_\lambda(k)\}), \tag{5.39}$$

where $g : [0, \infty) \to \mathbb{R}$ is some measurable mapping. For example, ϕ_g may yield the accumulated waiting time per cycle. Observe that ϕ_g depends on λ only through $\{\tau_\lambda(k)\}$ and β_λ.

Notice that $\{W(k)\}$ depends on $\{\tau_\lambda(k)\}$ only through the interarrival times, see (5.37). We have assumed that $\{\tau_\lambda(k+1) - \tau_\lambda(k)\}$ constitutes an i.i.d. sequence. Hence, we may as well assume that the initial customer arrives at time zero and set $W(1) = B(0)$. In other words, we shift the arrival process by $\sigma_0(1)$ to the left so that $\tau_\lambda(1) = 0$ a.s. The arrival process thus describes only the additional customers and the cycle performance becomes $\phi_g(\{\tau_\lambda(k)\}) \stackrel{\text{def}}{=} \phi_g(\{0\} \cup \{\tau_\lambda(k)\})$.

In the following we consider ϕ_g for the truncated arrival processes. Suppose that the initial customer is the only customer who arrives, that is, $\beta_\lambda = 0$ and

$\{\tau_\lambda(k) > 0 : 1 \leq k \leq \beta_\lambda\} = \emptyset$. In this case, the cycle performance equals

$$\phi_g(\emptyset) = g(W(1)) = g(B(0)), \qquad (5.40)$$

which stems from the fact that it takes the initial customer $B_j(0)$ time units until beginning of her/his service at station j. If the arrival stream contains in addition to the initial one an extra customer, and if this additional customer arrives at time $\tau > 0$, we set

$$W(2;\tau) \overset{\text{def}}{=} A(1) \otimes C(\tau) \otimes W(1) \oplus B(1).$$

For vectors x and y, write $x \rhd y$ if $x_i \geq y_i$ for all elements and if there exists at least one element j such that $x_j > y_j$. With this notation, we obtain

$$\phi_g(\tau) = g(B(0)) + 1_{W(2;\tau) \rhd B(1)} g(W(2;\tau)),$$

where $g(B(0))$ refers to the initial customer. The indicator mapping in the above equation expresses the fact that $W(2;\tau)$ only contributes to the cycle if $W(2;\tau) \neq B(1)$.

More generally, suppose customers arrive from the outside at time epochs $\tau_1, \ldots, \tau_k$, with $0 < \tau_1 < \tau_2 < \cdots < \tau_k < \infty$ and $k \geq 1$, then the waiting time of the customer arriving at τ_k is given by

$$W(k+1;\tau_1,\ldots,\tau_k) = A(k) \otimes C(\tau_{k+1} - \tau_k) \otimes W(k;\tau_1,\ldots,\tau_{k-1}) \oplus B(k)$$

and it holds that

$$\phi_g(\tau_1,\ldots,\tau_k) = g(B(0)) + \sum_{i=1}^{k} g(W(i+1;\tau_1,\ldots,\tau_i)) \prod_{j=1}^{i} 1_{W(j+1;\tau_1,\ldots,\tau_j) \rhd B(j)}.$$

Example 5.4.1 *Consider the M/G/1 queue. For this system we have $B(0) = 0$ and $\mathbb{E}[g(W(1))] = g(0)$; the arrival times of customers are given by the Poisson process $\{\tau_\lambda(k)\}$ and $\tau_\lambda(k+1) - \tau_\lambda(k)$ follows thus an exponential distribution with rate λ. Assume that the service time distribution has support $[0,\infty)$ and denote its density by f^S. The values $\mathbb{E}[W(2;\tau_\lambda(1))]$, $\mathbb{E}[W(3;\tau_\lambda(1),\tau_\lambda(2))]$ and $\mathbb{E}[W(4;\tau(1),\tau(2),\tau(3))]$ are easily computed with the help of the explicit formulae in Section 5.1.3.1. To see this, recall that we have assumed that the interarrival times are i.i.d. exponentially distributed with mean $1/\lambda$. Hence,*

$$\mathbb{E}[\phi_g(\tau_\lambda(1))] = g(0) + \mathbb{E}[1_{W(2;\tau_\lambda(1))>0} g(W(2;\tau_\lambda(1)))]$$

$$= g(0) + \int_0^\infty \int_a^\infty g(s-a) f^S(s) \lambda e^{-\lambda a} ds\, da, \qquad (5.41)$$

$$\mathbb{E}[\phi_g(\tau_\lambda(1),\tau_\lambda(2))]$$

$$= g(0) + \mathbb{E}\Big[1_{W(2;\tau_\lambda(1))>0} g(W(2;\tau(1)))\Big]$$

$$\qquad + \mathbb{E}\Big[1_{W(3;\tau_\lambda(1),\tau_\lambda(2))>0} 1_{W(2;\tau_\lambda(1))>0} g(W(3;\tau_\lambda(1),\tau_\lambda(2)))\Big]$$

$$= g(0) + \lambda^2 \int_0^\infty \int_0^\infty \int_{a_1+a_2}^\infty \int_0^\infty (g(s_1+s_2-a_1-a_2) + g(s_1-a_1))$$

$$\qquad \times f^S(s_2) f^S(s_1) e^{-\lambda(a_2+a_1)} ds_2\, ds_1\, da_2\, da_1$$

$$+\lambda^2 \int_0^\infty \int_0^\infty \int_{a_1}^{a_1+a_2} \int_{a+1+a_2-s_1}^\infty \big(g(s_1 + s_2 - a_1 - a_2) + g(s_1 - a_1)\big)$$
$$\times f^S(s_2)\, f^S(s_1)\, e^{-\lambda(a_2+a_1)}\, ds_2\, ds_1\, da_2\, da_1 \qquad (5.42)$$

and

$$\mathbb{E}\big[\phi_g(\tau_\lambda(1), \tau_\lambda(2), \tau_\lambda(3))\big] - g(0)$$
$$= \mathbb{E}\Big[\, 1_{W(2;\tau_\lambda(1))>0}\, g(W(2, \tau_\lambda(1)))$$
$$+ 1_{W(2;\tau_\lambda(1))>0}\, 1_{W(3;\tau_\lambda(1),\tau_\lambda(2))>0}\, g(W(3; \tau_\lambda(1), \tau_\lambda(2)))$$
$$+ 1_{W(2;\tau_\lambda(1))>0}\, 1_{W(3;\tau_\lambda(1),\tau_\lambda(2))>0}\, 1_{W(4;\tau_\lambda(1),\tau_\lambda(2),\tau_\lambda(3))>0}$$
$$\times g(W(4; \tau_\lambda(1), \tau_\lambda(2), \tau_\lambda(3)))\Big]$$

$$= \int_0^\infty \int_0^\infty \int_0^\infty \int_{a_1+a_2+a_3}^\infty$$
$$\int_0^\infty$$
$$\int_0^\infty h(s_1, s_2, s_3, a_1, a_2, a_3)\, ds_3\, ds_2\, ds_1\, da_3\, da_2\, da_1$$

$$+ \int_0^\infty \int_0^\infty \int_0^\infty \int_{a_1}^{a_1+a_2}$$
$$\int_{a_1+a_2+a_3-s_1}^\infty$$
$$\int_0^\infty h(s_1, s_2, s_3, a_1, a_2, a_3)\, ds_3\, ds_2\, ds_1\, da_3\, da_2\, da_1$$

$$+ \int_0^\infty \int_0^\infty \int_0^\infty \int_{a_1}^{a_1+a_2}$$
$$\int_{a_1+a_2-s_1}^{a_1+a_2+a_3-s_1}$$
$$\int_{a_1+a_2+a_3-s_1-s_2}^\infty$$
$$h(s_1, s_2, s_3, a_1, a_2, a_3)\, ds_3\, ds_2\, ds_1\, da_3\, da_2\, da_1$$

$$+ \int_0^\infty \int_0^\infty \int_0^\infty \int_{a_1+a_2}^{a_1+a_2+a_3}$$
$$\int_{a_1+a_2+a_3-s_1}^\infty$$
$$\int_0^\infty h_{i_1}(s_1, s_2, s_3, a_1, a_2, a_3)\, ds_3\, ds_2\, ds_1\, da_3\, da_2\, da_1$$

$$+ \int_0^\infty \int_0^\infty \int_0^\infty \int_{a_1+a_2}^{a_1+a_2+a_3} \int_0^{a_1+a_2+a_3-s_1} \int_{a_1+a_2+a_3-s_1-s_2}^\infty$$

$$\times h_{i_1}(s_1, s_2, s_3, a_1, a_2, a_3) \, ds_3 \, ds_2 \, ds_1 \, da_3 \, da_2 \, da_1 \,,$$

$$\tag{5.43}$$

with

$$h(s_1, s_2, s_3, a_1, a_2, a_3) = \big(g(s_1 + s_2 + s_3 - a_1 - a_2 - a_3) +$$
$$+ g(s_1 + s_2 - a_1 - a_2) + g(s_1 - a_1)\big)$$
$$\times f^S(s_3) f^S(s_2) f^S(s_1) \lambda^3 e^{-\lambda(a_1+a_2+a_3)} \,.$$

A basic property of Poisson processes is that a Poisson process with rate $0 < \lambda < \lambda_0$ can be obtained from a Poisson-λ_0-process through thinning the Poisson-λ_0-process in an appropriate way. Specifically, let $\{\tau_{\lambda_0}(k)\}$ denote a Poisson-λ_0-process and define $\{\tau_\lambda(k)\}$ as follows: with probability λ/λ_0 an element in $\{\tau_{\lambda_0}(k)\}$ is accepted and with probability $1 - \lambda/\lambda_0$ the element is rejected/deleted. Then, $\{\tau_\lambda(k)\}$ is a Poisson-λ-process. In order to make use of this property, we introduce an i.i.d. sequence $\{Y_\lambda(k)\}$ as follows

$$P(Y_\lambda(k) = 1) = \mu_\lambda(1) = \frac{\lambda}{\lambda_0} = 1 - P(Y_\lambda(k) = 0) = \mu_\lambda(0) \,.$$

Given $\{Y_\lambda(k)\}$, let $\{\tau_{\lambda_0}(k)|Y_\lambda(k)\}$ denote the subsequence of $\{\tau_{\lambda_0}(k)\}$ constituted of those $\tau_{\lambda_0}(k)$ for which $Y_\lambda(k) = 1$, that is, the mth element of $\{\tau_{\lambda_0}(k)|Y_\lambda(k)\}$ is given by

$$\tau_{\lambda_0}(n) \quad \text{if} \quad n = \inf\left\{ k \geq 1 : \sum_{l=1}^k Y_\lambda(l) = m \right\} \,,$$

and set

$$\phi_g(\{\tau_{\lambda_0}(k)\}, \{Y_\lambda(k)\}) = \phi_g(\{\tau_{\lambda_0}(k)|Y_\lambda(k)\}) \,.$$

By (5.39), ϕ_g depends on $\{Y_\lambda(k)\}$ only through the first β_{λ_0} elements:

$$\phi_g(\{\tau_{\lambda_0}(k)\}, \{Y_\lambda(k)\}) = \phi_g(\{\tau_{\lambda_0}(k)\}, (Y_\lambda(1), \ldots, Y_\lambda(\beta_{\lambda_0}))) \,.$$

Remark 5.4.1 *Notice that 'thinning' only affects additional customers. The reason for this is that with positive probability all customers of a cycle may be rejected. Obviously, a cycle has to contain at least one customer and we guarantee that the 'λ' version of the cycle obtained by thinning contains at least one customer through excluding the initial customer form thinning.*

*The resulting cycle is a legitimate sample of the first cycle under λ. To see this, notice that $W(k)$ depends on τ_λ only through the interarrival times and that the thinning decisions are i.i.d. by (**W3**)'.*

Thinning the Poisson-λ_0-processes according to $\{Y_\lambda(k)\}$ yields

$$\mathbb{E}[\phi_g(\{\tau_\lambda(k)\}) \,|\, \beta_{\lambda_0}]$$
$$= \mathbb{E}[\phi_g(\{\tau_{\lambda_0}(k)\}, (Y_\lambda(1), \ldots, Y_\lambda(\beta_{\lambda_0}))) \,|\, \beta_{\lambda_0}]$$
$$= \mathbb{E}\left[\sum_{\eta \in \{0,1\}^{\beta_{\lambda_0}}} \phi_g(\{\tau_{\lambda_0}(k)\}, \eta) \right.$$
$$\left. \times \left(\frac{\lambda}{\lambda_0}\right)^{\sum \eta_k} \left(1 - \frac{\lambda}{\lambda_0}\right)^{\beta_{\lambda_0} - \sum \eta_k} \,\Bigg|\, \beta_{\lambda_0} \right]$$
$$= \mathbb{E}\left[\sum_{\eta \in \{0,1\}^{\beta_{\lambda_0}}} \phi_g(\{\tau_{\lambda_0}(k)\}, \eta) \times \mu_\lambda^{\beta_{\lambda_0}}(\eta_1, \ldots, \eta_{\beta_{\lambda_0}}) \,\Bigg|\, \beta_{\lambda_0} \right],$$

where

$$\mu_\lambda^h(\eta_1, \ldots, \eta_h) = \prod_{k=1}^{h} \mu_\lambda(\eta_k),$$

for $\eta \in \{0,1\}^h$.

The measure μ_λ is ∞-times $\mathbb{R}^2$-differentiable, with $\mathbb{R}^2$-derivative

$$\mu_\lambda^{(1)} = \left(\frac{1}{\lambda_0}, \delta_1, \delta_0 \right) \tag{5.44}$$

and no higher order $\mathbb{R}^2$-derivative is significant, see Example 4.1.2. Moreover, μ_λ is $\mathbb{R}^2$-analytic on $[0, \lambda_0]$, see Example 4.6.2, and our product rule implies that $\prod_{m=1}^{h} \mu_\lambda$ is $\mathbb{R}^2$-analytic on $[0, \lambda_0]$ as well.

For the following we need an additional technical assumption:

(W4) Constants $c_0, c_1, p \in [0, \infty)$ exist such that

$$\left| \phi_g(\{\tau_{\lambda_0}(k)\}) \right| \leq c_0 + c_1 \left(\beta_{\lambda_0} \, \tau_{\lambda_0}(\beta_{\lambda_0}) \right)^p \quad \text{a.s.}$$

Example 5.4.2 *Denote by ϕ_{id} the accumulated waiting time per cycle in a $G/G/1$ queue, then ϕ_{id} satisfies (W4) for $p = 1$. Indeed, any waiting time has to be smaller than the cycle length $\tau_\lambda(\beta_\lambda)$ and there are β_λ non-zero waiting times in a cycle; hence,*

$$|\phi_{id}(W(1), \ldots, W(\beta_\lambda + 1))| \leq \beta_\lambda \, \tau_\lambda(\beta_\lambda) \quad \text{a.s.}$$

and because

$$\tau_{\beta_\lambda}(\beta_\lambda) \leq \tau_{\beta_{\lambda_0}}(\beta_{\lambda_0}) \quad \text{and} \quad \beta_\lambda \leq \beta_{\lambda_0} \quad \text{a.s.}$$

condition (W4) is satisfied for ϕ_{id}.

For ease of exposition, we assume in the following that $c_0 = 0$ and $c_1 = 1$. Applying the product rule of $\mathcal{D}$-differentiation, see Lemma 4.2.1, yields for any $m \geq n$

$$\frac{d^n}{d\lambda^n} \mathbb{E}\left[\phi_g(\{\tau_\lambda(k)\}) \middle| \beta_{\lambda_0} = m\right]$$

$$= \frac{d^n}{d\lambda^n} \mathbb{E}\left[\sum_{\eta \in \{0,1\}^{\beta_{\lambda_0}}} \phi_g(\{\tau_{\lambda_0}(k)\}, \eta)\mu_\lambda^{\beta_{\lambda_0}}(\eta_1, \ldots, \eta_{\beta_{\lambda_0}}) \middle| \beta_{\lambda_0} = m\right]$$

$$= \mathbb{E}\left[\sum_{l \in \mathcal{L}[1,\beta_{\lambda_0};n]} \sum_{i \in \mathcal{I}[l]} \frac{1}{\lambda_0^n} \frac{n!}{l_1! \cdots l_{\beta_{\lambda_0}}!} \sum_{\eta \in \{0,1\}^{\beta_{\lambda_0}}} \phi_g(\{\tau_{\lambda_0}(k)\}, \eta) \right.$$
$$\left. \times \prod_{k=1}^{\beta_{\lambda_0}} \left(\mu_\lambda^{(l_k, i_k)}(\eta_k) - \mu_\lambda^{(l_k, i_k^-)}(\eta_k)\right) \middle| \beta_{\lambda_0} = m\right]$$

$$= \mathbb{E}\left[\sum_{l \in \mathcal{L}[1,\beta_{\lambda_0};n]} \sum_{i \in \mathcal{I}[l]} \frac{n!}{\lambda_0^n} \sum_{\eta \in \{0,1\}^{\beta_{\lambda_0}}} \phi_g(\{\tau_{\lambda_0}(k)\}, \eta) \right.$$
$$\left. \times \prod_{k=1}^{\beta_{\lambda_0}} \left(\mu_\lambda^{(l_k, i_k)}(\eta_k) - \mu_\lambda^{(l_k, i_k^-)}(\eta_k)\right) \middle| \beta_{\lambda_0} = m\right],$$

where the last equality stems from the fact that only the first order derivative of μ_λ is significant, that is, $l_k \in \{0,1\}$. For $m < n$, the above n^{th} order derivative equals zero.

Remark 5.4.2 *Because only the first order derivative of μ_λ is significant, $\mathcal{L}[1, \beta_{\lambda_0}; n] = \emptyset$ if $\beta_{\lambda_0} < n$. Indeed, the set $\mathcal{L}[1, \beta_{\lambda_0}; n]$ contains those $l \in \{0,1\}^{\beta_{\lambda_0}}$ that satisfy $\sum_k l_k = n$, which already implies that l has at least n elements. In words, the derivative only contributes on the event $\{\beta_{\lambda_0} \geq n\}$ and is otherwise zero.*

For any $l \in \mathcal{L}[1, \beta_{\lambda_0}; n]$ and $i \in \mathcal{I}[l]$, the measures $\prod_{k=1}^{\beta_{\lambda_0}} \mu_\lambda^{(l_k, i_k)}$ and $\prod_{k=1}^{\beta_{\lambda_0}} \mu_\lambda^{(l_k, i_k^-)}$ shift the mass of the vectors η. More specifically, let $l \in \mathcal{L}[1, \beta_{\lambda_0}; n]$ and $i \in \mathcal{I}[l]$, if $l_k = 0$, then the kth point of the Poisson-λ_0-process is accepted with probability λ/λ_0 and rejected with probability $1 - \lambda/\lambda_0$; whereas if $l_k = 1$, then the kth point is *always* accepted if $i_k = 1$ or $i_k^- = 1$ and the point is *always* rejected if $i_k = -1$ or $i_k^- = -1$. Observe that for $l \in \mathcal{L}[1, \beta_{\lambda_0}; n]$ and $i \in \mathcal{I}[l]$ the measures $\prod_{k=1}^{\beta_{\lambda_0}} \mu_\lambda^{(l_k, i_k)}$ and $\prod_{k=1}^{\beta_{\lambda_0}} \mu_\lambda^{(l_k, i_k^-)}$ can accept at most β_{λ_0} points. By **(W4)**, it holds for any $\eta \in \{0,1\}^{\beta_{\lambda_0}}$

$$\left|\phi_g(\{\tau_\lambda(k)\}, \eta)\right| \leq \left(\beta_{\lambda_0} \tau_{\lambda_0}(\beta_{\lambda_0})\right)^p$$

and we obtain for the absolute value of the nth order derivative:

$$
\left| \frac{d^n}{d\lambda^n} \mathbb{E}\left[\phi_g(\{\tau_\lambda(k)\}) \Big| \beta_{\lambda_0} = m \right] \right|
$$

$$
= \left| \frac{1}{\lambda_0^n} \mathbb{E}\left[\sum_{l \in \mathcal{L}[1,\beta_{\lambda_0};n]} \sum_{i \in \mathcal{I}[l]} n! \sum_{\eta \in \{0,1\}^{\beta_{\lambda_0}}} \phi_g(\{\tau_{\lambda_0}(k)\}, \eta) \right.\right.
$$

$$
\left.\left. \times \prod_{k=1}^{\beta_{\lambda_0}} \left(\mu_\lambda^{(l_k, i_k)}(\eta_k) - \mu_\lambda^{(l_k, i_k^-)}(\eta_k) \right) \Big| \beta_{\lambda_0} = m \right] \right|
$$

$$
\leq \frac{2}{\lambda_0^n} \mathbb{E}\left[\sum_{l \in \mathcal{L}[1,\beta_{\lambda_0};n]} \sum_{i \in \mathcal{I}[l]} n! \left(\beta_{\lambda_0} \tau_{\lambda_0}(\beta_{\lambda_0}) \right)^p \Big| \beta_{\lambda_0} = m \right].
$$

The set $\mathcal{I}[l]$ has at most 2^{n-1} elements and

$$
\sum_{l \in \mathcal{L}[1,\beta_{\lambda_0};n]} n! \leq (\beta_{\lambda_0})^n ,
$$

see Section G.5 in the Appendix. Hence,

$$
\left| \frac{d^n}{d\lambda^n} \mathbb{E}\left[\phi_g(\{\tau_\lambda(k)\}) \Big| \beta_{\lambda_0} = m \right] \right|
$$

$$
\leq \frac{2^n}{\lambda_0^n} \mathbb{E}\left[\beta_{\lambda_0}^{n+p} \tau_{\lambda_0}^p(\beta_{\lambda_0}) \Big| \beta_{\lambda_0} = m \right]. \tag{5.45}
$$

From the above we conclude that

$$
\mathbb{E}\left[\beta_{\lambda_0}^{n+p} \tau_{\lambda_0}^p(\beta_{\lambda_0}) \right] < \infty
$$

is a sufficient condition for

$$
\sum_{m=1}^{\infty} \left| \frac{d^n}{d\lambda^n} \mathbb{E}\left[\phi_g(\{\tau_\lambda(k)\}) \,\Big|\, \beta_{\lambda_0} = m \right] \right| P(\beta_{\lambda_0} = m)
$$

to be finite, and applying the dominated convergence theorem yields

$$
\frac{d^n}{d\lambda^n} \mathbb{E}\left[\phi_g(\{\tau_\lambda(k)\}) \right]
$$

$$
= \frac{d^n}{d\lambda^n} \sum_{m=1}^{\infty} \mathbb{E}\left[\phi_g(\{\tau_\lambda(k)\}) \Big| \beta_{\lambda_0} = m \right] P(\beta_{\lambda_0} = m)
$$

$$
= \sum_{m=1}^{\infty} \frac{d^n}{d\lambda^n} \mathbb{E}\left[\phi_g(\{\tau_\lambda(k)\}) \Big| \beta_{\lambda_0} = m \right] P(\beta_{\lambda_0} = m) , \tag{5.46}
$$

inserting the explicit representation for the nth order derivative gives

$$
= \sum_{m=1}^{\infty} \mathbb{E}\Bigg[\sum_{l \in \mathcal{L}[1,\beta_{\lambda_0};n]} \sum_{i \in \mathcal{I}[l]} \frac{n!}{\lambda_0^n} \sum_{\eta \in \{0,1\}^{\beta_{\lambda_0}}} \phi_g(\{\tau_{\lambda_0}(k)\}, \eta)
$$
$$
\times \prod_{k=1}^{\beta_{\lambda_0}} \left(\mu_\lambda^{(l_k, i_k)}(\eta_k) - \mu_\lambda^{(l_k, i_k^-)}(\eta_k) \right) \Bigg| \beta_{\lambda_0} = m \Bigg] P(\beta_{\lambda_0} = m) \,,
$$

basic probability calculus now gives

$$
= \sum_{m=1}^{\infty} \mathbb{E}\Bigg[\sum_{l \in \mathcal{L}[1,\beta_{\lambda_0};n]} \sum_{i \in \mathcal{I}[l]} \frac{n!}{\lambda_0^n} \sum_{\eta \in \{0,1\}^{\beta_{\lambda_0}}} \phi_g(\{\tau_{\lambda_0}(k)\}, \eta)
$$
$$
\times 1_{\beta_{\lambda_0} = m} \prod_{k=1}^{\beta_{\lambda_0}} \left(\mu_\lambda^{(l_k, i_k)}(\eta_k) - \mu_\lambda^{(l_k, i_k^-)}(\eta_k) \right) \Bigg]
$$

$$
= \mathbb{E}\Bigg[\sum_{l \in \mathcal{L}[1,\beta_{\lambda_0};n]} \sum_{i \in \mathcal{I}[l]} \frac{n!}{\lambda_0^n} \sum_{\eta \in \{0,1\}^{\beta_{\lambda_0}}} \phi_g(\{\tau_{\lambda_0}(k)\}, \eta)
$$
$$
\times 1_{\beta_{\lambda_0} \geq n} \prod_{k=1}^{\beta_{\lambda_0}} \left(\mu_\lambda^{(l_k, i_k)}(\eta_k) - \mu_\lambda^{(l_k, i_k^-)}(\eta_k) \right) \Bigg]
$$

$$
= \mathbb{E}\Bigg[\sum_{l \in \mathcal{L}[1,\beta_{\lambda_0};n]} \sum_{i \in \mathcal{I}[l]} \frac{n!}{\lambda_0^n} \sum_{\eta \in \{0,1\}^{\beta_{\lambda_0}}} \phi_g(\{\tau_{\lambda_0}(k)\}, \eta)
$$
$$
\times \prod_{k=1}^{\beta_{\lambda_0}} \left(\mu_\lambda^{(l_k, i_k)}(\eta_k) - \mu_\lambda^{(l_k, i_k^-)}(\eta_k) \right) \Bigg| \beta_{\lambda_0} \geq n \Bigg] P(\beta_{\lambda_0} \geq n) \,,
$$

where the last but one equality follows from the fact that the expression for the derivative only contributes on the event $\{\beta_{\lambda_0} \geq n\}$, see Remark 5.4.2.

The above analysis leads to (the first part of) the following theorem.

Theorem 5.4.1 *Under assumption* **(W1)** − **(W4)**, *suppose that for* $n \in \mathbb{N}$ *it holds that*
$$
\mathbb{E}\left[\beta_{\lambda_0}^{n+p} \tau_{\lambda_0}^p(\beta_{\lambda_0}) \right] < \infty \,.
$$
Then, for any λ *with* $0 < \lambda < \lambda_0$, $\mathbb{E}[\phi_g(\{\tau_\lambda(k)\})]$ *is finite and it holds that*

$$
\frac{d^n}{d\lambda^n} \mathbb{E}\left[\phi_g(\{\tau_\lambda(k)\}) \right]
$$
$$
= \mathbb{E}\Bigg[\sum_{l \in \mathcal{L}[1,\beta_{\lambda_0};n]} \sum_{i \in \mathcal{I}[l]} \frac{n!}{\lambda_0^n} \sum_{\eta \in \{0,1\}^{\beta_{\lambda_0}}} \phi_g(\{\tau_{\lambda_0}(k)\}, \eta)
$$
$$
\times \prod_{k=1}^{\beta_{\lambda_0}} \left(\mu_\lambda^{(l_k, i_k)}(\eta_k) - \mu_\lambda^{(l_k, i_k^-)}(\eta_k) \right) \Bigg] \,.
$$

If, for $0 < \lambda_0$, a number $r_{\lambda_0} \geq p$ exists such that

$$\mathbb{E}\left[\tau_{\lambda_0}^p(\beta_{\lambda_0})\, e^{r_{\lambda_0}\beta_{\lambda_0}}\right] < \infty,$$

then the Taylor series for $\mathbb{E}\left[\phi_g(\{\tau_\lambda(k)\})\right]$ exists at any point $\lambda \in (0, \lambda_0)$ and the radius of convergence of the series is at least $\frac{1}{2}\lambda_0(r_{\lambda_0} - p)$.

Proof: Finiteness of $\mathbb{E}[\phi_g(\{\tau_\lambda(k)\})]$ follows from (**W4**). It holds that

$$\mathbb{E}_\lambda\left[\phi_g(\{\tau_\lambda(k)\})\right] = \sum_{m=1}^\infty \mathbb{E}\left[\phi_g(\{\tau_\lambda(k)\})\mid \beta_{\lambda_0} = m\right] P(\beta_{\lambda_0} = m).$$

As a first step of the proof, we show that $\mathbb{E}\left[\phi_g(\{(\tau_\lambda(k)\})\mid \beta_{\lambda_0} = m\right]$ is analytic. Writing $\mathbb{E}[\phi_g(\{\tau_\lambda(k)\})\mid \beta_{\lambda_0} = m]$ as a Taylor series at λ, with $0 < \lambda < \lambda_0$, gives:

$$\sum_{n=0}^\infty \frac{1}{n!}|\Delta|^n \left|\frac{d^n}{d\lambda^n}\mathbb{E}\left[\phi_g(\{\tau_\lambda(k)\})\mid \beta_{\lambda_0} = m\right]\right|$$

$$\overset{(5.45)}{\leq} \sum_{n=0}^\infty \frac{1}{n!}|\Delta|^n 2^n \mathbb{E}\left[\frac{m^{n+p}}{\lambda_0^n}\tau_{\lambda_0}^p(m)\,\bigg|\,\beta_{\lambda_0} = m\right]$$

$$= \mathbb{E}\left[\sum_{n=0}^\infty \frac{1}{n!}|\Delta|^n 2^n \frac{m^{n+p}}{\lambda_0^n}\tau_{\lambda_0}^p(m)\,\bigg|\,\beta_{\lambda_0} = m\right]$$

$$= \mathbb{E}\left[\tau_{\lambda_0}^p(m)\, m^p \sum_{n=0}^\infty \frac{1}{n!}\left(\frac{2|\Delta|m}{\lambda_0}\right)^n\,\bigg|\,\beta_{\lambda_0} = m\right]$$

$$= \mathbb{E}\left[\tau_{\lambda_0}^p(m)\, m^p\, e^{\frac{2|\Delta|m}{\lambda_0}}\,\bigg|\,\beta_{\lambda_0} = m\right]$$

$$\leq \mathbb{E}\left[\tau_{\lambda_0}^p(m)\, e^{\left(p + \frac{2|\Delta|}{\lambda_0}\right)m}\,\bigg|\,\beta_{\lambda_0} = m\right], \tag{5.47}$$

where the last inequality stems from the fact that for $m \geq 1$ it holds that $m^p = e^{\ln(m)p} \leq e^{mp}$. By assumption, the expression on the right-hand side of the above series of inequalities is finite provided that

$$p + \frac{2|\Delta|}{\lambda_0} \leq r_{\lambda_0},$$

or, equivalently, if

$$|\Delta| \leq \frac{1}{2}\lambda_0(r_{\lambda_0} - p).$$

Hence, $\mathbb{E}\left[\phi_g(\{\tau_\lambda(k)\})\mid \beta_{\lambda_0} = m\right]$ can be written as a Taylor series at any $\lambda < \lambda_0$. The domain of convergence of the Taylor series is (at least) the entire interval $\left(\lambda_0 - \frac{1}{2}\lambda_0(r_{\lambda_0} - p),\ \lambda_0 + \frac{1}{2}\lambda_0(r_{\lambda_0} - p)\right)$.

For the second step of the proof, we sum the Taylor series expansions for m. For $|\Delta| \leq \frac{1}{2}\lambda_0(r_{\lambda_0} - p)$, the bound for the Taylor series conditioned on the event $\{\beta_0 = m\}$ in (5.47) satisfies

$$\sum_{m=0}^{\infty} \mathbb{E}\left[\tau_{\lambda_0}^p(m)\, e^{\left(p + \frac{2\Delta}{\lambda_0}\right)m}\,\middle|\, \beta_{\lambda_0} = m\right]\, P(\beta_{\lambda_0} = m) < \infty\,,$$

and interchanging the order of summation and differentiation is justified in the following row of equations, see Theorem G.4.1 in Section G.4 in the Appendix:

$$\mathbb{E}_{\lambda}\left[\phi_g(\{\tau_{\lambda}(k)\})\right]$$

$$= \sum_{m=1}^{\infty} \mathbb{E}_{\lambda}\left[\phi_g(\{\tau_{\lambda}(k)\})\,\middle|\, \beta_{\lambda_0} = m\right] P(\beta_{\lambda_0} = m)$$

$$= \sum_{m=1}^{\infty} \sum_{n=0}^{\infty} \frac{1}{n!}\Delta^n \frac{d^n}{d\lambda^n}\mathbb{E}\left[\phi_g(\{\tau_{\lambda}(k)\})\,\middle|\, \beta_{\lambda_0} = m\right] P(\beta_{\lambda_0} = m)$$

$$= \sum_{n=0}^{\infty} \frac{1}{n!}\Delta^n \sum_{m=1}^{\infty} \frac{d^n}{d\lambda^n}\mathbb{E}\left[\phi_g(\{\tau_{\lambda}(k)\})\,\middle|\, \beta_{\lambda_0} = m\right] P(\beta_{\lambda_0} = m)$$

$$= \sum_{n=0}^{\infty} \frac{1}{n!}\Delta^n \frac{d^n}{d\lambda^n}\mathbb{E}\left[\phi_g(\{\tau_{\lambda}(k)\})\right]\,,$$

where the last equality follows from (5.46). This concludes the proof. $\square$

We now turn to the application of the above results to waiting times. Let π_{λ} denote the stationary distribution of $W(k)$ provided that the arrival stream has intensity λ. We write $\mathbb{E}_{\pi_{\lambda}}$ to indicate that the expectation is taken with respect to π_{λ}. Under conditions $(\mathbf{W1}) - (\mathbf{W3})'$, a sufficient condition for the existence of a unique stationary distribution is that $\lambda < \mathbf{a}$, see Theorem 2.3.1. It follows from renewal theory that

$$\mathbb{E}_{\pi_{\lambda}}[g(W)] = \frac{1}{\mathbb{E}[\beta_{\lambda} + 1]}\mathbb{E}\left[\sum_{k=1}^{\beta_{\lambda}+1} g(W(k))\right]\,, \tag{5.48}$$

where

$$\sum_{k=1}^{\beta_{\lambda}+1} g(W(k)) = \phi_g(\{\tau(k)\})\,. \tag{5.49}$$

Theorem 5.4.1 provides sufficient conditions for differentiability of $\mathbb{E}\left[\sum_{k=1}^{\beta_{\lambda}+1} g(W(k))\right]$. Moreover, setting

$$\beta_{\lambda} = \phi_1(\{\tau_{\lambda}(k)\})\,, \tag{5.50}$$

we obtain sufficient conditions for differentiability of $\mathbb{E}[\beta_{\lambda}]$ as well. If Theorem 5.4.1 applies, we can, in principle, expand the left-hand side of (5.48), that is, the stationary waiting time, into a Taylor series. Unfortunately, due to the fact that $\mathbb{E}_{\pi_\lambda}[g(W)]$ is given through a fraction whose numerator and denominator both depend on λ, higher-order derivatives are too complex to be of any practical use. However, when a light traffic approximation is considered, the individual derivatives have a surprisingly simple representation.

5.4.2 Light Traffic Approximation

In the previous section we studied (higher order) derivatives evaluated at a point λ that had to lie between 0 and a predefined reference point λ_0. Instead of derivatives we could have considered left sided derivatives in the above analysis and in the Taylor series expansion we would then have replaced higher order derivatives by their left sided counterparts. The resulting theorem is stated below.

Theorem 5.4.2 *(Theorem 5.4.1 revisited) Assume that assumptions* (**W1**) − (**W4**) *are satisfied, and denote the maximal Lyapunov exponent of* $\{A(k)\}$ *by* **a**. *For any* λ *with* $\mathbf{a} > \lambda > 0$ *the following holds: If a number* $r_\lambda > p$ *exists such that*

$$\mathbb{E}\left[\tau_\lambda^p(\beta_\lambda)\,e^{r_\lambda\beta_\lambda}\right] < \infty\,,$$

then

$$\mathbb{E}\left[\phi_g(\{\tau_{\lambda+\Delta}(k)\})\right]$$
$$= \sum_{n=0}^{h} \frac{d^n}{d\lambda^n}\mathbb{E}\left[\phi_g(\{\tau_\lambda(k)\})\right]\frac{\Delta^n}{n!}$$
$$+ \frac{1}{h!}\int_\lambda^{\lambda+\Delta}(\lambda+\Delta-t)^h\left.\frac{d^{h+1}}{d\lambda^{h+1}}\right|_{\lambda=t}\mathbb{E}\left[\phi_g(\{\tau_\lambda(k)\})\right]dt\,,$$

where $|\Delta| < \frac{1}{2}\lambda(r_\lambda - p)$.

The expected stationary waiting time can be expressed via expected values taken over the first cycle of waiting times, see (5.48). Theorem 5.4.2 applies to the numerator and the denominator appearing on the right-hand side of (5.48). Hence, the Taylor series for the stationary waiting at λ exists and letting λ tend to zero in the Taylor series yields a so called *light traffic approximation* of the stationary waiting time. Light traffic approximations of stationary waiting times in open max-plus linear queuing systems have been intensively studied in the literature. Let W_i denote the ith component of the vector of stationary waiting times. The pioneering paper on light traffic expansions for $\mathbb{E}[W_i]$ is [17], where sufficient conditions for the existence of the light traffic approximation for $\mathbb{E}[W_i]$ are established and the (first) elements of the Taylor series are computed analytically. These results have been extended in [16] to $\mathbb{E}[f(W_i)]$, where f belongs to the class of performance measures $\mathcal{F}$, where $h \in \mathcal{F}$ if $h : [0,\infty) \to [0,\infty)$ and $h(x) \leq c\,x^\nu$ for $x \geq 0$ and $\nu \in \mathbb{N}$. In [5] expansions are obtained for $\mathbb{E}[f(W_i, W_j)]$, for $f : [0, \infty)^2 \to [0, \infty)$ with $f(x, y) \leq cx^{\nu_1} x^{\nu_2}$ for $x, y \geq 0$ and $\nu_1, \nu_2 \in \mathbb{N}$. In [3, 4], explicit expressions are given for the moments, Laplace transform and tail probability of the waiting time of the nth customer. Furthermore, starting with these exact expressions for transient waiting times, exact expressions for moments, Laplace transform and tail probability of stationary waiting times in a certain class of max-plus linear systems with deterministic service are computed.

In the remainder of this section we will provide a heuristic approach to light traffic approximations. To begin with, we will discuss light-traffic approximations of the cycle performance. As it will turn out, the elements of the light traffic approximation are closely related to the variables $W[m; i_1, i_2, \ldots, i_k]$ introduced in Section 5.1.3.1.

According to Theorem 5.4.1, we have to compute

$$
\lim_{\lambda \downarrow 0} \frac{d^n}{d\lambda^n} \mathbb{E}_\lambda \left[\sum_{k=1}^{\beta_\lambda+1} g(W(k)) \right]
$$

$$
= \lim_{\lambda \downarrow 0} \frac{n!}{\lambda^n} \mathbb{E} \left[\sum_{l \in \mathcal{L}[1,\beta_\lambda;n]} \sum_{i \in \mathcal{I}[l]} \sum_{\eta \in \{0,1\}^{\beta_\lambda}} \phi_g(\{\tau_\lambda(k)\}, \eta) \right.
$$

$$
\left. \times \prod_{k=1}^{\beta_\lambda} \left(\mu_\lambda^{(l_k, i_k)}(\eta_k) - \mu_\lambda^{(l_k, i_k^-)}(\eta_k) \right) \right],
$$

for ϕ_g as in (5.49). The random variable in the above expression for the nth order derivative only contributes if $\beta_\lambda \geq n$, which stems from the fact that $\mathcal{L}[1, \beta_\lambda; n] = \emptyset$ for $\beta_\lambda < n$ (see Remark 5.4.2). Letting λ tend to zero implies that β_λ tends to zero. Hence, the measures $\prod_{k=1}^{\beta_\lambda} \mu_\lambda^{(l_k, i_k)}(\eta_k)$ and $\prod_{k=1}^{\beta_\lambda} \mu_\lambda^{(l_k, i_k^-)}(\eta_k)$ converge to point masses as λ tends to 0. Specifically, $\mu_\lambda^{(1,1)}$ becomes the Dirac measure in 1 and $\mu_\lambda^{(1,-1)}$ becomes the Dirac measure in 0, see (5.44). Note that $\mathcal{L}[1, n; n] = \{(1, 1, \ldots, 1)\}$ and let $\mathbf{1}$ denote the vector in $\mathbb{R}^n$ with all elements equal to one. For λ sufficiently small, it holds

$$
\sum_{l \in \mathcal{L}[1,\beta_\lambda;n]} \sum_{i \in \mathcal{I}[l]} \prod_{k=1}^{\beta_\lambda} \mu_\lambda^{(l_k, i_k)}(\cdot) \approx \sum_{i \in \mathcal{I}[1]} \prod_{k=1}^{n} \mu_\lambda^{(1, i_k)}(\cdot)
$$

$$
= \sum_{i \in \mathcal{I}[1]} \prod_{k=1}^{n} \left(1_{i_k=1} \delta_1(\cdot) + 1_{i_k=-1} \delta_0(\cdot) \right)
$$

and

$$
\sum_{l \in \mathcal{L}[1,\beta_\lambda;n]} \sum_{i \in \mathcal{I}[l]} \prod_{k=1}^{\beta_\lambda} \mu_\lambda^{(l_k, i_k^-)}(\cdot) \approx \sum_{i \in \mathcal{I}[1]} \prod_{k=1}^{n} \mu_\lambda^{(1, i_k^-)}(\cdot)
$$

$$
= \sum_{i \in \mathcal{I}[1]} \prod_{k=1}^{n} \left(1_{i_k^-=1} \delta_1(\cdot) + 1_{i_k^-=-1} \delta_0(\cdot) \right).
$$

In words, the measures degenerate in the limit to point masses. For $i \in \mathcal{I}[1]$, set

$$\eta_k(i) = \begin{cases} 1 & \text{if } i_k = 1, \\ 0 & \text{if } i_k = -1, \end{cases}$$

for $1 \leq k \leq n$, and

$$\mathcal{I}^+[n] = \bigcup_{i \in \mathcal{I}[1]} \{\eta(i)\} \quad \text{and} \quad \mathcal{I}^-[n] = \bigcup_{i \in \mathcal{I}^-[1]} \{\eta(i)\}.$$

Hence, for λ sufficiently small,

$$\lim_{\lambda \downarrow 0} \frac{d^n}{d\lambda^n} \mathbb{E}_\lambda \left[\sum_{k=1}^{\beta_\lambda+1} g(W(k)) \right]$$

$$\approx \frac{d^n}{d\lambda^n} \mathbb{E}_\lambda \left[\sum_{k=1}^{n+1} g(W(k)) \right]$$

$$= \frac{n!}{\lambda^n} \mathbb{E} \left[\sum_{\eta \in \mathcal{I}^+[n]} \phi_g(\{\tau_\lambda(k)\}, \eta) - \sum_{\eta \in \mathcal{I}^-[n]} \phi_g(\{\tau_\lambda(k)\}, \eta) \right].$$

$$(5.51)$$

For example, for $n = 1$, we have $\mathcal{I}^+[1] = \{1\}$ and $\mathcal{I}^-[1] = \{0\}$, which implies that

$$\frac{d}{d\lambda}\bigg|_{\lambda=0} \mathbb{E} \left[\sum_{k=1}^{\beta_\lambda+1} g(W(k)) \right] \approx \frac{1}{\lambda} \mathbb{E} \left[\phi_g(\tau_\lambda(1)) - \phi_g(\emptyset) \right],$$

for λ sufficiently small, where $\phi_g(\emptyset)$ evaluates a sample path where the cycle consists only of the initial customer and it holds that $\phi_g(\emptyset) = g(B(0))$, see equation (5.40). It thus remains to calculate the term $\mathbb{E}[\phi_g(\{\tau_\lambda(1)\})]$. This term describes the following experiment. At time zero an initial customer enters the system and an additional customer arrives at time $\tau_\lambda(1)$. This quantity has already been computed in the previous section, where it was denoted by $W(2; \tau(1))$. For λ sufficiently small it holds that

$$\frac{d}{d\lambda}\bigg|_{\lambda=0} \mathbb{E} \left[\sum_{k=1}^{\beta_\lambda+1} g(W(k)) \right] \approx \frac{1}{\lambda} \mathbb{E} \left[\phi_g(\tau_\lambda(1)) - \phi_g(\emptyset) \right]$$

$$= \frac{1}{\lambda} \mathbb{E} \left[g(B(0)) + 1_{W(2;\tau_\lambda(1))>0} g(W(2; \tau_\lambda(1))) - g(B(0)) \right]$$

$$\stackrel{\text{def}}{=} \mathbf{V}_\lambda^g(1).$$

Notice that the event $\{W(2; \tau_\lambda(1)) > 0\}$ in the above equation describes the event that the cycle contains $W(1)$ and $W(2)$. We set

$$\lim_{\lambda \downarrow 0} \mathbf{V}_\lambda^g(1) = \mathbf{V}^g(1),$$

provided the limit exists.

Example 5.4.3 *For the M/G/1 queue, we obtain*

$$\mathbf{V}_\lambda^g(1) \stackrel{(5.41)}{=} \int_0^\infty \int_a^\infty g(s-a) \, f^S(s) e^{-\lambda a} ds \, da \, .$$

In particular, if the service times are exponentially distributed with rate μ and $g \in C_p([0,\infty))$ for some p, then

$$\mathbf{V}^g(1) = \int_0^\infty \int_a^\infty g(s-a) \, \mu \, e^{-\mu s} \, ds \, da \, .$$

We now turn to the second order derivative. For $n = 2$, we have $\mathcal{I}^+[2] = \{(1,1),(0,0)\}$ and $\mathcal{I}^-[2] = \{(1,0),(0,1)\}$, which implies that the second order derivative is approximated by

$$\frac{d^2}{d\lambda^2}\bigg|_{\lambda=0} \mathbb{E}\left[\sum_{k=1}^{\beta_\lambda+1} g(W(k))\right]$$

$$\approx \frac{d^2}{d\lambda^2} \mathbb{E}\left[\sum_{k=1}^3 g(W(k))\right]$$

$$= \frac{2}{\lambda^2} \mathbb{E}\left[\phi_g(\tau_\lambda(1),\tau_\lambda(2)) + \phi_g(\emptyset) - \phi_g(\tau_\lambda(1)) - \phi_g(\tau_\lambda(2))\right]$$

$$= \frac{2}{\lambda^2} \mathbb{E}\left[1_{W(2;\tau_\lambda(1))>0}1_{W(3;\tau_\lambda(1),\tau_\lambda(2))>0}\, g\big(W(3;\tau_\lambda(1),\tau_\lambda(2))\big)\right.$$

$$\left. -1_{W(2;\tau_\lambda(2))>0}\, g\big(W(2;\tau_\lambda(2))\big)\right]$$

$$\stackrel{\text{def}}{=} \mathbf{V}_\lambda^g(2) \, ,$$

for λ sufficiently small.

Example 5.4.4 *For the M/G/1 queue, a closed form expression for*

$$\mathbb{E}\left[1_{W(2;\tau_\lambda(1))>0}1_{W(3;\tau_\lambda(1),\tau_\lambda(2))>0}\, g\big(W(3;\tau_\lambda(1),\tau_\lambda(2))\big)\right]$$

can be obtained via (5.42). We address computing

$$\mathbb{E}\left[1_{W(2;\tau_\lambda(2))>0}\, g\big(W(2;\tau_\lambda(2))\big)\right] \, .$$

Recall that the sum of two independent exponentially distributed random variables with mean $1/\lambda$ is governed by a Gamma-$(2,\lambda)$-distribution with density $\lambda^2 x e^{-\lambda x}$ for $x \geq 0$. Hence, following (5.41),

$$\mathbb{E}\left[1_{W(2;\tau_\lambda(2))>0}\, g\big(W(2;\tau(2))\big)\right] = \lambda^2 \int_0^\infty \int_a^\infty g(s-a) \, a \, f^S(s) \, e^{-\lambda a} \, ds \, da \, .$$

In particular, if the service times are exponentially distributed with rate μ and $g \in C_p([0,\infty))$ for some p, then

$$\lim_{\lambda\downarrow 0} \mathbf{V}_\lambda^g(2) = \mathbf{V}^g(2)$$

exists, and it holds that

$$
\mathbf{V}^g(2) = 2 \int_0^\infty \int_0^\infty \int_{a_1+a_2}^\infty \int_0^\infty g(s_1 + s_2 - a_1 - a_2)
$$
$$
\times \mu^2\, e^{-\mu s_1}\, e^{-\mu s_1}\, ds_2\, ds_1\, da_2\, da_1
$$
$$
+ 2 \int_0^\infty \int_0^\infty \int_{a_1}^{a_1+a_2} \int_{a_1+a_2-s_1}^\infty g(s_1 + s_2 - a_1 - a_2)
$$
$$
\times \mu^2\, e^{-\mu s_1}\, e^{-\mu s_1}\, ds_2\, ds_1\, da_2\, da_1
$$
$$
- 2 \int_0^\infty \int_a^\infty g(s - a)\, a\, \mu\, e^{-\mu s}\, ds\, da\, .
$$

For the third order derivative, first observe that

$$
\mathcal{I}^+[3] = \{(1,1,1),(1,0,0),(0,1,0),(0,0,1)\}
$$

and

$$
\mathcal{I}^-[3] = \{(1,1,0),(1,0,1),(0,1,1),(0,0,0)\}\,.
$$

The third order derivative is evaluated through

$$
\frac{d^3}{d\lambda^3}\bigg|_{\lambda=0} \mathbb{E}\left[\sum_{k=1}^{\beta_\lambda+1} g(W(k))\right]
$$
$$
\approx \frac{6}{\lambda^3} \mathbb{E}\Big[\phi_g\big(\tau_\lambda(1),\tau_\lambda(2),\tau_\lambda(3)\big) + \phi_g\big(\tau_\lambda(1)\big) + \phi_g\big(\tau_\lambda(2)\big) + \phi_g\big(\tau_\lambda(3)\big)\}
$$
$$
- \phi_g\big(\tau_\lambda(1),\tau_\lambda(2)\big) - \phi_g\big(\tau_\lambda(1),\tau_\lambda(3)\big) - \phi_g\big(\tau_\lambda(2),\tau_\lambda(3)\big) - \phi_g(\emptyset)\Big]
$$
$$
\approx \frac{6}{\lambda^3} \mathbb{E}\Big[1_{W(2;\tau_\lambda(1))>0}\, 1_{W(3;\tau_\lambda(1),\tau_\lambda(2))>0}\, 1_{W(4;\tau_\lambda(1),\tau_\lambda(2),\tau_\lambda(3))>0}
$$
$$
\times g\big(W(4;\tau_\lambda(1),\tau_\lambda(2),\tau_\lambda(3))\big)
$$
$$
1_{W(2;\tau_\lambda(1))>0}\, 1_{W(3;\tau_\lambda(1),\tau_\lambda(2))>0}\, g\big(W(3;\tau_\lambda(1),\tau_\lambda(2))\big)
$$
$$
+ 2\cdot 1_{W(2;\tau_\lambda(1))>0}\, g\big(W(2;\tau_\lambda(1))\big)
$$
$$
+ 1_{W(2;\tau_\lambda(2))>0}\, g\big(W(2;\tau_\lambda(2))\big)
$$
$$
+ 1_{W(2;\tau_\lambda(3))>0}\, g\big(W(2;\tau_\lambda(3))\big)
$$
$$
- 2\cdot 1_{W(2;\tau_\lambda(1))>0}\, g\big(W(2;\tau_\lambda(1))\big)
$$
$$
- 1_{W(2;\tau_\lambda(2))>0}\, g\big(W(2;\tau_\lambda(2))\big)
$$
$$
- 1_{W(2;\tau_\lambda(1))>0}\, 1_{W(3;\tau_\lambda(1),\tau_\lambda(2))>0}\, g\big(W(3;\tau_\lambda(1),\tau_\lambda(2))\big)
$$
$$
- 1_{W(2;\tau_\lambda(1))>0}\, 1_{W(3;\tau_\lambda(1),\tau_\lambda(3))>0}\, g\big(W(3;\tau_\lambda(1),\tau_\lambda(3))\big)
$$
$$
- 1_{W(2;\tau_\lambda(2))>0}\, 1_{W(3;\tau_\lambda(2),\tau_\lambda(3))>0}\, g\big(W(3;\tau_\lambda(2),\tau_\lambda(3))\big)\Big]
$$

$$\approx \frac{6}{\lambda^3} \mathbb{E}\Big[1_{W(2;\tau_\lambda(1))>0} \, 1_{W(3;\tau_\lambda(1),\tau_\lambda(2))>0} \, 1_{W(4;\tau_\lambda(1),\tau_\lambda(2),\tau_\lambda(3))>0}$$
$$\times g\big(W(4;\tau_\lambda(1),\tau_\lambda(2),\tau_\lambda(3))\big)$$
$$+ 1_{W(2;\tau_\lambda(3))>0} \, g\big(W(2;\tau_\lambda(3))\big)$$
$$- 1_{W(2;\tau_\lambda(1))>0} \, 1_{W(3;\tau_\lambda(1),\tau_\lambda(3))>0} \, g\big(W(3;\tau_\lambda(1),\tau_\lambda(3))\big)$$
$$- 1_{W(2;\tau_\lambda(2))>0} \, 1_{W(3;\tau_\lambda(2),\tau_\lambda(3))>0} \, g\big(W(3;\tau_\lambda(2),\tau_\lambda(3))\big)\Big]$$
$$\stackrel{\text{def}}{=} \mathbf{V}_\lambda^g(3).$$

Example 5.4.5 *The term*

$$\mathbb{E}[1_{W(2,\tau_\lambda(1))>0} \, 1_{W(3;\tau_\lambda(1),\tau_\lambda(2))>0} \, 1_{W(4;\tau_\lambda(1),\tau_\lambda(2),\tau_\lambda(3))>0}$$
$$\times g(W(4;\tau_\lambda(1),\tau_\lambda(2),\tau_\lambda(3)))]$$

in the above expression is easily evaluated via (5.43), and, choosing the densities appropriately, the other terms can be evaluated via (5.41) and (5.42), respectively. In particular, if the service times are exponentially distributed with rate μ and $g \in C_p([0,\infty))$ for some p, then the limit for $\mathbf{V}_\lambda^g(3)$ as λ tends to zero, denoted by $\mathbf{V}^g(3)$, exists, and can be computed in the same vein as $\mathbf{V}^g(2)$ and $\mathbf{V}^g(1)$.

Generally, we set

$$\mathbf{V}_\lambda^g(n) \stackrel{\text{def}}{=} \lim_{\lambda \downarrow 0} \frac{d^n}{d\lambda^n} \mathbb{E}\left[\sum_{k=1}^{\beta_\lambda+1} g(W(k))\right]$$

provided that the limit exists. We summarize our discussion in the following scheme.

Light Traffic Approximation Scheme: Assume that assumptions (**W1**) − (**W4**) are satisfied. Suppose that for some $\lambda_0 > 0$ it holds that

$$\mathbb{E}_{\lambda_0}[\tau \, e^{r_{\lambda_0}\beta}] < \infty$$

for $r_{\lambda_0} > 0$ sufficiently small. Provided that the limit $\mathbf{V}_\lambda^g(n)$ exits for $n \leq h$, the following light traffic approximation exists for $\mathbb{E}_{\pi_\lambda}[g(W)]$:

$$\mathbb{E}\left[\sum_{k=1}^{\beta_\lambda+1} g(W(k))\right] = \sum_{n=1}^{h} \frac{\lambda^n}{n!}\mathbf{V}^g(n) + r_h(\lambda),$$

with $r_h(\lambda) \to 0$ as λ tend to zero.

We illustrate the above light traffic approximation by the following numerical example.

Example 5.4.6 *Consider a $M/M/1$ queueing system with arrival rate λ and service rate μ, and assume that $\rho \stackrel{\text{def}}{=} \lambda/\mu < 1$. For this system the expected accumulated waiting time per cycle is equal to*

$$\frac{\rho}{\mu(1-\rho)^2} \, . \tag{5.52}$$

To see this, note that the expected stationary waiting time is equal to

$$\frac{\rho}{\mu \, (1-\rho)} \, .$$

Let $\{X_n\}$ be the Markov chain describing the queue length in a $M/M/1$ queue. Notice that the arrival of a customer triggers an upward jump of X_n. Start $\{X_n\}$ in 0 and denote the expected number of upward jumps of $\{X_n\}$ until returns to state 0 by C, then

$$\rho = \frac{2C-1}{2C} \, ,$$

which gives

$$C = \frac{1}{1-\rho} \, ,$$

and using (5.48) equation (5.52) follows.

The accumulated waiting time is described through the functional ϕ_{id}, that is, we take $g(x) = x$ in the previous section. Recall that ϕ_{id} satisfies condition (**W4**), see Example 5.4.2. Inserting $\mu e^{-\mu x}$ for $f^S(x)$ in the formulae provided in Example 5.4.3 to Example 5.4.5, the first three elements of the light traffic approximation are explicitly given by $\mathbf{V}^g(n)$, for $n = 1, 2, 3$.

The light traffic approximation is given by

$$\mathbb{E}\left[\sum_{k=1}^{\beta_\lambda + 1} g(W(k))\right] \approx \sum_{n=1}^{h} \frac{\lambda^n}{n!} \mathbf{V}^g(n) \, .$$

For the numerical experiments we set $\mu = 1$. Figure 5.15 shows a light traffic approximation of degree $h = 3$ and Figure 5.16 shows a light traffic approximation corresponding to $h = 5$. In both figures, the thin line represents the true expected accumulated waiting time and the thick line represents the Taylor series approximation. It is worth noting that the light traffic approximations are fairly accurate up to $\rho \approx 0.3$ for $h = 3$ and $\rho \approx 0.4$ for $h = 5$.

We now turn to light-traffic approximations for stationary waiting times. Under conditions (**W1**) $-$ (**W3**)$'$, a sufficient condition for the existence of a unique stationary distribution is that $\lambda < \mathbf{a}$, see Theorem 2.3.1. It follows from renewal theory that

$$\mathbb{E}_{\pi_\lambda}[g(W)] = \frac{1}{\mathbb{E}[\beta_\lambda + 1]} \mathbb{E}\left[\sum_{k=1}^{\beta_\lambda + 1} g(W(k))\right] \, . \tag{5.53}$$

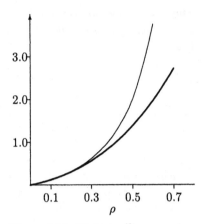

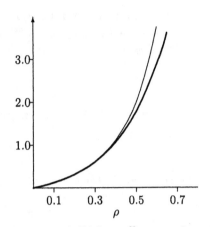

Figure 5.15: Light traffic approximation of degree $h = 3$ for the accumulated waiting time per cycle in a
M/M/1 queue.

Figure 5.16: Light traffic approximation of degree $h = 5$ for the accumulated waiting time per cycle in a
M/M/1 queue.

Recall that for $g(x) = 1$, we can deduce expressions for higher-order derivatives
of $\mathbb{E}[\beta_\lambda + 1]$ from the Taylor series expansion for $\mathbb{E}\left[\sum_{k=1}^{\beta_\lambda+1} g(W(k))\right]$.

   Notice that

$$\lim_{\lambda\downarrow 0} \mathbb{E}\left[\beta_\lambda + 1\right] = 1,\tag{5.54}$$

and, provided that $g(B(0)) = 0$,

$$\lim_{\lambda\downarrow 0} \mathbb{E}\left[\sum_{k=1}^{\beta_\lambda+1} g(W(k))\right] = 0.\tag{5.55}$$

We thus obtain for the derivative of $\mathbb{E}_{\pi_\lambda}[g(W)]$

$$\lim_{\lambda\downarrow 0} \frac{d}{d\lambda}\mathbb{E}_{\pi_\lambda}[g(W)] = \lim_{\lambda\downarrow 0} \frac{d}{d\lambda}\mathbb{E}\left[\sum_{k=1}^{\beta_\lambda+1} g(W(k))\right]$$
$$= \lim_{\lambda\downarrow 0} \mathbf{V}_\lambda^g(1).$$

Higher-order derivatives are obtained just as easy, where we make use of (5.54)
and (5.55) to simplify the expressions for the derivatives. We conclude the section with a numerical example.

Example 5.4.7 *The situation is as in Example 5.4.6 and we consider the expected stationary waiting time as performance measure of interest; this quantity
can be computed through*

$$\frac{\rho}{\mu(1-\rho)},$$

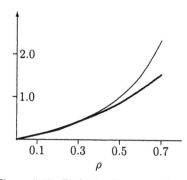

Figure 5.17: Light traffic approximation of degree $h = 3$ for the stationary waiting time in a M/M/1 queue.

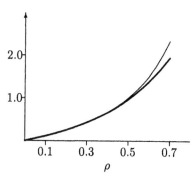

Figure 5.18: Light traffic approximation of degree $h = 5$ for the stationary waiting time in a M/M/1 queue.

for $\rho = \lambda/\mu < 1$. For the numerical experiments we set $\mu = 1$.

Figure 5.17 shows a light traffic approximation of degree $h = 3$ and Figure 5.18 shows a light traffic approximation corresponding to $h = 5$. In both
figures, the thin line represents the true expected stationary waiting time and
the thick line represents the Taylor series approximation. Notice that the light
traffic approximation is fairly accurate up to $\rho \approx 0.35$ for $h = 3$ and $\rho \approx 0.55$
for $h = 5$.

Appendix A

Basic Algebra

Let R be a non-empty set equipped with a binary mapping $\odot$. The mapping $\odot$ is called *associative* if

$$\forall a, b, c \in R : \quad a \odot (b \odot c) = (a \odot b) \odot c.$$

The mapping is called *commutative* if

$$\forall a, b \in R : \quad a \odot b = b \odot a.$$

An element $z \in R$ is called *neutral element,* or, *identity* for $\odot$ if

$$\forall a \in R : \quad a \odot z = z \odot a = a.$$

If $\odot$ represents 'addition,' then z is also called a *zero element* of $\odot$ and if $\odot$ represents 'multiplication,' then z is also called a *unity element* of $\odot$.

Let $\odot'$ be another binary mapping on R. We say that $\odot$ is *right distributive over* $\odot'$ if

$$\forall a, b, c \in R : \quad (a \odot' b) \odot c = (a \odot c) \odot' (b \odot c)$$

and $\odot$ is called *left distributive over* $\odot'$ if

$$\forall a, b, c \in R : \quad a \odot (b \odot' c) = (a \odot c) \odot' (b \odot c).$$

An element $u \in R$ is called *absorbing* for $\odot$ if

$$\forall a \in R : \quad a \odot u = u.$$

Appendix B

A Network with Breakdowns

In this section, we derive the sample-path dynamic for the model with breakdowns introduced in Example 1.5.5.

For the first beginning of service[1] at the single-server station to take place, two conditions have to be satisfied: the customer initially in service has to leave the station, which happens at time $x_2(1)$, and a new customer has to arrive at the single-server station. This happens at time $x_4(1)$ because the first customer arriving at the single-server station is the first customer who leaves the multiserver station. In formula,

$$x_1(1) = x_2(1) \oplus x_4(1)$$

and, by finite induction,

$$x_1(k) = x_2(k) \oplus x_4(k), \quad k \geq 1.$$

The first departure from the single-server station takes place at time σ and the second departure takes place σ time units after the first beginning of service. In formula,

$$x_2(1) = \sigma \quad \text{and} \quad x_2(2) = x_1(1) \otimes \sigma.$$

Letting $x_2(0) = 0$, finite induction yields:

$$x_2(k+1) = x_1(k) \otimes \sigma, \quad k \geq 0.$$

We now turn to the multi-server station. Following the same line of argument as for the single-server station, the departure times at the multi-server station follow

$$x_4(k+1) = x_3(k) \otimes \sigma', \quad k \geq 0,$$

---

[1] The first beginning of service is triggered by the first customer arriving at the station. The initial customer is not considered as an arrival.

where we set $x_3(0) = 0$.

Consider the multi-server station with no breakdown. The first beginning of service occurs at time $x_3(1) = 0$ and for the second beginning of service the following two conditions have to be satisfied: the first departure from the multi-server takes place and a new customer arrives from the single-server station. In formula,

$$x_3(1) = 0$$

and

$$x_3(2) = x_2(1) \oplus x_4(1) . \tag{B.1}$$

By finite induction,

$$x_3(k + 1) = x_2(k) \oplus x_4(k) , \quad k \geq 0 ,$$

where we set $x_4(0) = 0$. The sample-path dynamic of the network with no breakdown is thus given by

$$
\begin{aligned}
x_1(k + 1) &= x_2(k + 1) \oplus x_4(k + 1) \\
x_2(k + 1) &= x_1(k) \otimes \sigma \\
x_3(k + 1) &= x_2(k) \oplus x_4(k) \\
x_4(k + 1) &= x_3(k) \otimes \sigma' ,
\end{aligned}
$$

for $k \geq 0$. Replacing $x_2(k+1)$ and $x_4(k+1)$ in the first equation by the expression on the right-hand side of equations two and four above, respectively, yields

$$x_1(k + 1) = (x_1(k) \otimes \sigma) \oplus (x_3(k) \otimes \sigma') .$$

Hence, for $k \geq 0$,

$$
\begin{aligned}
x_1(k + 1) &= (x_1(k) \otimes \sigma) \oplus (x_3(k) \otimes \sigma') \\
x_2(k + 1) &= x_1(k) \otimes \sigma \\
x_3(k + 1) &= x_2(k) \oplus x_4(k) \\
x_4(k + 1) &= x_3(k) \otimes \sigma' ,
\end{aligned}
$$

which reads in matrix-vector notation:

$$
x(k + 1) = \begin{pmatrix}
\sigma & \varepsilon & \sigma' & \varepsilon \\
\sigma & \varepsilon & \varepsilon & \varepsilon \\
\varepsilon & e & \varepsilon & e \\
\varepsilon & \varepsilon & \sigma' & \varepsilon
\end{pmatrix} \otimes x(k) ,
$$

with $k \geq 0$.

In case a breakdown at the multi-server station has occurred, the first beginning of service at the multi-server station takes place upon the departure of the customer initially in service:

$$x_3(1) = x_4(1) .$$

The second beginning of service depends on two conditions: the second departure from the multi-server station takes place and a new customer arrives from the single-server station. In formula,

$$x_3(2) = x_2(1) \oplus x_4(2),$$

compare with (B.1). By finite induction,

$$x_3(k+1) = x_2(k) \oplus x_4(k+1), \quad k \geq 0.$$

The sample-path dynamic of the network with breakdown is therefore given by

$$\begin{aligned}
x_1(k+1) &= x_2(k+1) \oplus x_4(k+1) \\
x_2(k+1) &= x_1(k) \otimes \sigma \\
x_3(k+1) &= x_2(k) \oplus x_4(k+1) \\
x_4(k+1) &= x_3(k) \otimes \sigma',
\end{aligned}$$

for $k \geq 0$. As we have already explained, the above set of equations implies that

$$x_1(k+1) = (x_1(k) \otimes \sigma) \oplus (x_3(k) \otimes \sigma').$$

Furthermore, replacing $x_4(k+1)$ on the right-hand side of the equation for $x_3(k+1)$ by $x_3(k) \otimes \sigma'$ yields

$$x_3(k+1) = x_2(k) \oplus (x_3(k) \otimes \sigma').$$

Hence, for $k \geq 0$,

$$\begin{aligned}
x_1(k+1) &= (x_1(k) \otimes \sigma) \oplus (x_3(k) \otimes \sigma') \\
x_2(k+1) &= x_1(k) \otimes \sigma \\
x_3(k+1) &= x_2(k) \oplus (x_3(k) \otimes \sigma') \\
x_4(k+1) &= x_3(k) \otimes \sigma',
\end{aligned}$$

which reads in matrix-vector notation:

$$x(k+1) = \begin{pmatrix} \sigma & \varepsilon & \sigma' & \varepsilon \\ \sigma & \varepsilon & \varepsilon & \varepsilon \\ \varepsilon & e & \sigma' & \varepsilon \\ \varepsilon & \varepsilon & \sigma' & \varepsilon \end{pmatrix} \otimes x(k),$$

with $k \geq 0$.

Appendix C

Bounds on the Moments of the Binomial Distribution

For $p \in (0, 1)$ it holds that

$$\sum_{n=0}^{\infty} p^n = \frac{1}{1-p} .$$

Taking the derivative with respect to p implies

$$\sum_{n=1}^{\infty} n p^{n-1} = \frac{1}{(1-p)^2} .$$

Multiplying both sides by $(1-p)$ yields

$$\sum_{n=1}^{\infty} n p^{n-1}(1-p) = \frac{1}{(1-p)} ,$$

which is noticeably the first moment of the Binomial distribution.

For higher moments we derive an upper bound. Starting point is the following equation:

$$\frac{d^m}{dp^m} \sum_{n=0}^{\infty} p^n = \frac{d^m}{dp^m} \frac{1}{1-p} . \tag{C.1}$$

Note that

$$\frac{d^m}{dp^m} \frac{1}{1-p} = \frac{m!}{(1-p)^{m+1}} \tag{C.2}$$

and

$$\frac{d^m}{dp^m} \sum_{n=0}^{\infty} p^n = \sum_{n=0}^{\infty} (n+1)(n+2) \cdots (n+m) \, p^n$$

$$\geq \sum_{n=0}^{\infty} n^m \, p^n . \tag{C.3}$$

Inserting (C.2) and (C.3) into (C.1) yields

$$\sum_{n=0}^{\infty} n^m p^n \leq \frac{m!}{(1-p)^{m+1}} \, ,$$

which implies

$$\sum_{n=0}^{\infty} n^m p^n (1-p) \leq \frac{m!}{(1-p)^m} \, .$$

Hence, the mth moment of the Binomial distribution is bounded by

$$\frac{m!}{(1-p)^m} \, .$$

Appendix D

The Shifted Negative Binomial Distribution

Perform a series of mutually independent experiments each of which has probability of success $0 < q \leq 1$ and let $\beta_n = k$ describe the event that the nth success occurred at the kth experiment. Then, the distribution of $\beta_n - n$, that is, the number of failures until the nth success is called *negative binomial distribution*. The random variable β_n is thus governed by a *shifted negative binomial distribution*.

In the following we will compute $\mathbb{E}[\,(\beta_n)^l\,]$, for $l \geq 1$. The basic equation for the shifted binomial distribution reads

$$\sum_{k=n}^{\infty} \binom{k-1}{n-1} (1-q)^{k-n}\, q^n = 1\,,$$

that is,

$$P(\beta_n = k) = \binom{k-1}{n-1} (1-q)^{k-n}\, q^n\,, \quad k \geq n\,,$$

which implies

$$\sum_{k=n}^{\infty} \binom{k-1}{n-1} (1-q)^k = \frac{(1-q)^n}{q^n}$$

and

$$\sum_{k=n}^{\infty} \binom{k-1}{n-1} (1-q)^{k+l} = \frac{(1-q)^{l+n}}{q^n}\,,$$

for $l \geq 1$. Taking the lth derivative with respect to q yields

$$(-1)^l \sum_{k=n}^{\infty} (k+l)\,(k+l-1)\cdots(k+1) \binom{k-1}{n-1} (1-q)^k = \frac{d^l}{dq^l}\left(\frac{(1-q)^{l+n}}{q^n}\right)\,.$$

Hence,

$$\sum_{k=n}^{\infty} (k+l)\,(k+l-1)\cdots(k+1)\,P(\beta_n = k) = (-1)^l \, \frac{q^n}{(1-q)^n} \, \frac{d^l}{dq^l}\left(\frac{(1-q)^{l+n}}{q^n}\right),$$

which implies

$$\mathbb{E}[(\beta_n)^l] \leq (-1)^l \frac{q^n}{(1-q)^n} \, \frac{d^l}{dq^l}\left(\frac{(1-q)^{l+n}}{q^n}\right).$$

Appendix E

Probability Theory

E.1 Measures

Let $S \neq \emptyset$ be a set. A *σ-field* $\mathcal{S}$ on S is a collection of subsets of S with the following properties: (i) $S \in \mathcal{S}$, (ii) if $A \in \mathcal{S}$, then $A^c \in \mathcal{S}$, where $A^c = \{s \in S : s \notin A\}$, and (iii) if $A_i \in \mathcal{S}$, for $i \in \mathbb{N}$, then $\bigcup_{i \in \mathbb{N}} A_i \in \mathcal{S}$. Let $\mathcal{A}$ denote a collection of subsets of S. We denote by $\sigma(\mathcal{A})$ the σ-field generated by $\mathcal{A}$, that is, the smallest σ-field that contains $\mathcal{A}$. Let $(S, \mathcal{T})$ be a topological space. The *Borel field* of S, denoted by $\mathcal{B}$, is the σ-field generated by the collection of open sets $\mathcal{T}$, in formula: $\mathcal{B} = \sigma(\mathcal{T})$.

The pair $(S, \mathcal{S})$, where $\mathcal{S}$ is a σ-field on S, is called a *measurable space*. A *measure* μ on a measurable space $(S, \mathcal{S})$ is a mapping $\mu : \mathcal{S} \to \mathbb{R} \cup \{-\infty, \infty\}$ such that for any sequence $\{A_n\}$ of mutually disjoint elements of $\mathcal{S}$ it holds that

$$\mu\left(\sum_{n=1}^{\infty} A_n\right) = \sum_{n=1}^{\infty} \mu\left(A_n\right).$$

The measure m on $(\mathbb{R}, \mathcal{B})$, where $\mathcal{B}$ denotes the Borel field on $\mathbb{R}$, assigning $m((a, b]) = b - a$ to an interval $(a, b]$ is called *Lebesgue measure*. It generalizes the notion of length in geometry and is the case closest to everyday intuition.

The collection $(S, \mathcal{S}, \mu)$ is called *measure space*. A measure μ is called *signed* if $\mu(A) < 0$ for some $A \in \mathcal{S}$ and otherwise it is called *non-negative*. Furthermore, a measure μ is called *finite* if $\mu(A) \in \mathbb{R}$ for any $A \in \mathcal{S}$. We denote the set of signed measures on $(S, \mathcal{S})$ by $\mathcal{M}$. A non-negative measure μ is called *σ-finite* if there exist countably many sets A_i in $\mathcal{S}$ such that $\mu(A_i) < \infty$ and $\bigcup_i A_i = S$.

Let $(S, \mathcal{S})$ and $(R, \mathcal{R})$ be two measurable spaces. A mapping $g : S \to R$ is said to be *measurable* if for any $A \in \mathcal{R}$ it holds true that $\{s \in S : g(s) \in A\} \in \mathcal{S}$. A measurable mapping is also called *random variable*.

Let $\mu \in \mathcal{M}$ be non-negative. Then for any measurable mapping $g : S \to \mathbb{R}$ the *μ-integral* of g, denoted by

$$< g, \mu > = \int_S g(s)\, \mu(ds),$$

is defined although it may take values in $\{-\infty, \infty\}$. In particular, for any $A \in \mathcal{S}$,

$$< 1_A, \mu > = \mu(A) ,$$

where $1_A : S \to \mathbb{R}$ is defined by $1_A(s) = 1$ for $s \in A$ and $1_A(s) = 0$ otherwise.

For any signed measure $\mu \in \mathcal{M}$ a measurable set S_μ^+ exists such that, for any $A \in \mathcal{S}$, it holds that $\mu(A \cap S_\mu^+) \geq 0$, whereas $\mu(A \cap (S \setminus S_\mu^+)) \leq 0$, see, for example, Proposition IV.1.1 in [86] for a proof. The positive part of μ is defined by

$$[\mu]^+(A) = \mu(A \cap S_\mu^+), \quad A \in \mathcal{S}$$

and the negative part by

$$[\mu]^-(A) = -\mu(A \cap (S \setminus S_\mu^+)), \quad A \in \mathcal{S} .$$

The pair $([\mu]^+, [\mu]^-)$ is called *Hahn-Jordan decomposition*. The absolute measure $|\mu|$ is defined by $|\mu| = [\mu]^+ + [\mu]^-$. Integration with respect to a signed measure is defined by

$$< g, \mu > = < g, [\mu]^+ > - < g, [\mu]^- >$$

and integration with respect to an absolute measure is defined by

$$< g, |\mu| > = < g, [\mu]^+ > + < g, [\mu]^- > , \tag{E.1}$$

provided that the terms on the right-hand side of the above formulas are finite. The Hahn-Jordan decomposition is unique in the sense that if $\hat{G}$ is another set, such that $\mu(A \cap \hat{G}) \geq 0$ and $\mu(A \cap \hat{G}^c) \leq 0$ for any $A \in \mathcal{S}$, then $\mu(A \cap G) = \mu(A \cap S_\mu^+)$ for any $A \in \mathcal{S}$. A signed measure $\mu \in \mathcal{M}$ is finite if $[\mu]^+(S)$ and $[\mu]^-(S)$ are finite.

A *probability measure* μ is a non-negative measure such that $\mu(S) = 1$ (which already implies that $\mu(\emptyset) = 0$). If μ is a probability measure on $(S, \mathcal{S})$, then the collection $(S, \mathcal{S}, \mu)$ is called *probability space*.

Consider σ-finite measures μ and ν on a measurable space $(S, \mathcal{S})$. μ is said to be *absolutely continuous* with respect to ν if $\nu(A) = 0$, for $A \in \mathcal{S}$, implies $\mu(A) = 0$. If μ is absolutely continuous with respect to ν, then a measurable mapping $d\mu/d\nu : S \to \mathbb{R}$ exists such that

$$\mu(A) = \int_A \frac{d\mu}{d\nu}(s) \, \nu(ds) , \quad A \in \mathcal{S} .$$

The mapping $d\mu/d\nu$ is called ν-*density of* μ, or *Radon-Nikodym derivative*.

Let $(S, \mathcal{S}, \mu)$ and $(T, \mathcal{T}, \nu)$ be probability spaces. The product of μ and ν on $S \times T$, denoted by $\mu \times \nu$, is a measure such that

$$\forall A \in \mathcal{S}, B \in \mathcal{T} : \quad (\mu \times \nu)(A \times B) = \mu(A) \nu(B)$$

and *Fubini's theorem* states that

$$\int_{S \times T} f(s,t) \, (\mu \times \nu)(ds, dt) = \int_T \left(\int_S f(s,t) \, \mu(ds) \right) \nu(dt)$$

$$= \int_S \left(\int_T f(s,t) \, \nu(dt) \right) \mu(ds) ,$$

for any measurable mapping $f : S \times T \to \mathbb{R}$.

Let $(S, \mathcal{S}, \mu)$ be a measure space and $g : S \to \mathbb{R}$ a measurable mapping from $(S, \mathcal{S})$ to $(\mathbb{R}, \mathcal{R})$, where $\mathcal{R}$ is a σ-field over $\mathbb{R}$. The *induced measure* of g, denoted by μ^g, is defined as follows

$$\mu^g(A) = \mu\Big(\{s \in S : g(s) \in A)\}\Big), \quad A \in \mathcal{R}.$$

The cumulative distribution function (c.d.f.) of a real-valued random variable X defined on a probability space $(S, \mathcal{S}, \mu)$ is the function $F : [-\infty, \infty] \to [0, 1]$, where

$$F(x) = \mu^X((-\infty, x]), \quad -\infty \le x \le \infty.$$

We take the domain $[-\infty, \infty]$ since it is natural to assign the values 0 and 1 to $F(-\infty)$ and $F(\infty)$. A c.d.f. has the decomposition $F(x) = F'(x) + F''(x)$, where $F'(x)$ is positive only on a set of Lebesgue measure zero, and $F''(x)$ is absolutely continuous with respect to the Lebesgue measure. The Radon-Nikodym derivative of F'' with respect to the Lebesgue measure exists and is called probability density function (p.d.f.). If f is the p.d.f. of the c.d.f. F, then it holds that $f(x) = dF(x)/dx$ except for a set of Lebesgue measure zero.

Let μ be a finite measure on $(S, \mathcal{S})$, where S is a locally compact Hausdorff space, see any book on functional analysis for definitions, and $\mathcal{S}$ contains the Borel field on S. The measure μ is called *regular* if

$$\mu(A) = \inf\{\mu(U) : U \text{ open in } S, A \subset U\}, \quad A \in \mathcal{S},$$

and for any open set $U \subset S$ it holds

$$\mu(U) = \sup\{\mu(F) : F \text{ is compact in } S, F \subset U\}.$$

E.2 Polish Spaces

Let S be a nonempty set with zero element 0_S. A *norm* is a mapping $\|\cdot\| : S \to [0, \infty)$ having the properties (i) $0 < \|x\| < \infty$ for $x \neq 0_S$ and $\|0_S\| = 0$, (ii) $\|\alpha x\| = |\alpha| \|x\|$ for $\alpha \in \mathbb{R}$ and (iii) $\|x + y\| \le \|x\| + \|y\|$ (triangle inequality), for any $x, y \in S$.

A *metric* is a mapping $d : S \times S \to [0, \infty)$ having the properties (i) $d(x, y) = d(y, x)$, (ii) $d(x, y) = 0 \Leftrightarrow x = y$, and (iii) $d(x, z) \le d(x, y) + d(y, z)$ (triangle inequality), for any $x, y, z \in S$. If (i) and (iii) hold but $d(x, y) = 0$ is possible when $x \neq y$, we call d a *pseudo metric*. A *metric space* (S, d) is a set S paired with metric d.

An *open set* of (S, d) is a set $A \subset S$ such that, for each $s \in A$, $\delta > 0$ exists such that $\{x \in S : d(x, s) < \delta\} \subset A$. The collection of open subsets of S is denoted by $\mathcal{T}(d)$. Hence, $(S, \mathcal{T}(d))$ is a *topological space*. The Borel field on a metric space (S, d) is the σ-field generated by $\mathcal{T}(d)$.

A metric is said to be *complete* if the metric space (S, d) is complete, that is, if the limiting point of any Cauchy sequence in S lies in S. If there is a countable collection of open subsets of $T(d)$ such that any open subset of S can be written as union of these sets, then $T(d)$ is said to have a *countable basis*. A topological space (S, T) is called a *Polish space* if (i) its topology is defined by a complete metric (that is, there exists a metric d such that $T = T(d)$) and (ii) T has a countable basis.

E.3   The Shift-Operator

Many stochastic concepts, such as stationarity or coupling, can be expressed through the shift-operator in a very elegant manner. Let $(\Omega, \mathcal{F}, \mathcal{P})$ be a probability space. We call the mapping $\theta : \Omega \to \Omega$ *shift-operator* if

- the mapping θ is a bijective and measurable mapping from Ω onto itself,

- the law $\mathcal{P}$ is left invariant by θ, namely $\mathbb{E}[X] = \mathbb{E}[X \circ \theta]$ for any measurable and integrable random variable.

For any $n, m \in \mathbb{Z}$, we set $\theta^n \circ \theta^m = \theta^{n+m}$. In particular, θ^0 is the identity and $(\theta^n)^{-1} = \theta^{-n}$. By convention, the composition operator '$\circ$' has highest priority in all formulae, that is, $X \circ \theta Y$ means $(X \circ \theta) Y$.

The shift operator allows to define sequences of random variables. To see this, let X be a measurable mapping defined on $(\Omega, \mathcal{F})$ and set $X(n, \omega) = X(\theta^n \omega)$, for $n \in T \subset \mathbb{Z}$. Because the law $\mathcal{P}$ is invariant, the distribution of $X(n)$ is independent of n. This motivates the following definition. We call $\{X(t) : t \in T\}$, with $X(t)$ a $\mathbb{R}$-valued random variable defined on $(\Omega, \mathcal{F})$ and $T \subset \mathbb{Z}$, θ-*stationary* if

$$X(t; \omega) = X(0, \theta^t \omega), \quad \omega \in \Omega, \qquad (E.2)$$

for any t. We call a sequence $\mathbf{X} = \{X(t) : t \in T\}$ *compatible with shift operator* θ if a version of $\mathbf{X}$ exists satisfying (E.2). Moreover, we call $\mathbf{X}$ *stationary* if $\mathbf{X}$ is compatible with shift operator θ so that $\mathbf{X}$ is θ-stationary.

The shift θ is called *ergodic* if

$$\lim_{n \to \infty} \frac{1}{n} \sum_{k=1}^{n} X \circ \theta^k = \mathbb{E}[X] \quad \text{a.s.},$$

for any measurable and integrable function $X : \Omega \to \mathbb{R}$. We call a sequence $\mathbf{X} = \{X(t) : t \in T\}$ *ergodic* if $\mathbf{X}$ is compatible with an ergodic shift operator.

An event $A \in \mathcal{F}$ is called *invariant* if $P(A) = P(\theta^t A)$ for any t, where $\theta^t A = \{\theta^t \omega : \omega \in A\}$. Ergodicity of a shift operator is characterized by Birkhoff's pointwise ergodic theorem: the shift operator θ is ergodic if (and only if) the only events in $\mathcal{F}$ that are invariant are Ω and $\emptyset$, see [20].

Let $\mathbf{X} = \{X(t) : t \in T\}$ be a sequence of random elements on a state space S. For $m \geq 1$, let $\alpha \in S^m$ be a sequence of states such that

$$(X(t + m - 1), X(t + m - 2), \ldots, X(t)) = \alpha$$

with positive probability. Define the sequence of hitting times of $\mathbf{X}$ on α as follows:

$$T_0 = \inf\{t \geq 0 : (X(t+m-1), X(t+m-2), \ldots, X(t)) = \alpha\}$$

and, for $k > 0$,

$$T_{k+1} = \inf\{t > T_k + m : (X(t+m-1), X(t+m-2), \ldots, X(k)) = \alpha\}.$$

Result: (Theorem 2.10 in [11]) If $\mathbf{X}$ is a stationary and ergodic sequence compatible with shift operator θ, then it holds that (i) $T_k < \infty$ with probability one for all k, and (ii) $\lim_{k\to\infty} T_k = \infty$ with probability one.

E.4   Types of convergence

Let X, X_n, $n \geq 0$, be real-valued random variables defined on a common probability space $(\Omega, \mathcal{F}, P)$ with state space S and let S be equipped with the Borel field.

E.4.1   Almost Sure Convergence

The sequence $\{X_n\}$ converges *almost surely* to X as n tends to ∞ if for any $\delta > 0$

$$\lim_{n\to\infty} P\left(\sup_{m \geq n} |X_m - X| > \delta \right) = 0,$$

or, equivalently,

$$P\left(\lim_{n\to\infty} \sup_{m \geq n} |X_m - X| > \delta \right) = 0$$

and yet another equivalent condition is that the event

$$\left\{ \lim_{n\to\infty} X_n = X \right\}$$

has probability one.

E.4.2   Convergence in Probability

The sequence $\{X_n\}$ converges *in probability* to X as n tends to ∞ if for any $\delta > 0$

$$\lim_{n\to\infty} P(|X_n - X| \geq \delta) = 0,$$

or, equivalently,

$$\lim_{n\to\infty} P\left(|X_n - X| > \delta \right) = 0.$$

Result: Almost sure convergence of $\{X_n\}$ to X implies convergence in probability of $\{X_n\}$ to X. On the other hand, convergence in probability of $\{X_n\}$ to X implies a.s. convergence of a subsequence of $\{X_n\}$ to X.

E.4.3   Convergence in Distribution (Weak Convergence)

Let $C_b(\mathbb{R})$ denote the set of bounded continuous mapping from S onto $\mathbb{R}$. A sequence $\{\mu_n\}$ of measures on S is said to *converge weakly* to a distribution μ if

$$\lim_{n\to\infty} \int_S f\, d\mu_n = \int_S f\, d\mu\,, \quad f \in C_b(\mathbb{R})\,.$$

Let μ_n denote the distribution of X_n and μ the distribution of X. If $\{\mu_n\}$ converges weakly to μ as n tends to ∞, then we say that $\{X_n\}$ *converges in distribution* to X.

Result: Convergence in probability implies convergence in distribution but the converse is not true.

E.4.4   Convergence in Total Variation

The *total variation norm* of a (signed) measure μ on S is defined by

$$||\mu||_{tv} = \sup_{\substack{f \in C_b(\mathbb{R}) \\ |f| \le 1}} \left| \int_S f\, d\mu \right|\,.$$

In particular, weak convergence of a sequence $\{\mu_n\}$ of measures on S towards a distribution μ is equivalent to

$$\lim_{n\to\infty} ||\mu_n - \mu||_{tv} = 0\,.$$

Let again μ_n denote the distribution of X_n and μ the distribution of X. If $\{\mu_n\}$ converges in total variation to μ as n tends to ∞, then we say that $\{X_n\}$ *converges in total variation* to X. The convergence in total variation of $\{X_n\}$ to X can be expressed equivalently by

$$\lim_{n\to\infty} \sup_{A \in S} |P(X_n \in A) - P(X \in A)| = 0\,.$$

Result: Convergence in total variation implies convergence in distribution (or, weak convergence) but the converse is not true.

E.4.5   Weak Convergence and Transformations

With the notation of the previous section we now state the *continuous mapping theorem*. Let $h : \mathbb{R} \to \mathbb{R}$ be measurable with discontinuity points confined to a set D_h, where $\mu(D_h) = 0$. If μ_n converges weakly towards μ as n tends to ∞, then μ_n^h tends to μ^h as n tends to ∞, or, equivalently,

$$\lim_{n\to\infty} \int f(h(x))\, \mu_n(dx) = \int f(h(x))\, \mu(dx)\,, \quad f \in C_b(\mathbb{R})\,.$$

Hence, if $\{X_n\}$ converges weakly and h is continuous, then $\{h(X_n)\}$ converges weakly.

E.5 Weak Convergence and Norm Convergence

Let (S, d) be a separable metric space and denote the set of continuous real-valued mappings on S by $C(S)$. Let $v : S \to \mathbb{R}$ be a measurable mapping such that

$$\inf_{s \in S} v(s) \geq 1.$$

The set of mappings from S to $\mathbb{R}$ can be equipped with the so-called v-norm introduced presently. For $g : S \to \mathbb{R}$, the v-norm of g, denoted by $\|g\|_v$, is defined by

$$\|g\|_v \overset{\text{def}}{=} \sup_{s \in S} \frac{|g(s)|}{v(s)},$$

see, for example, [64] for the use of the v-norm in the theory of measure-valued differentiation of Markov chains. If g has finite v-norm, then $|g(s)| \leq c\,v(s)$ for any $s \in S$ and some finite constant c. For example, the set of real, continuous v-dominated functions, defined by

$$\mathcal{D}_v(S) \overset{\text{def}}{=} \{ g \in C(S) \, | \exists c > 0 : |g(s)| \leq cv(s), \forall s \in S \}, \tag{E.3}$$

can be characterized as the set of all continuous mappings $g : S \to \mathbb{R}$ having finite v-norm. Note that $C^b(S)$ is a particular $\mathcal{D}_v(S)$-space, obtained for $v = const$. Moreover, the condition that $\inf_{s \in S} v(s) \geq 1$ implies that $C^b(S) \subset \mathcal{D}_v(S)$ for any choice of v.

The v-norm of a measure μ on $(S, \mathcal{S})$, with $\mathcal{S}$ the Borel-field with respect to the metric d, is defined through

$$\|\mu\|_v \overset{\text{def}}{=} \sup_{\|g\|_v \leq 1} \left| \int_S g(s)\, \mu(ds) \right|,$$

or, more explicitly,

$$\|\mu\|_v = \sup_{|g| \leq v} \left| \int_S g(s)\, \mu(ds) \right|.$$

In particular, it holds that

$$\|\mu\|_v = \int v(s)\, |\mu|(ds), \tag{E.4}$$

see (E.1). Let $\{\mu_n\}$ be a sequence of measures on $(S, \mathcal{S})$ and let μ be a measure on $(S, \mathcal{S})$. We say that μ_n converges in v-norm towards μ if

$$\lim_{n \to \infty} \|\mu_n - \mu\|_v = 0.$$

It can be shown that the set $\mathcal{D}_v(S)$ endowed with the v-norm is a Banach space. This last remark indicates the following fact: For each measure μ with $\int v(s)\mu(ds)$ finite, the mapping $T_\mu : \mathcal{D}_v(S) \to \mathbb{R}$ defined through

$$T_\mu(g) \overset{\text{def}}{=} \int g\, d\mu,$$

is a continuous linear functional on the Banach space $\mathcal{D}_v(S)$ and the operator norm of T_μ satisfies $\|T_\mu\| = \|\mu\|_v$. The Cauchy-Schwartz inequality thus holds for v-norms, i.e.,

$$\left| \int g(s)\mu(ds) \right| \leq \|g\|_v \cdot \|\mu\|_v,$$

for all $g \in \mathcal{D}_v(S)$ and μ such that v is μ-integrable.

Let $\{\mu_n\}$ be a sequence of measures on $(S, \mathcal{S})$ and let μ be a measure on $(S, \mathcal{S})$. We say that μ_n converges weakly in $\mathcal{D}_v(S)$-sense towards μ if

$$\lim_{n \to \infty} \int_S g(s)\, \mu_n(ds) = \int_S g(s)\, \mu(ds),$$

for all $g \in \mathcal{D}_v(S)$; in symbols $\mu_n \overset{\mathcal{D}_v(S)}{\Longrightarrow} \mu$.

Remark E.5.1 *Note that v-norm convergence implies $\mathcal{D}_v(S)$-convergence. This can be seen as follows. According to Cauchy-Schwartz Inequality, for each $g \in \mathcal{D}_v(S)$ it holds that:*

$$\left| \int g(s)\mu_n(ds) - \int g(s)\mu(ds) \right| = \left| \int g(s)(\mu_n - \mu)(ds) \right| \leq \|g\|_v \cdot \|\mu_n - \mu\|_v.$$

Hence, $\|\mu_n - \mu\|_v \to 0$ implies that the left-hand side in the above relation converges to 0 as $n \to \infty$.

For a $\mathcal{D}_v(S)$-differentiable measure μ_θ, $\mathcal{D}_v(S)$-convergence of the measure $(\mu_{\theta+\Lambda} - \mu_\theta)/\Delta$ as Δ tends to zero implies v-norm continuity of μ_θ. The precise statement is given in the following theorem, where $\mathcal{M}^v(S)$ denotes the set of all measures μ on $(S, \mathcal{S})$ such that $\int v\,d\mu$ exists and is finite.

Theorem E.5.1 *Let $\{\mu_\theta\}_{\theta \in \Theta} \subset \mathcal{M}^v(S)$ be $\mathcal{D}_v(S)$-differentiable at $\theta \in \Theta$. Then $\mu_{\theta+h}$ converges in v-norm to μ_θ, as $h \to 0$. In symbols: $\lim_{h \to 0} \|\mu_{\theta+h} - \mu_\theta\|_v = 0$.*

Proof: Assume without loss of generality that θ is an interior point of Θ. Thus, we can choose $\Delta > 0$ such that $[\theta - \Delta, \theta + \Delta] \subset \Theta$. Denote by $T(h)$ the linear continuous functional on $\mathcal{D}_v(S)$ defined as ($h \neq 0$):

$$(T(h))(g) = \frac{1}{h} \int g(s)(\mu_{\theta+h} - \mu_\theta),$$

for all $g \in \mathcal{D}_v(S)$. The operator norm of $T(h)$ satisfies:

$$\|T(h)\|_v = \frac{\|\mu_{\theta+h} - \mu_\theta\|_v}{|h|},$$

for all $h \neq 0$ such that $|h| \leq \Delta$. By $\mathcal{D}_v(S)$-differentiability of μ_θ, for each $g \in \mathcal{D}_v(S)$:

$$\sup_{|h| \leq \Delta} \left| (T(h))(g) \right| < \infty,$$

and the Banach-Steinhaus Theorem[1] yields

$$\sup_{|h| \leq \Delta} \|T(h)\|_v = M < \infty.$$

Thus,

$$\|\mu_{\theta+h} - \mu_\theta\|_v \leq M \cdot |h|,$$

for $|h| \leq \Delta$. Letting $h \to 0$, concludes the proof. $\qquad\square$

Recall that we introduced the following notation in Chapter 3. The set of all probability measures on $(S, \mathcal{S})$ is denoted by $\mathcal{M}_1 = \mathcal{M}_1(S)$. Moreover, $\mathcal{L}^1(\mu_\theta)$ denotes the set of continuous absolutely integrable mappings with respect to μ_θ and

$$\mathcal{L}^1(\mu_\theta : \theta \in \Theta) = \bigcap_{\theta \in \Theta} \mathcal{L}^1(\mu_\theta)$$

denotes the set of continuous absolutely integrable mappings with respect to μ_θ for any $\theta \in \Theta$; see the section 'List of Symbols.'

We assume that S, Z are two separable complete metric spaces endowed with the mappings $v : S \to [1, \infty)$ and $u : Z \to [1, \infty)$, respectively. Let $w : S \times Z \to [1, \infty)$ be defined as $w(s, z) \stackrel{\text{def}}{=} v(s)u(z)$. Then we can consider the space of continuous mappings on $S \times Z$ bounded by w up to a multiplicative constant, denoted by $\mathcal{D}_w(S \times Z)$, see (E.3). Furthermore, if $\mu : \Theta \to \mathcal{M}_1(S)$ and $\nu : \Theta \to \mathcal{M}_1(Z)$, then $\mathcal{D}_v(S) \subset \mathcal{L}^1(\mu_\theta : \theta \in \Theta)$ and $\mathcal{D}_u(Z) \subset \mathcal{L}^1(\nu_\theta : \theta \in \Theta)$ implies $\mathcal{D}_w(S \times Z) \subset \mathcal{L}^1((\mu \times \nu)_\theta : \theta \in \Theta)$; for a proof use Fubini's Theorem.

Lemma E.5.1 *Let $\{\mu_n\}_{n \geq 1} \subset \mathcal{M}^v(S)$ and $\{\nu_n\}_{n \geq 1} \subset \mathcal{M}^u(Z)$. If μ_n converges in v-norm to μ and $\nu_n \stackrel{\mathcal{D}_u(Z)}{\Longrightarrow} \nu$, then*

$$\mu_n \times \nu_n \stackrel{\mathcal{D}_w(S \times Z)}{\Longrightarrow} \mu \times \nu \quad \text{and} \quad \nu_n \times \mu_n \stackrel{\mathcal{D}_w(S \times Z)}{\Longrightarrow} \nu \times \mu.$$

Proof: Note that $\nu_n \stackrel{\mathcal{D}_u(Z)}{\Longrightarrow} \nu$ implies

$$(\nu_n - \nu) \stackrel{\mathcal{D}_u(Z)}{\Longrightarrow} \mathcal{O}, \qquad (E.5)$$

where $\mathcal{O}$ denotes the null-measure assigning value 0 to any measurable set. Hence, in order to prove the first part of lemma, we may assume without loss of generality that $\nu_n \stackrel{\mathcal{D}_u(Z)}{\Longrightarrow} \mathcal{O}$ and we have to show that

$$\mu_n \times \nu_n \stackrel{\mathcal{D}_w(S \times Z)}{\Longrightarrow} \mathcal{O}.$$

Let $g \in \mathcal{D}_w(S \times Z)$, i.e.,

$$|g(s, z)| \leq \|g\|_w \, v(s) \, u(z). \qquad (E.6)$$

---

[1]The result is also known in the literature as *The Principle of Uniform Boundedness* and it basically asserts that weak and strong boundedness are equivalent. The precise statement is as follows. Let $(X, \| \cdot \|)$ be a Banach space and let $\mathcal{K}$ be a family of continuous linear functionals from X to $\mathbb{R}$. If for each $x \in X$ the set $\{T(x) : T \in \mathcal{K}\}$ is bounded in $\mathbb{R}$, then $\sup\{\|T\| : T \in \mathcal{K}\} < \infty$.

Then we have:

$$\int g(s,z)(\mu_n \times \nu_n)(ds, dz) = \int \int g(s,z)\mu_n(ds)\nu_n(dz) = \int h_n(z)\nu_n(dz), \quad \text{(E.7)}$$

where $h_n(z) = \int g(s,z)\mu_n(ds)$, for $n \geq 1$. Set $h(z) = \int g(s,z)\mu(ds)$. Then,

$$|h(z)| \leq \int |g(s,z)| \cdot |\mu|(ds) \leq \|g\|_w \int v(s)u(z)|\mu|(ds)$$

$$= \|g\|_w \underbrace{\left(\int v(s)|\mu|(ds) \right)}_{=\|\mu\|_v} u(z),$$

for all $z \in Z$, where the inequality follows from (E.6). Thus, $\|h\|_u \leq \|g\|_w \|\mu\|_v$, which implies $h \in \mathcal{D}_u(Z)$. Consequently, by (E.5), $\int h(z)\nu_n(dz) \to 0$. To prove the lemma it now suffices to show that $\int (h_n(z) - h(z))\nu_n(dz) \to 0$. To this end, we note that for all $z \in Z$:

$$|h_n(z) - h(z)| \leq \int |g(s,z)| \cdot |\mu_n - \mu|(ds)$$

$$\leq \|g\|_w \cdot u(z) \int v(s)|\mu_n - \mu|(ds)$$

$$= \|g\|_w \cdot \|\mu_n - \mu\|_v \cdot u(z),$$

where the last equality follows from (E.4). This yields

$$\left| \int (h_n(z) - h(z))\nu_n(dz) \right| \leq \int |h_n(z) - h(z)| \cdot |\nu_n|(dz) \leq \|g\|_w \cdot \|\mu_n - \mu\|_v \cdot \|\nu_n\|_u.$$

We have $\int u(z)|\nu_n|(dz) = \|\nu_n\|_u < \infty$ for all n, and an immediate application of the Banach-Steinhaus theorem yields

$$\sup_{n \geq 1} \|\nu_n\|_u < \infty.$$

Now, the fact that μ_n converges in v-norm to μ concludes the proof of the first part of the lemma.

For the proof of the second part of the lemma we apply Fubini's theorem in order to reverse the order of integration in (E.7). The proof of the second part of the lemma then follows from the same line of argument as the proof of the first part. $\qquad \square$

In order to apply Lemma E.5.1, one has to assume that any $g \in \mathcal{D}_w(S \times Z)$ is continuous. However, it is possible to slightly deviate from the continuity assumption. If g is bounded by some $h \in \mathcal{D}_w(S \times Z)$ and if the set of discontinuities, denoted by D_g, satisfies $(\mu \times \nu)(D_g) = 0$ (resp. $(\nu \times \mu)(D_g) = 0$), then Lemma E.5.1 applies to g as well.

E.6   Coupling

E.6.1   Coupling Convergence

We say that there is *coupling convergence in finite time* (or, merely coupling) of a sequence $\{X_n\}$ to a stationary sequence $\{Y \circ \theta^n\}$ if

$$\lim_{n \to \infty} P\left(\forall k : X_{n+k} = Y \circ \theta^{n+k}\right) = 1,$$

or, equivalently, there exists an a.s. finite random variable N such that

$$X_{N+k} = Y \circ \theta^{N+k}, \quad k \geq 0.$$

Result: Coupling (convergence) implies total variation convergence.

E.6.2   Strong Coupling Convergence and Goldstein's Maximal Coupling

We say that there is *strong coupling convergence in finite time* (or, merely strong coupling) of a sequence $\{X_n\}$ to a stationary sequence $\{Y \circ \theta^n\}$ if

$$N^0 = \inf\{n \geq 0 \,|\, \forall k \geq 0 : X_{n+k} \circ \theta^{-n-k} = Y\}$$

is finite with probability one.

Result: Strong coupling convergence implies coupling convergence but the converse is not true.

We illustrate this with the following example. Let ξ_m, with $\xi_m \in \mathbb{Z}$ and $\mathbb{E}[\xi_1] = \infty$, be an i.i.d. sequence and define X_n, for $n \geq 1$, as follows

$$X_n = \begin{cases} \xi_0 & \text{for } X_{n-1} = 0, \\ X_{n-1} - 1 & \text{for } X_{n-1} \geq 2, \\ X_n & \text{for } X_{n-1} = 1, \end{cases}$$

where $X_0 = 0$. It is easily checked that $\{X_n\}$ couples with the constant sequence 1 after $\xi_0 - 1$ transitions. To see that $\{X_n\}$ fails to converge in strong coupling, observe that the shift operator applies to the 'stochastic noise' ξ_m as well. Specifically, for $k \geq 0$,

$$X_n \circ \theta^{-k} = \begin{cases} \xi_0 \circ \theta^{-k} & \text{for } X_{n-1} \circ \theta^{-k} = 0, \\ X_{n-1} \circ \theta^{-k} - 1 & \text{for } X_{n-1} \circ \theta^{-k} \geq 2, \\ X_n \circ \theta^{-k} & \text{for } X_{n-1} \circ \theta^{-k} = 1; \end{cases}$$

where $X_0 \circ \theta^{-k} = 0$, and $\xi_{-k} = \xi_0 \circ \theta^{-k}$. This implies

$$\begin{aligned} N^0 &= \inf\{n \geq 0 \,|\, \forall k \geq 0 : X_{n+k} \circ \theta^{-n-k} = 1\} \\ &= \inf\{n \geq 0 \,|\, \forall k \geq 0 : \xi_{n+k} - 1 \leq n\} \\ &= \infty \quad \text{a.s.} \end{aligned}$$

Result (Goldstein's maximal coupling [80]): Let $\{X_n\}$ and Y be defined on a Polish state space. If $\{X(n)\}$ converges with coupling to Y, then a version $\{\tilde{X}(n)\}$ of $\{X(n)\}$ and a version $\tilde{Y}$ of Y defined on the same probability space exists such that $\{\tilde{X}(n)\}$ converges with strong coupling to $\tilde{Y}$.

E.6.3 δ-Coupling

Coupling and strong coupling, as introduced above, are related to total variation convergence. We now state the definition of δ-coupling which is related to weak convergence. (The classical terminology is ε-coupling. We have changed it to δ-coupling to avoid confusion with the notation $\varepsilon = -\infty$ for the max-plus semiring.)

Consider a metric space (E, d) and two sequences $\{X_n\}$ and $\{Y_n\}$ defined on E. We say that there is δ-*coupling* of these two sequences if

- for each $\delta > 0$, versions of $\{X_n\}$ and $\{Y_n\}$ exist defined on a common probability space, and

- an a.s. finite random variable η_δ exists such that, for $n \geq \eta_\delta$, it holds that $d(X_n, Y_n) \leq \delta$.

Result: Consider a sequence $\{X_n\}$ and a stationary sequence $\{Y_n\}$ defined on a metric space E. If there is δ-coupling of the two sequences, then $\{X_n\}$ converges weakly to Y.

E.7   The Dominated Convergence Theorem

Let $(S, \mathcal{S}, \mu)$ be a probability space. Let $f_n : S \to \mathbb{R}$, for $n \in \mathbb{N}$, be measurable and assume that $f, g : S \to \mathbb{R}$ are measurable mappings such that, for any $n \in \mathbb{N}$, the set of points $s \in S$ with

$$|f_n(s)| \leq g(s)$$

and

$$\lim_{n \to \infty} f_n(s) = f(s)$$

has μ-measure one. If

$$\int_S |g(s)| \mu(ds) < \infty,$$

then

$$\lim_{n \to \infty} \int_S f_n(s) \mu(ds) = \int_S f(s) \mu(ds).$$

E.8   Wald's Equality

Let $\{X(n)\}$ be an i.i.d. sequence such that $\mathbb{E}[X(1)]$ is finite. Furthermore, let η be a non-negative integer-valued random variable with finite mean. If for all $m \geq 0$ the event $\{\eta = m\}$ is independent of $\{X(m+n) : n \geq 1\}$, then

$$\mathbb{E}\left[\sum_{i=1}^{\eta} X(i)\right] = \mathbb{E}[\eta]\,\mathbb{E}[X(1)] < \infty.$$

E.9   Regenerative Processes

Let $\{X(n)\}$ denote a stochastic process with state space $(S, \mathcal{S})$. A random time τ_k is called *stopping time* if the occurrence or non-occurrence of τ_k at time t is known from $\{X(n) : n \leq t\}$, that is, the event $\{\tau_k = t\}$ lies in $\sigma(\{X(n) : n \leq t\})$, for any t. The process $\{X(n)\}$ is called *classical regenerative*, or, *regenerative* if there exists a sequence of stopping times $\{\tau_k\}$ such that

- $\{\tau_{k+1} - \tau_k, \; k \geq 0\}$ is an i.i.d. sequence;

- for every sequence of times $0 < t_1 < t_2 < \cdots < t_n$ and every $k \geq 0$, the random vectors $(X(t_1), X(t_2), \ldots, X(t_n))$ and $(X(\tau_k + t_1), X(\tau_k + t_2), \ldots, X(\tau_k+t_n))$ have the same distributions, and the processes $\{X(n) : n \leq \tau_k\}$ and $\{X(n) : n > \tau_k\}$ are independent.

Thus, in a regenerative process, the regeneration points $\{\tau_k : k \geq 0\}$ cut the process into independent and identically distributed *cycles* of the form $\{X(n) : \tau_k \leq n < \tau_{k+1}\}$. A distribution function is called *lattice* if it assigns probability one to a set of the form $\{0, \delta, 2\delta, \ldots\}$, for some $\delta > 0$, and it is called *non-lattice* otherwise.

Result: Let $\{X(n)\}$ be a regenerative process such that the distribution of $\tau_{k+1} - \tau_k$ is non-lattice. If, for a measurable mapping $f : S \to \mathbb{R}$, $\mathbb{E}[\sum_{n=\tau_1}^{\tau_2-1} f(X(n))]$ is finite, then

$$\lim_{N \to \infty} \frac{1}{N} \sum_{n=1}^{N} f(X(n)) = \frac{\mathbb{E}\left[\sum_{n=\tau_1}^{\tau_2-1} f(X(n))\right]}{\mathbb{E}[\tau_2 - \tau_1]} \quad \text{a.s.}$$

Appendix F

Markov Chains

Let $(S, \mathcal{S})$ denote a Polish state space, where $\mathcal{S}$ denotes the Borel field of S. The mapping $P : S \times \mathcal{S} \to [0, 1]$ is a *Markov kernel* (on $(S, \mathcal{S})$) if

(a) $P(s; \cdot)$ is a probability measure on $(S, \mathcal{S})$, for all $s \in S$; and

(b) $P(\cdot; B)$ is $\mathcal{S}$ measurable for all $B \in \mathcal{S}$.

The product of Markov kernels is again a Markov kernel. Specifically, let P, Q be two Markov kernels on $(S, \mathcal{S})$, then the product of P and Q is defined as follows: for $s \in S$ and $B \in \mathcal{S}$ set $P Q(s; B) = (P \circ Q)(s, B) = \int_S P(s; dz) Q(z; B)$. Moreover, write $P^n(s; \cdot)$ for the measure obtained by the n fold product of P in the above way.

When an initial distribution μ is given, P defines a Markov chain $\{X(n)\}$ with state space $(S, \mathcal{S})$:

$$\mathbb{P}(X(n) \in B) = \int \mu(ds) P^n(s; B),$$

where $\mathbb{P}$ denotes the underlying probability measure on $(S, \mathcal{S})$.

Let ϕ be a σ-finite measure on $(S, \mathcal{S})$. A Markov chain with transition kernel $P(x; B)$ is *ϕ-irreducible* if

$$\mathbb{P}\left(\bigcup_{n=1}^{\infty} X(n) \in B \,\middle|\, X(0) = x \right) > 0, \quad x \in S, \ B \in \mathcal{S},$$

whenever $\phi(B) > 0$.

A Markov chain $\{X(n)\}$ is called *uniformly ϕ-recurrent* if there exists a non-trivial measure ϕ on $(S, \mathcal{S})$ such that for each $A \in \mathcal{S}$, with $\phi(A) > 0$,

$$\lim_{k \to \infty} \sum_{m=1}^{k} {}_A P^m(x, A) = 1$$

uniformly in x, where $_A P^m(x, A)$ is the *taboo probability* defined by

$$_A P^m(x, A) = \mathbb{P}(\{X(m) \in A, X(0) = x, X_i \notin A, 1 \le i \le m - 1\}), \quad A \in \mathcal{S}.$$

A uniformly ϕ-recurrent Markov chain is also called *Harris recurrent*.

A *d-cycle* of a ϕ-irreducible chain $\{X(n)\}$ is a collection $\{S_1, \ldots, S_d\}$ of disjoint subsets of S such that $\phi(S - \cup_{i=1}^d S_i) = 0$ and $P(s; S_{i+1}) = 1$ for $s \in S_i$ and $1 \le i \le d$ (take $S_{i+1} = S_1$ when $i = d$). At least one d-cycle exists for a ϕ-irreducible chain and the *period* of the chain is the smallest d for which a d-cycle exists. The chain is called *aperiodic* if it is ϕ-irreducible and $d = 1$; otherwise it is called *periodic*. Observe that aperiodicity of a chain already implies its ϕ-irreducibility. A uniformly ϕ-recurrent and aperiodic Markov chain is also called *Harris-ergodic*.

Result: A uniformly ϕ-recurrent and aperiodic (resp. Harris ergodic) Markov chain converges, for any initial distribution, weakly towards a unique stationary regime π. Moreover, for any measurable mapping $f : S \to \mathbb{R}$, with $\int_S f(s) \pi(ds)$ finite, it holds that

$$\lim_{N \to \infty} \frac{1}{N} \sum_{n=1}^N f(X(n)) = \int_S f(s)\pi(ds) \quad \text{a.s.}$$

Let $\{X(n)\}$ be a Harris ergodic Markov chain. For $B \in \mathcal{S}$, let $\tau_n \stackrel{\text{def}}{=} \tau_B(n)$ denote the nth hitting time of $X(n)$ on B, where we set $\tau_n = \infty$ if $X(n)$ doesn't visit B for at least n times. Hence, τ_n is a stopping time. A set $B \in \mathcal{S}$ is called a *regeneration set* if, with probability one, $\tau_n < \infty$, for any $n \in \mathbb{N}$, and with probability one:

$$\lim_{n \to \infty} \tau_n = \infty.$$

A regeneration set B is called *atom* if the regeneration points $\{\tau_k : k \ge 0\}$ cut the Markov chain into independent and identically distributed cycles of the form $\{X(n) : \tau_k \le n < \tau_{k+1}\}$. Thus, whenever $X(n)$ hits B it starts independent from the past. In particular, if we consider two versions of $X(n)$, where one version is started according to an initial distribution μ and the other according to an initial distribution ν, then both versions couple when they simultaneously hit B, which occurs after a.s. finitely many transitions.

Result: A Harris ergodic Markov chain $\{X(n)\}$ with atom converges, for any initial distribution, in strong coupling to its unique stationary regime. In addition to that, let B denote an atom of $\{X(n)\}$, then it holds that

$$\int_S f(s) \pi(ds) = \frac{\mathbb{E}\left[\sum_{n=1}^{\tau_1 - 1} f(X(n)) \,\Big|\, X(0) \in B\right]}{\mathbb{E}[\tau_1 \,|\, X(0) \in B]},$$

for any measurable mapping $f : S \to \mathbb{R}$ such that $\int_S f(s) \pi(ds)$ is finite, where τ_1 denotes the first hitting time of $X(n)$ on B.

Appendix G

Tools from analysis

G.1 Cesàro limits

A real-valued sequence $\{x_n\}$ is called *Cesàro-summable* if

$$\lim_{n \to \infty} \frac{1}{n} \sum_{m=1}^{n} x_m$$

exists. If

$$\lim_{n \to \infty} x_n = x$$

exists, then

$$\lim_{n \to \infty} \frac{1}{n} \sum_{m=1}^{n} x_m = x \,.$$

In words, any convergent sequence is Cesàro-summable. The converse is, however, not true. To see this, consider the sequence $x_n = (-1)^n$, $n \in \mathbb{N}$.

G.2 Lipschitz and Uniform Continuity

Let $X \subset \mathbb{R}$ be a compact set. A mapping $f : X \to \mathbb{R}$ is called *Lipschitz continuous* if $K \in \mathbb{R}$ exists such that for any $x, x + \Delta \in X$ is holds that

$$|f(x) - f(x + \Delta)| \leq K \, |\Delta| \,.$$

The constant K is called *Lipschitz constant*.

Result (Mean-Value Theorem): For $X = [a, b] \subset \mathbb{R}$, let $f : X \to \mathbb{R}$ be continuous on $[a, b]$ and differentiable on $]a, b[$. Then $\xi \in]a, b[$ exists such that

$$\frac{f(b) - f(a)}{b - a} = f'(\xi) \,,$$

where f' denotes the derivative of f.

For differentiable mappings, a sufficient condition for Lipschitz continuity can be found by the Mean-Value Theorem. The precise statement is given in the following.

Result: For $X = [a, b] \subset \mathbb{R}$, let $f : X \to \mathbb{R}$ be continuous on $[a, b]$ and differentiable on $]a, b[$. If

$$\sup_{x \in]a,b[} |f'(x)| \stackrel{\text{def}}{=} K < \infty ,$$

then f is Lipschitz continuous on (a, b) with Lipschitz constant K.

G.3   Interchanging Limit and Differentiation

Let $\mathcal{F}$ denote a set of mappings from $X = [a, b] \subset \mathbb{R}$ to $\mathbb{R}$. $\mathcal{F}$ is called *uniformly bounded* if $M \in [0, \infty)$ exists such that, for any $f \in \mathcal{F}$,

$$|f(x)| \leq M , \quad x \in X .$$

A mapping $f : X = [a, b] \subset \mathbb{R}$ is called *uniformly continuous* if for any $\eta > 0$ a $\delta > 0$ exists such that, for any $x_1, x_2 \in X = [a, b] \subset \mathbb{R}$ with $|x_1 - x_2| < \delta$, it holds that

$$|f(x_1) - f(x_2)| < \eta .$$

The set $\mathcal{F}$ is called *uniformly continuous* if for any $\eta > 0$ a $\delta > 0$ exists such that, for any $x_1, x_2 \in X = [a, b] \subset \mathbb{R}$ with $|x_1 - x_2| < \delta$ and any $f \in \mathcal{F}$, it holds that

$$|f(x_1) - f(x_2)| < \eta .$$

Result (Arzela-Ascoli): Let $\mathcal{F}$ be a (at least) countable set of mappings from $X = [a, b] \subset \mathbb{R}$ to $\mathbb{R}$. If $\mathcal{F}$ is uniformly bounded and uniformly continuous, then one can choose a uniformly convergent sequence out of $\mathcal{F}$.

Result: For $n \in \mathbb{N}$, let f_n be a continuously differentiable mapping from $X = [a, b] \subset \mathbb{R}$ to $\mathbb{R}$. If

  (i) f_n converges pointwise to f on X,

  (ii) the sequence of derivatives f'_n converges uniformly on X,

then f is differentiable and it holds that

$$\lim_{n \to \infty} f'_n(x) = f'(x) .$$

We combine the above results to the following statement.

Theorem G.3.1 *For $n \in \mathbb{N}$, let f_n be a twice differentiable mapping from $X = [a, b] \subset \mathbb{R}$ to $\mathbb{R}$. If*

(i) f_n converges pointwise to f on X ,

(ii) f'_n converges pointwise on X ,

(iii) a constant M exists, such that

$$\max \left(\sup_{n \in \mathbb{N}} \sup_{x \in X} |f'_n(x)| , \sup_{n \in \mathbb{N}} \sup_{x \in X} |f''_n(x)| \right) \leq M ,$$

where f''_n denotes the second order derivative of f ,

then f is differentiable and it holds that

$$\lim_{n \to \infty} f'_n(x) = f'(x) .$$

Proof: By the foregoing result, it remains to be shown that f'_n converges uniformly on X . Set

$$\mathcal{F} = \{f'_n : n \in \mathbb{N}\} .$$

By assumption, $\mathcal{F}$ is uniformly bounded. Uniform continuity follows from the fact that f'_n is Lipschitz continuous with Lipschitz constant M . See Section G.2. Since the Lipschitz constant is independent of n , $\mathcal{F}$ is uniformly continuous. Hence, according to the Arzela-Ascoli Theorem we may choose a sequence $\{f_{n_m} : m \in \mathbb{N}\}$ out of $\{f_n\}$ that converges uniformly on X . This yields

$$\lim_{n \to \infty} f'_n(x) = \lim_{m \to \infty} f'_{n_m}(x)$$
$$= f'(x) ,$$

which concludes the proof of the theorem. $\square$

G.4 Taylor Series Expansions

Fix $x_0 \in \mathbb{R}$ and $\Delta > 0$, let $f : [x_0, x_0 + \Delta] \to \mathbb{R}$ be an $(n+1)$ times continuously differentiable mapping on $[x_0, x_0 + \Delta]$ (where differentiability at the boundary has to interpreted as one-sided differentiability). Then it holds that

$$f(x_0 + \Delta) = \sum_{m=0}^{n} \frac{\Delta^m}{m!} \frac{d^m}{dx^m} \bigg|_{x=x_0} f(x) + R_{n+1}(x_0) \tag{G.1}$$

and $\nu \in (0,1)$ exists such that

$$R_{n+1}(x_0) = \frac{\Delta^{n+1}}{(n+1)!} \frac{d^{n+1}}{dx^{n+1}} \bigg|_{x=x_0+\nu\Delta} f(x) .$$

The above remainder term is called *Lagrange* remainder. An alternative way of expressing R_{n+1} is the *Cauchy* remainder:

$$R_{n+1}(x_0) = \frac{\Delta^{n+1}}{n!} (1 - \nu')^n \frac{d^{n+1}}{dx^{n+1}} \bigg|_{x=x_0+\nu'\Delta} f(x) ,$$

where ν' is again a number in $(0, 1)$. In addition to that, R_{n+1} can be expressed using integration as follows:

$$R_{n+1}(x_0) = \frac{1}{n!} \int_{x_0}^{x_0+\Delta} (x_0 + \Delta - t)^n \left. \frac{d^{n+1}}{dx^{n+1}} \right|_{x=t} f(x)\, dt . \qquad \text{(G.2)}$$

The expression on the right-hand side of (G.1) is called a *Taylor polynomial for f of degree n at x_0*. If

$$\sum_{m=0}^{\infty} \frac{\Delta^m}{m!} \left. \frac{d^m}{dx^m} \right|_{x=x_0} f(x) \qquad \text{(G.3)}$$

exists, for given x_0 and Δ, then this series is called *Taylor series* or *Taylor series expansion for f at x_0* evaluated at Δ. In the particular case $x_0 = 0$, (G.3) is also called *MacLaurin series*. The radius of convergence of a Taylor series, denoted by $r(x_0)$, is the largest Δ such that the sum in (G.3) exists and is finite. Because Taylor series are power series, they converge absolutely if they converge at all. Hence, if the radius of convergence of the Taylor series expansion for f at x_0 is $r(x_0) > 0$, then the series converges for any $x_0 + \Delta$, with $|\Delta| \leq r(x_0)$. The radius of convergence of the Taylor series for f at x_0 is given by the formula of Cauchy-Hadamard:

$$\frac{1}{r(x_0)} = \limsup \left(\frac{1}{n!} \left| \left. \frac{d^n}{dx^n} \right|_{x=x_0} f(x) \right| \right)^{\frac{1}{n}} , \qquad \text{(G.4)}$$

where $r(x_0) = 0$ if the lim sup equals ∞ and $r(x_0) = \infty$ if the lim sup equals 0.

A real-valued mapping $f : U \to \mathbb{R}$, with $U \subset \mathbb{R}$, is called *analytic* if, for any $x_0 \in U$, a $r(x_0) > 0$ exists such that the Taylor series for f at x_0 equals f for any $|\Delta| \leq r(x_0)$. For U open, it can be shown that analyticity of f on U is equivalent to the existence of a holomorphic extension of f to the complex plane, which explains the term 'analytic.'

Let the Taylor series expansion for f at x_0 have radius of convergence $r(x_0) > 0$ such that the Taylor series for f at x_0 equals f for any $|\Delta| \leq r(x_0)$. This implies that the expression in (G.1) converges to f for at least those $x_0 + \Delta$, with $|\Delta| \leq r(x_0)$. In other words, the remainder term $R_{n+1}(x_0)$ tends to 0 as n tends to ∞ for all $|\Delta| \leq r(x_0)$. Does this mean that increasing the degree of the Taylor polynomial improves the accuracy of the approximation? To answer this question, note that, for $n \in \mathbb{N}$,

$$\left| f(x_0 + \Delta) - \sum_{m=0}^{n} \frac{\Delta^m}{m!} \left. \frac{d^m}{dx^m} \right|_{x=x_0} f(x) \right| = \left| \sum_{m=n+1}^{\infty} \frac{\Delta^m}{m!} \left. \frac{d^m}{dx^m} \right|_{x=x_0} f(x) \right|$$

$$\leq \sum_{m=n+1}^{\infty} \frac{|\Delta|^m}{m!} \left| \left. \frac{d^m}{dx^m} \right|_{x=x_0} f(x) \right|$$

$$\overset{\text{def}}{=} H_{n+1} .$$

Existence of the Taylor series implies that $H_n \searrow 0$. Observe that $R_n \leq H_n$. Hence, the error in predicting $f(x_0 + \Delta)$ by a Taylor polynomial of degree n is at most H_n, where H_n is a monotone decreasing sequence. Thus, with growing n, R_{n+1} eventually decreases. Unfortunately, this does not imply that increasing the order of a Taylor polynomial from n to $n+1$ will improve the quality of the approximation. To see this, let f be given such that $d^m f/dx^m = 0$ for $m \geq N$, for some finite N. In this case, it is not ruled out that the accuracy of the approximation decreases with increasing the degree of the Taylor polynomial provided the degree is smaller than N. For example, it can happen that

$$R_n < R_{n+1} \quad \text{for } n \leq N - 1$$

and $R_{n+1} = 0$ for $n \geq N$. In such a case, increasing the degree of the Taylor polynomial may even decrease the quality of the approximation.

Let $\{f_n\}$ be a sequence of functions that converges point-wise to a function f. Under appropriate conditions the limit of the Taylor series for f_n will converge to the Taylor series for f. The exact statement is given in the following theorem.

Theorem G.4.1 *Consider $X = [x_0, x_0 + \Delta] \subset \mathbb{R}$ and let $\{f_n\}$ be a sequence of mappings such that*

(i) f_n converges pointwise to a mapping f on X,

(ii) $d^k f_n/dx^k$ converges pointwise on X as n tends to ∞,

(iii) on X, the Taylor series for f_n exists and converges to the true value of f_n,

(iv) a sequence $\{M_k\}$ exists, whereby

(a)

$$\sup_{n \in \mathbb{N}} \sup_{x \in X} \left| \frac{d^k}{dx^k} f_n(x) \right| \leq M_k, \quad k \in \mathbb{N},$$

and

(b)

$$\sum_{k=0}^{\infty} \frac{\Delta^k}{k!} M_k < \infty,$$

then it holds that

$$f(x_0 + \Delta) = \sum_{k=0}^{\infty} \frac{\Delta^k}{k!} \frac{d^k}{dx^k}\bigg|_{x=x_0} f(x) = \lim_{n \to \infty} \sum_{k=0}^{\infty} \frac{\Delta^k}{k!} \frac{d^k}{dx^k}\bigg|_{x=x_0} f_n(x).$$

Proof: Repeated application of Theorem G.3.1 yields, for any k,

$$\lim_{n \to \infty} \frac{d^k}{dx^k} f_n(x) = \frac{d^k}{dx^k} f(x) \tag{G.5}$$

on X, where differentiability at the boundary of X has to be understood as one-sided differentiability. Assumption (iv)(a) implies that, for any $n \in \mathbb{N}$ and any $x \in X$,

$$\left| \frac{d^k}{dx^k} f_n(x) \right| < M_k.$$

Together with assumption (iv)(b), the dominated convergence theorem can be applied. This yields

$$\lim_{n\to\infty} \sum_{k=0}^{\infty} \frac{\Delta^k}{k!} \frac{d^k}{dx^k} f_n(x) = \sum_{k=0}^{\infty} \frac{\Delta^k}{k!} \lim_{n\to\infty} \frac{d^k}{dx^k} f_n(x)$$

and inserting (G.5) gives

$$\sum_{k=0}^{\infty} \frac{\Delta^k}{k!} \lim_{n\to\infty} \frac{d^k}{dx^k} f_n(x) = \sum_{k=0}^{\infty} \frac{\Delta^k}{k!} \frac{d^k}{dx^k} f(x) \,,$$

which concludes the proof. □

G.5 Combinatorial Aspects of Derivatives

Let $\{f_k\}$ be a sequence of n times differentiable mappings from $\mathbb{R}$ to $\mathbb{R}$ and denote the nth order derivative of f_k by $f_k^{(n)}$, and let $f_k^{(0)} = f_k$. We denote the argument of f_k by x. The first order derivative of the product $\prod_{i=1}^{m} f_i$ is given by

$$\frac{d}{dx} \prod_{i=1}^{m} f_i = \sum_{k=1}^{m} \prod_{i=1}^{m} f_i^{(1_{k=i})} \,,$$

where $1_{k=i} = 1$ if $k = i$ and zero otherwise. Generally, the nth order derivative of the product of m mappings is given by

$$\frac{d^n}{dx^n} \prod_{i=1}^{m} f_j = \sum_{k_1=1}^{m} \sum_{k_2=1}^{m} \cdots \sum_{k_n=1}^{m} \prod_{i=1}^{m} f_i^{\left(\sum_{j=1}^{n} 1_{k_j=i}\right)} \,.$$

Obviously, the above expression has m^n elements. However, some elements occur more than once. Specifically, let

$$\mathcal{L}[1, m; n] = \left\{ (l_1, \ldots, l_m) \in \{0, \ldots, n\}^m \,\middle|\, \sum_{k=1}^{m} l_k = n \right\}$$

and interpret $l = (l_1, \ldots, l_m)$ as 'taking the l_k^{th} order derivative of the kth element of an m fold product,' then the combination of higher-order derivatives corresponding to $l \in \mathcal{L}[1, m; n]$ occurs

$$\frac{n!}{l_1! \cdots l_m!}$$

times in the nth order derivative. Indeed, there are $n!/l_1! \cdots l_m!$ possibilities of placing n balls in m urns such that finally urn k contains l_k balls. Hence,

$$\frac{d^n}{dx^n} \prod_{i=1}^{m} f_i = \sum_{l \in \mathcal{L}[1,m;n]} \frac{n!}{l_1! \cdots l_m!} \prod_{i=1}^{m} f_i^{(l_i)} \,.$$

The above sum has

$$\binom{m+n-1}{n}$$

elements, which stems from the fact that this is exactly the number of possibilities of placing n balls in m urns. Denoting the number of elements of a set $A \subset \mathbb{N}^n$ by $|A|$, this can be written by

$$\left| \mathcal{L}[1, m; n] \right| = \binom{m+n-1}{n}.$$

Recall that the overall derivative is built out of m^n elementary expressions and thus

$$\left| \sum_{l \in \mathcal{L}[1,m;n]} \frac{n!}{l_1! \cdots l_m!} \right| = m^n.$$

For $l \in \mathcal{L}[1, m; n]$ introduce the set

$$\mathcal{I}[l] =$$
$$\left\{ (i_1, \ldots, i_m) \,\middle|\, i_k \in \{0, +1, -1\}, i_k = 0 \text{ iff } l_k = 0 \text{ and } \prod_{\substack{i_1, \ldots, i_m \\ i_k \neq 0}} i_k = +1 \right\}.$$

The set $\mathcal{I}[l]$ has at most 2^{n-1} elements, that is,

$$\forall l \in \mathcal{L}[1, m; n] : \quad \left| \mathcal{I}[l] \right| \leq 2^{n-1}.$$

This can be seen as follows. Any $l \in \mathcal{L}[1, m; n]$ has at most n entries different from zero. We can place any possible allocation of '+1' and '−1' on $n - 1$ places. The nth place is completely determined by this allocation because we have to chose this element so that

$$\prod_{\substack{i_1, \ldots, i_m \\ i_k \neq 0}} i_k = +1.$$

Appendix H

Appendix to Section 5.1.3.2

Table H.1 lists values for the Taylor polynomial of degree $h = 2$ and $h = 3$, respectively, for various traffic loads of the M/M/1 queue. Specifically, the upper values refer to the Taylor polynomial of degree $h = 2$, the values in the second row are those for the Taylor polynomial of degree $h = 3$, and the values in brackets are the 'true' values (which stem from intensive simulation). For the naive approximation, the values $V_{id}(2;0,1)$ are the upper values and the values $V_{id}(3;0,1,2)$ are listed on the second row. Eventually, the table list the stationary expected waiting for the various traffic loads.

Table H.2 lists values for the Taylor polynomial of degree $h = 2$ and $h = 3$, respectively, for various traffic loads of the D/M/1 queue. Specifically, the upper values refer to the Taylor polynomial of degree $h = 2$, the values in the second row are those for the Taylor polynomial of degree $h = 3$, and the values in brackets are the 'true' values (which stem from intensive simulation). For the naive approximation, the values $V_{id}(2;0,1)$ are the upper values and the values $V_{id}(3;0,1,2)$ are listed on the second row.

Table H.1: Approximating the expected m^{th} waiting time in a M/M/1 queue via a Taylor polynomial of degree 2 and 3, respectively.

m	$\rho = 0.1$	$\rho = 0.3$	$\rho = 0.5$	$\rho = 0.7$	$\rho = 0.9$
	0.1066	0.3598	0.7893	1.3080	1.4534
5	0.1101	0.4010	0.8025	1.1374	1.5364
	[0.1110]	[0.4082]	[0.7804]	[1.1697]	[1.5371]
	0.1066	0.3600	0.8444	1.9948	2.4537
10	0.1100	0.4036	0.9802	1.6228	1.5537
	[0.1110]	[0.4264]	[0.9235]	[1.5894]	[2.3303]
	0.1066	0.3600	0.8457	2.2934	6.0912
20	0.1100	0.4036	0.9963	2.6459	-5.7325
	[0.1110]	[0.4283]	[0.9866]	[1.9661]	[3.3583]
	0.1066	0.3600	0.8457	2.3077	14.7404
50	0.1100	0.4036	0.9963	2.8626	-6.1387
	[0.1115]	[0.4283]	[0.9994]	[2.2617]	[5.0376]
naive	0.1066	0.3249	0.5185	0.6810	0.8161
	0.1100	0.3707	0.6378	0.8813	1.0931
analytic $(n = \infty)$	0.1111	0.4285	1	2.3333	9.0000

Table H.2: Approximating the expected m^{th} waiting time in a D/M/1 queue via a Taylor polynomial of degree 2 and 3, respectively.

m	$\rho = 0.1$	$\rho = 0.3$	$\rho = 0.5$	$\rho = 0.7$	$\rho = 0.9$
	0.00004542	0.04241	0.2443	0.6171	0.8955
5	0.00004542	0.04259	0.2412	0.5918	1.0584
	[0.00004601]	[0.04266]	[0.2410]	[0.5981]	[1.0225]
	0.00004542	0.04241	0.2127	0.8314	1.1456
10	0.00004542	0.04261	0.2533	0.7115	1.7005
	[0.00004602]	[0.04268]	[0.2535]	[0.7460]	[1.5102]
	0.00004542	0.04241	0.2536	0.9246	2.2680
20	0.00004542	0.04261	0.2560	0.8606	0.3134
	[0.00004615]	[0.04268]	[0.2552]	[0.8387]	[2.1053]
	0.00004542	0.04241	0.2536	0.9292	5.7393
50	0.00004542	0.04261	0.2560	0.8957	-4.4469
	[0.00004607]	[0.04268]	[0.2552]	[0.8747]	[2.9735]
naive	0.00004542	0.04118	0.1902	0.3791	0.5579
	0.00004542	0.04229	0.2175	0.4743	0.7389

Bibliography

[1] Altman, E., B. Gaujal, and A. Hordijk. Admission control in stochastic event graphs. *IEEE Transactions on Automatic Control*, 45:854–867, 2000.

[2] Altman, E., B. Gaujal, and A. Hordijk. *Discrete-Event Control of Stochastic Networks: Multimodularity and Regularity*. Lecture Notes in Mathematics, vol. 1829. Springer, Berlin, 2003.

[3] Ayhan, H., and D. Seo. Tail probability of transient and stationary waiting times in (max,+)-linear systems. *IEEE Transactions on Automatic Control*, 47:151–157, 2000.

[4] Ayhan, H., and D. Seo. Laplace transform and moments of waiting times in Poisson driven (max,+)-linear systems. *Queueing Systems – Theory and Applications*, 37:405–436, 2001.

[5] Ayhan, H., and F. Baccelli. Expansions for joint Laplace transforms for stationary waiting times in (max,+)-linear systems with Poisson input. *Queueing Systems – Theory and Applications*, 37:291–328, 2001.

[6] Baccelli, F. Ergodic theory of stochastic Petri networks. *Annals of Probability*, 20:375–396, 1992.

[7] Baccelli, F., and D. Hong. Analytic expansions of (max,+) Lyapunov exponents. *Annals of Applied Probability*, 10:779–827, 2000.

[8] Baccelli, F., and D. Hong. Analyticity of iterates of random non–expansive maps. *Advances in Applied Probability*, 32:193–220, 2000.

[9] Baccelli, F., E. Gelenbe, and B. Plateau. An end–to–end approach to the resequencing problem. *Journal of the Association for Computing Machinery*, 31:474–485, 1984.

[10] Baccelli, F., G. Cohen, G.J. Olsder, and J.P. Quadrat. *Synchronization and Linearity*. John Wiley and Sons, (this book is out of print and can be accessed via the max-plus web portal at http://maxplus.org), 1992.

[11] Baccelli, F., and J. Mairesse. *Ergodic Theorems for Stochastic Operators and Discrete Event Networks.* In *Idempotency* (editor J. Gunawardena), vol. 11 of *Publications of the Newton Institute.* Cambridge University Press, 1998.

[12] Baccelli, F., and M. Canales. Parallel simulation of stochastic Petri nets using recurrence equations. *ACM Transactions on Modeling and Computer Simulation*, 3:20–41, 1993.

[13] Baccelli, F., and P. Brémaud. *Elements of Queueing Theory.* Springer, Berlin, 1984.

[14] Baccelli, F., and P. Konstantopoulos. Estimates of cycle times in stochastic Petri nets. In *Lecture Notes in Control and Information Science 177*, pages 1–20. Springer, Berlin, 1992.

[15] Baccelli, F., S. Hasenfuß, and V. Schmidt. Transient and stationary waiting times in (max,+)–linear systems with Poisson input. *Queueing Systems – Theory and Applications*, 26:301–342, 1997.

[16] Baccelli, F., S. Hasenfuß, and V. Schmidt. Expansions for steady–state characteristics of (max,+)–linear systems. *Stochastic Models*, 14:1–24, 1998.

[17] Baccelli, F., and V. Schmidt. Taylor series expansions for Poisson–driven (max,+)–linear systems. *Annals of Applied Probability*, 6:138–185, 1996.

[18] Baccelli, F., and Z. Liu. Comparison properties of stochastic decision free Petri nets. *IEEE Transactions on Automatic Control*, 37:1905–1920, 1992.

[19] Baccelli, F., and Z. Liu. On a class of stochastic evolution equations. *Annals of Probability*, 20:350–374, 1992.

[20] Billingsley, P. *Ergodic Theory and Information.* Wiley, New York, 1968.

[21] Blondel, V., S. Gaubert, and J. Tsitsiklis. Approximating the spectral radius of sets of matrices in the max-plus algebra is NP hard. *IEEE Transactions on Automatic Control*, 45:1762–1765, 2000.

[22] Borovkov, A. *Ergodicity and Stability of Stochastic Processes.* Probability and Statistics. Wiley, Chichester, 1998.

[23] Borovkov, A., and S. Foss. Stochastically recursive sequences and their generalizations. *Siberian Advances in Mathematics*, 2:16–81, 1992.

[24] Bougerol, P., and J. Lacroix. *Products of Random Matrices with Applications to Schrödinger Operators.* Birkhäuser, Boston, 1985.

[25] Bouillard, A., and B. Gaujal. Coupling time of a (max,plus) matrix. In *Proceedings of the workshop on (max,+)-algebra and applications*, pages 235–239. Prague, Czech Republic, August 2001, 1991.

[26] Brauer, A. On a problem of partitions. *American Journal of Mathematics*, 64:299–312, 1942.

[27] Brémaud, P. Maximal coupling and rare perturbation analysis. *Queueing Systems – Theory and Applications*, 11:307–333, 1992.

[28] Brilman, M., and J. Vincent. Dynamics of synchronized parallel systems. *Stochastic Models*, 13:605–617, 1997.

[29] Brilman, M., and J. Vincent. On the estimation of throughput for a class of stochastic resources sharing systems. *Mathematics of Operations Research*, 23:305–321, 1998.

[30] Cao, X.R. The MacLaurin series for performance functions of Markov chains. *Advances in Applied Probability*, 30:676–692, 1998.

[31] Cheng, D. Tandem queues with general blocking: a unified model and comparison results. *Journal of Discrete Event Dynamic Systems*, 2:207–234, 1993.

[32] Cochet-Terrasson, J., G. Cohen, S. Gaubert, M. Mc Gettrick, and J.P. Quadrat. Numerical computation of spectral elements in max-plus-algebra. In *Proceedings of the IFAC conference on Systems Structure and Control*, pages 699–706. Nantes, France, July 1998, 1998.

[33] Cohen, G., D. Dubois, J.P. Quadrat, and M. Viot. Analyse du comportement périodique de systèmes de production par la théorie des dioïdes. *INRIA Research Report No. 191, INRIA Rocquencourt, 78153 Le Chesnay, France*, 1983.

[34] Cohen, G., D. Dubois, J.P. Quadrat, and M. Viot. A linear system-theoretic view of discrete event processes and its use for performance evaluation in manufacturing. *IEEE Transactions on Automatic Control*, 30:210–220, 1985.

[35] Cohen, J. Subadditivity, generalized products of random matrices and operations research. *SIAM Reviews*, 30:69–86, 1988.

[36] Cohen, J. Erratum "Subadditivity, generalized products of random matrices and Operations Research". *SIAM Reviews*, 35:124, 1993.

[37] Cuninghame-Green, R.A. *Minimax algebra*. vol. 166 of Lecture Notes in Economics and Mathematical Systems. Springer, Berlin, 1979.

[38] Cuninghame-Green, R.A. Maxpolynomial equations. *Fuzzy Sets and Systems*, 75:179–187, 1995.

[39] Cuninghame-Green, R.A. *Minimax algebra and its applications*. Advances in Imaging and Electron Physics. Vol. 90. Academic Press, New York, 1995.

[40] Daduna, H. Exchangeable items in repair systems: delay times. *Operations Research*, 38:349–354, 1990.

[41] de Vries, R.E. *On the Asymptotic Behavior of Discrete Event Systems.* PhD thesis, Faculty of Technical Mathematics and Informatics, University of Technology, Delft, The Netherlands, 1992.

[42] Dumas, Y., and P. Robert. On the throughput of a resource sharing model. *Mathematics of Operations Research*, 26:163–173, 2001.

[43] Gunawardena, J. (editor). *Idempotency.* Publications of the Newton Institute, Cambirgde University Press, 1998.

[44] Fu, M., and J.Q. Hu. *Conditional Monte Carlo: Gradient Estimation and Optimization Applications.* Kluwer, Boston, 1997.

[45] Furstenberg, H. Noncommuting random products. *Transactions of the American Mathematical Society*, 108:377–428, 1995.

[46] Gaubert, S. Performance evaluation of (max,+) automata. *IEEE Transactions on Automatic Control*, 40:2014–2025, 1995.

[47] Gaubert, S. Methods and applications of (max,+)–linear algebra. In *Proceedings of the STACS'1997, Lecture Notes in Computer Science, vol 1200*. Springer (this report can be accessed via the WEB at http://www.inria.fr/RRRT/RR-3088.html), 1997.

[48] Gaubert, S., and D. Hong. Series expansions of Lyapunov exponents and forgetful monoids. *INRIA Research Report No. 3971*, 2000.

[49] Gaubert, S., and J. Gunawardena. The duality theorem for min-max functions. *Comptes Rendus de l'Academie des Sciences, Série I, Mathématique, Paris*, t. 326, Série I:699–706, 1998.

[50] Gaubert, S., and J. Mairesse. Asymptotic analysis of heaps of pieces and application to timed Petri nets. In *Proceedings of the 8th International Workshop on Petri Nets and Performance Models (PNPM'99)*. Zaragoza, Spain, 1999.

[51] Gaubert, S., and J. Mairesse. Modeling and analysis of timed Petri nets using heaps of pieces. *IEEE Transactions on Automatic Control*, 44:683–698, 1999.

[52] Glasserman, P. Structural conditions for perturbation analysis derivative estimation finite–time performance indices. *Operations Research*, 39(5):724–738, 1991.

[53] Glasserman, P., and D. Yao. Stochastic vector difference equations with stationary coefficients. *Journal of Applied Probability*, 32:851–866, 1995.

[54] Glasserman, P., and D. Yao. Structured buffer-allocation problems. *Journal of Discrete Event Dynamic Systems*, 6:9–41, 1996.

[55] Grigorescu, S., and G. Oprisan. Limit theorems for j–x processes with a general state space. *Zeitschrift für Wahrscheinlichkeitstheorie und Verwandte Gebiete*, 35:65–73, 1976.

[56] Gunawardena, J. Cycle times and fixed points of min-max functions. In *11th International Conference on Analysis and Optimization of Systems*, pages 266–272. Springer Lecture Notes in Control and Information Science 199, 1994.

[57] Gunawardena, J. Min-max functions. *Journal of Discrete Event Dynamic Systems*, 4:377–407, 1994.

[58] Gunawardena, J. From max-plus algebra to nonexpansive maps. *Theoretical Computer Science*, 293:141–167, 2003.

[59] Hajek, B. Extremal splittings of point processes. *Mathematics of Operations Research*, 10(4):543–556, 1985.

[60] Hartmann, M., and C. Arguelles. Transience bounds for long walks. *Mathematics of Operations Research*, pages 414–439, 1999.

[61] Heidergott, B. A characterization for (max,+)–linear queueing systems. *Queueing Systems – Theory and Applications*, 35:237–262, 2000.

[62] Heidergott, B. A differential calculus for random matrices with applications to (max,+)–linear stochastic systems. *Mathematics of Operations Research*, 26:679–699, 2001.

[63] Heidergott, B. Variability expansion for performance characteristics of (max,plus)-linear systems. In *Proceedings of the International Workshop on DES, Zaragoza, Spain*, pages 245–250. IEEE Computer Society, 2002.

[64] Heidergott, B., and A. Hordijk. Taylor series expansions for stationary Markov chains. *Advances in Applied Probability*, 23:1046–1070, 2003.

[65] Heidergott, B., G.J. Olsder, and J. van der Woude. *Max Plus at Work: Modeling and Analysis of Synchronized Systems*. Princeton University Press, Princeton, 2006.

[66] Hennion, B. Limit theorems for products of positive random matrices. *Annals of Applied Probability*, 25:1545–1587, 1997.

[67] Ho, Y.C., M. Euler, and T. Chien. A gradient technique for general buffer storage design in a serial production line. *International Journal of Production Research*, 17:557–580, 1979.

[68] Ho, Y.C., and X.R. Cao. *Perturbation Analysis of Discrete Event Systems*. Kluwer Academic, Boston, 1991.

[69] Hong, D. *Exposants de Lyapunov de Réseaux stochastiques max-plus linéaires.* PhD thesis, INRIA, 2000.

[70] Hong, D. Lyapunov exponents: When the top joins the bottom. Technical report no. 4198, INRIA Rocquencourt, 2001.

[71] Jean–Marie, A. Waiting time distributions in Poisson–driven deterministic systems. Technical report no. 3083, INRIA Sophia Antipolis, 1997.

[72] Jean-Marie, A., and G.J. Olsder. Analysis of stochastic min-max-plus systems: results and conjectures. *Mathematical Computing and Modelling*, 23:175–189, 1996.

[73] Karp, R. A characterization of the minimum cycle mean in a digraph. *Discrete Mathematics*, 23:309–311, 1978.

[74] Kingman, J.F.C. The ergodic theory of subadditive stochastic processes. *Journal of Royal Statistical Society*, 30:499–510, 1968.

[75] Kingman, J.F.C. Subadditve ergodic theory. *Annals of Probability*, 1:883–909, 1973.

[76] Knuth, D. *The Art of Computing, Vol. I.* Addison-Wesley, Massachusetts, 1997.

[77] Krivulin, N. A max-algebra approach to modeling and simulation of tandem queueing systems. *Mathematical Computing and Modelling*, 22:25–37, 1995.

[78] Krivulin, N. The max-plus algebra approach in modelling of queueing networks. In *Proc. 1996 Summer Computer Simulation Conference, Portland, July 21-25,1996*, pages 485–490. SCS, 1996.

[79] Le Boudec, J.Y., and P. Thiran. *Network Calculus: A Theory of Deterministic Queueing Systems for the Internet.* Springer, Lecture Notes in Computer Science, No. 2050, Berlin, 1998.

[80] Lindvall, T. *Lectures on the Coupling Method.* Wiley, 1992.

[81] Loynes, R. The stability of queues with non–independent inter-arrival and service times. *Proceedings of the Cambridge Philosophical Society*, 58:497–520, 1962.

[82] Mairesse, J. A graphical representation of matrices in the (max,+) algebra. *INRIA Technical Report PR-2078, Sophia Antipolis, France*, 1993.

[83] Mairesse, J. A graphical approach to the spectral theory in the (max,+) algebra. *IEEE Transactions on Automatic Control*, 40:1783–1789, 1995.

[84] Mairesse, J. Products of irreducible random matrices in the (max,+) algebra. *Advances of Applied Probability*, 29:444–477, 1997.

[85] McEneany W. *Max-Plus Methods for Nonlinear Control and Estimation.* Birkhäuser, Boston, 2006.

[86] Neveu, J. *Mathematical Foundations of the Calculus of Probability.* Holden–Day, San Francisco, 1965.

[87] Olsder, G.J. Analyse de systemès min-max. *Recherche Opérationelle (Operations Research)*, 30:17–30, 1996.

[88] Peres, Y. Domains of analytic continuation for the top Lyapunov exponent. *Annales de l'Institut Henry Poincaré Probabilités et Statistiques*, 28:131 – 148, 1992.

[89] Pflug, G. Derivatives of probability measures – concepts and applications to the optimization of stochastic systems. In *Discrete Event Systems: Models and Applications, Lecture Notes Control Information Sciences 103*, pages 252–274. IIASA, 1988.

[90] Pflug, G. *Optimization of Stochastic Models.* Kluwer Academic, Boston, 1996.

[91] Pólya, G., and G. Szegö. *Problems and Theorems in Analysis, Vol. 1.* Springer, New–York, 1976.

[92] Propp, J., and D. Wilson. Exact sampling with coupled Markov chains and applications to statistical mechanics. *Random Structures and Algorithms*, 9:223 – 252, 1996.

[93] Resing, J.A.C., R.E. de Vries, G. Hooghiemstra, M.S. Keane, and G.J. Olsder. Asymptotic behavior of random discrete event systems. *Stochastic Processes and their Applications*, 36:195–216, 1990.

[94] Rubinstein, R. *Monte Carlo Optimization, Simulation and Sensitivity Analysis of Queueing Networks.* Wiley, 1986.

[95] Rubinstein, R., and A. Shapiro. *Discrete Event Systems: Sensitivity Analysis and Optimization by the Score Function Method.* Wiley, 1993.

[96] Saheb, N. Concurrency measure in communication monoids. *Discrete Applied Mathematics*, 24:223–236, 1989.

[97] Seidel, W., K. von Kocemba, and K. Mitreiter. On Taylor series expansions for waiting times in tandem queues: an algorithm for calculating the coefficients and an investigation of the approximation error. *Performance Evaluation*, 38:153–171, 1999.

[98] Subiono and J. van der Woude. Power algorithms for (max,+)- and bipartite (min,max,+)-systems. *Journal of Discrete Event Dynamic Systems*, 10:369–389, 2000.

[99] van den Boom, T., B. De Schutter., and B. Heidergott. Complexity reduction in MPC for stochastic max-plus-linear systems by variability expansion. In *Proceedings of the 41st IEEE Conference on Decision and Control*, pages 3567–3572, Las Vegas, Nevada, December 2002.

[100] van den Boom, T., B. Heidergott, and B. De Schutter. Variability expansion for model predictive control. *Automatica*, (to appear).

[101] van der Woude, J. A simplex–like method to compute the eigenvalue of an irreducible (max,+)–system. *Linear Algebra and its Applications*, 330:67–87, 2001.

[102] Vincent, J. Some ergodic results on stochastic iterative discrete event systems. *Journal of Discrete Event Dynamic Systems*, 7:209–232, 1997.

[103] Wagneur, E. Moduloids and pseudomodules 1.: dimension theory. *Discrete Mathematics*, 98:57–73, 1991.

[104] Zazanis, M. Analyticity of Poisson–driven stochastic systems. *Advances in Applied Probability*, 24:532–541, 1992.

List of Symbols

The symbols are listed in order of their appearance.

$\mathbb{N} = \{0, 1, 2, \ldots\}$ the set of natural numbers

$\mathbb{R}$ the set of finite real numbers

$\mathbb{Z}$ the set of integers

$\frac{d^n}{d\theta^n}\big|_{\theta=\theta_0} f(\theta)$ the nth derivative of f evaluated at θ_0, page v

$\mathbb{E}_\theta$ the expected value evaluated at θ

$\mathbb{R}_{\max}$ the set $\mathbb{R} \cup \{-\infty\}$, page 4

$\mathcal{R}_{\max}$ the structure $(\mathbb{R}_{\max}, \oplus = \max, \otimes = +, \varepsilon = -\infty, e = 0)$, page 4

$\mathcal{R}_{\min}$ the structure $(\mathbb{R} \cup \{\infty\}, \oplus = \min, \otimes = +, \varepsilon = \infty, e = 0)$, page 4

$A^\top$ the transpose of A, page 4

$\mathcal{G}(A)$ the communication graph of matrix A, page 5

$\mathcal{D}(A)$ the set of edges in the communication graph of matrix A, page 5

$A^n = A^{\otimes n} = \bigotimes_{k=1}^{n} A(k)$, for $A(k) = A$, the nth power of A, page 6

A^* the $\oplus$-sum over all powers of A, page 13

$\|A\|_\oplus = \max_{1 \le i \le I} \max_{1 \le j \le J} 1_{A_{ij} \in \mathbb{R}} |A_{ij}|$, page 50

$\|A\|_{\min} = \min_{1 \le i \le I} \min_{1 \le j \le J} A_{ij}$, page 51

$\|A\|_{\max} = \max_{1 \le i \le I} \max_{1 \le j \le J} A_{ij}$, page 52

$\mathcal{G}^c(A)$ the critical graph of matrix A, page 62

$i\mathcal{R}j$ reachability relation between node i and j in a graph, page 62

$\mathbf{e}$ a vector with all entries equal to e, page 63

λ^{top} the top Lyapunov exponent, page 69

λ^{bot} the bottom Lyapunov exponent, page 69

$x \otimes \mathbf{e}$ a vector with all entries equal to some number x, page 110

η the backward coupling time, page 112

$\mathcal{M}_1 = \mathcal{M}_1(S)$ the set of probability measures on a set S, page 120

$\mathcal{L}^1(\mu_\theta : \theta \in \Theta)$ the set of functions that are continuous absolutely integrable with respect to μ_θ, for $\theta \in \Theta$, page 120

$C^b(S)$ the set of bounded continuous real-valued functions $g : S \mapsto \mathbb{R}$, page 120

$\mathcal{D}(S, Z)$ a space of mappings on the product space $S \times Z$, page 126

$C_p(S, || \cdot ||_S)$ the set of all functions $g : S \to \mathbb{R}$, such that $|g(x)| \leq c_1 + c_2 ||x||^p$, page 128

$\mathcal{C}(S, Z)$ a space of continuous mappings on the product space $S \times Z$, page 130

$M^{I \times J}$ an extended state space to accommodate $\mathcal{D}$-derivatives, page 136

$\equiv$ weak equivalence, $A \equiv B$ if for any g out of a specified set D it holds that $E[g(A)] = E[g(B)]$, page 138

$s(\mu_\theta)$ the order of the highest significant $\mathcal{D}$-derivative of probability measure μ_θ, page 152

$\mathcal{L}[m_1, m_2; n]$ set of multi-indices to describe higher-order derivatives, page 155

$\text{card}(H)$ the cardinality of set H, page 155

$\mathcal{I}[l]$ set of multi-indices to describe positive and negative parts of higher-order derivatives, page 155

$s(X_\theta)$ the order of the highest significant $\mathcal{D}$-derivative of random variable X_θ page 158

$||\mathcal{A}||_\oplus = \sup_{A \in \mathcal{A}} ||A||_\oplus$, page 168

$\mathbf{c}_{A(0)}$ normalizing constant for finite products, page 168

$B_{g,m,\{A(k)\}}(n, p)$ an upper bound for the nth order derivative over a finite product, page 168

1_A the indicator mapping for event A, page 192

$A_{\tilde{a}}(k)$ the halted system, page 202

$\tau_{\tilde{a}}^{(n)}$ the maximal length of the nth order derivative, page 209

$B_{g,\tau_{\tilde{a}},\{A(k)\}}(n, p)$ an upper bound for the nth order derivative over a random horizon product, page 211

$a(q, M, n, p)$ an upper bound, page 213

$b(q, M, n)$ an upper bound, page 218

$\mathbf{C}_{A(0)}$ supremum of a normalizing constant for finite products, page 220

$\eta_{\bar{a}}^{(n)}$ the maximal length of the nth order derivative, page 230

$B^0_{g, \eta_{\bar{a}}, \{A(k)\}}(n, p)$ an upper bound for the nth order derivative over a random product for the Lyapunov exponent, page 231

List of Assumptions

Index